Before Newton

Before Newton

The life and times
of Isaac Barrow

Edited by
MORDECHAI FEINGOLD

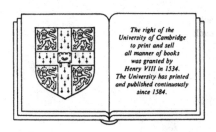

The right of the
University of Cambridge
to print and sell
all manner of books
was granted by
Henry VIII in 1534.
The University has printed
and published continuously
since 1584.

CAMBRIDGE UNIVERSITY PRESS

Cambridge
New York   Port Chester   Melbourne   Sydney

CAMBRIDGE UNIVERSITY PRESS
Cambridge, New York, Melbourne, Madrid, Cape Town, Singapore, São Paulo

Cambridge University Press
The Edinburgh Building, Cambridge CB2 8RU, UK

Published in the United States of America by Cambridge University Press, New York

www.cambridge.org
Information on this title: www.cambridge.org/9780521306942

© Cambridge University Press 1990

First published 1990
This digitally printed version 2008

A catalogue record for this publication is available from the British Library

Library of Congress Cataloguing in Publication data

Before Newton : the life and times of Isaac Barrow / edited by
Mordechai Feingold.

p. cm.

ISBN 0-521-30694-9

1. Barrow, Isaac. 1630–1677. 2. Mathematicians – England –
Biography. 3. Clergy – England – Biography. I. Feingold,
Mordechai.
QA29.B37B44 1990
510′.92 – dc20
[B]   89-38508
                                                        CIP

ISBN 978-0-521-30694-2 hardback
ISBN 978-0-521-06385-2 paperback

CONTENTS

CONTRIBUTORS

MORDECHAI FEINGOLD is Associate Professor of the History of Science at Virginia Polytechnic and State University. His publications include *The Mathematicians' Apprenticeship: Science, Universities and Society in England, 1560–1640* (1984). He is presently engaged in a study of Newtonianism.

JOHN GASCOIGNE is Senior Lecturer in the School of History, University of New South Wales, Sydney. He is author of *Cambridge in the Age of the Enlightenment* (1989) and is working on a study of Sir Joseph Banks and his intellectual milieu.

ANTHONY GRAFTON teaches history and history of science at Princeton University. His *Joseph Scaliger: A Study in the History of Classical Scholarship* was published in 1983, and at present he is completing the second volume of this intellectual biography of Scaliger.

MICHAEL S. MAHONEY is Professor of History of Science at Princeton University. He is the author of many studies on the development of mathematics and mechanics in the seventeenth century, including *The Mathematical Career of Pierre de Fermat* (1973). His current research is in the history of computing.

ALAN E. SHAPIRO is Professor of History of Science and Technology at the University of Minnesota. He has written widely on the history of optics and seventeenth-century science and is the editor of *The Optical Papers of Isaac Newton*.

IRÈNE SIMON, O.B.E., is Emeritus Professor at the University of Liège. She has written extensively on English literature and the history of ideas, including *Three Restoration Divines* (1967, 1976) and *Neo-Classical Criticism* (1971). At present, she is completing a study on Robert South's "Description of Poland" and is engaged in a study of the contemporary English novel.

EDITOR'S PREFACE

Our appraisal of historical scientific figures tends to be synonymous with their purported legacy to posterity – as adjudged by subsequent scholars; we rarely attempt to take into account their reputation among contemporaries or gauge their accomplishments within the context of their age. Obviously, a critical approach that equates historical "worth" with an appreciable contribution to future generations will – and must – be the primary focus of historians of science. Nonetheless, to use such a yardstick exclusively is to disregard the complexity of past events and to hamper seriously our ability to comprehend the precise character of the scientific enterprise; for all discoveries and breakthroughs in science, irrespective of the unique contribution of the individual who inaugurated them, cannot be considered in isolation, independent of a large community of teachers, fellow students, and scholars of the second order. The fact that the input of such persons is vague and amorphous, and cannot be assigned a simple letter in an equation, does not negate their contribution – it only makes more difficult our task of definition. And by failing to recognize this connectedness between individual genius and larger community, we not only distort the complexity and singularity of a historical moment, but also risk misconstruing the nature and magnitude of the innovation itself.

Isaac Barrow is a case in point. Esteemed by contemporaries as a profound mathematician as much as later generations esteemed him as a divine, his work had already, by the time of his death in 1677, been superseded by the discoveries of Newton and Leibniz. Nevertheless, his reputation as a mathematician persisted well into the nineteenth century, both because of and in spite of the role he played, posthumously, in the raucous calculus priority dispute. In a manner that has few, if any, parallels in the history of science, Barrow's reputation correlated inversely with the vicissitudes of the combatants. One of the tacks taken by both sides was to accuse and castigate the other as a beneficiary of Barrow's discoveries.

To deny even the least shadow of this charge, both protagonists refused even to allude to Barrow in any context – let alone acknowledge the slightest debt. This process of silent repudiation was hastened and made explicit in the present century with the advent of meticulous and rigorous new scholarship in the history of mathematics. Partly as a reaction to J. M. Child's reckless attempt in 1916 to credit Barrow with the invention of the calculus – an effort that resulted in a vehement rebuttal by Dietrich Mahnke a decade later – the research of Joseph Hofmann and D. T. Whiteside established, respectively, the independence of Leibniz's and Newton's discoveries, at the same time denying the debt of either man to Barrow, either in the form of personally transmitted knowledge or via his publications. The verdict of both scholars was that Barrow's mathematical knowledge was derivative, and hence extraneous to the intellectual development of either Newton or Leibniz. This new orthodoxy has rapidly been assimilated and perpetuated by the more recent literature, so much so that Barrow's reputation, to use A. Rupert Hall's eloquent phrase, "once so high (at least among English speakers), seems to be descending toward the status of an elegant codifier" (*Philosophers at War* [Cambridge, 1980], p. 8).

Unfortunately, none of these scholars ever embarked on a close and sustained study of Barrow, save for those aspects conspicuously pertinent to Newtonian and Leibnizian concerns. It is the purpose of this volume, therefore, to provide the requisite evaluation. By seeking to understand Barrow's life and work primarily in the confines of the pre-Newtonian scientific framework, we believe that a more balanced picture of the essence of the man and the intellectual issues that preoccupied him will emerge. Thus, the chapters devoted to Barrow's attainments in mathematics and optics entail a substantial rethinking of his scientific acumen, while the chapters dealing with Barrow's life and times endeavor to emphasize the interdependence of his science and his religious and philosophical world views. For this reason, a sizable portion of the book is devoted to the many facets of Barrow's interests and undertakings, particularly his lifelong and incessant commitment to theology, education, and the classics.

A collective enterprise of this sort risks duplication of material among the chapters. Propitiously, all authors were fully aware of the domains and responsibilities of the other contributors, and conscious efforts were made to avoid addressing issues discussed more fully elsewhere. Thus, despite the affinity between geometry and optics in Barrow's thought, both Mahoney (Chapter 3) and Shapiro (Chapter 2) have resisted the temptation to infringe upon each other's domain. In a like manner, Gascoigne (Chapter 4) has circumvented biographical material treated in the first chapter and aspects of Barrow's theology addressed by Irène Simon (Chapter 6). For his part, the editor has followed vigilantly the decision to provide an intellectual biography of Barrow, while leaving the discussion of the content of his work to the other contributors. However, no editorial policy to ensure uniformity of opinion has been employed, and, subsequently, variations in interpretation may occasionally – and perhaps wisely – be found in the respective chapters.

I wish to thank the master and fellows of Trinity College, Cambridge, for permission to use and quote material in their archives. It is also a duty, and pleasure, to acknowledge the kindness and perseverence with which the staff of the library – and in particular Dr. Timothy Hobbs – assisted our efforts. I also thank the keeper of Western manuscripts at the Bodleian Library, Oxford, for permission to provide a modern edition of Barrow's library catalogue in its possession, and to Lord Delamere for kind permission to quote from the Hartlib papers housed at Sheffield University Library. I am also grateful for information received from Mrs. Alison Sproston, assistant librarian of Gonville and Caius College, Cambridge, and Miss Suzanne Eward, librarian of Salisbury Cathedral Library. Among individuals who contributed in a variety of ways to the volume, the following friends and colleagues ought to be mentioned: Willy Applebaum, Roger Ariew, Ann Blair, I. Bernard Cohen, Michael Hunter, Elisabeth Leedham-Green, Maren Niehoff, Kristine Peterson, and Tim Wells. Last, but not least, I wish to thank the contributors to the volume, who shared my zeal for the subject matter of the book – and patiently awaited its completion – and Helen Wheeler and Mary Nevader, who lovingly attended the book while in press.

1

Isaac Barrow: divine, scholar, mathematician

MORDECHAI FEINGOLD

The education of a scholar

Isaac Barrow appears to have descended from an ancient Suffolk family, but his immediate ancestors were professional men from Cambridgeshire. His great-grandfather was Philip Barrough, author of the popular *Methode of physicke* (1583) and a licentee by Cambridge University to practice chirurgy and medicine in 1559 and 1572, respectively. Philip's brother, Isaac, was elected a scholar of Trinity College, Cambridge, in 1560 and fellow the following year, graduating M.D. in 1570. He is said to have been tutor to Robert Cecil, the future lord treasurer, to whose father William, Lord Burghley, Philip Barrough's *Methode of physicke* was dedicated. Two of Philip's sons, Samuel and Isaac, matriculated at Trinity College in 1575 and 1584, respectively, the latter becoming a justice of peace at Cambridgeshire. Of the children of the J.P., Thomas was the father of the master of Trinity College, while his brother – yet another Isaac – was to become the bishop of St. Asaph.[1]

In his old age, Thomas Barrow recollected that it was his father's severity that drove him to leave home and acquire a trade instead of following the footsteps of his ancestors into one of the three learned professions. Thus, "he came to London and was apprentice to a linnen-draper," rising in due course to become linen draper to Charles I. In 1624, Thomas married Ann, the daughter of William Buggin of North Cray, Kent, who gave birth in October 1630 to Isaac Barrow. It appears that Barrow was the only child of this union – certainly the only child to survive infancy. Ann died c. 1634, and the widowed father sent the lad to his grandfather, the Cambridgeshire J.P., who resided at Spinney Abbey. Within two years, however, Thomas remarried, perhaps – as Osmond speculated – in order to establish a new family home and thus remove young Isaac from the care of his grandfather. As Thomas Barrow told Aubrey years later, he "was

1

faine to force him away, for there he would have been good for nothing."[2] The new wife was Katherine Oxinden, sister of Henry Oxinden of Maydekin, Kent. Obviously, Thomas had married above his station. A somewhat condescending tone toward Thomas is evident in the *Oxinden Letters,* although the Oxindens were not above making use of the new brother-in-law in both private and business matters. Later, following the outbreak of the Civil War, Barrow and the Oxindens parted ways, for the latter adhered to Parliament side. Of this marriage, at least one daughter, Elizabeth (born 1641), is known to have survived the Restoration, having spent the period of the Civil War and Interregnum with another branch of the Oxinden family at Dean.[3]

What Thomas missed by defying his father's wishes and becoming a merchant he compensated for in his attitude toward his son, and from an early age the elder Barrow intended Isaac to become a scholar. To such an end, he made a special arrangement with Robert Brooke, headmaster of Charterhouse, according to which Brooke would be paid double the going rate of £2 p.a. in return for a promise to take special care of young Isaac. The choice of Brooke and of Charterhouse was reasonable, given the school's reputation for providing a solid classical education that would prepare a young man for university. (Two recent alumni included the poets Richard Crashaw and Richard Lovelace.) In addition, the headmaster was known for his stern Anglican beliefs, sentiments that would cost him his position following the Puritans' rise to power.[4] Quite possibly, an additional incentive was the fact that the school's captain, a kinsman, was expected to supervise Isaac's education. Such plans, however, backfired, for the above-mentioned kinsman informed Thomas that Brooke was scandalously neglectful of Isaac's education; if anything, the lad was distinguishing himself as the school ruffian, "being much given to fighting, and promoting it in others." So much so that his father "often solemnly wished, that if it pleased God to take away any of his children it might be his son Isaac"![5] Thus, after two or three years, Isaac was removed from Charterhouse and transferred to Felsted School in Essex, under the care of its noted schoolmaster, Martin Holbeach.[6]

Sir John Bramston, commenting in his autobiography on the upbringing of the regicide Sir Henry Mildmay, stated that "from his mother's woombe and his master's tuition [Sir Henry] suckt in such principles of disloyaltie and rebellion" that he later became noted for. The master that Bramston was referring to was none other than Martin Holbeach, who, continued the judge, "scarce bred any man that was loyall to his prince." Such was obviously a biased opinion. True, Holbeach was noted for his Puritan sympathies. He had once "whiscked to safety" the celebrated Thomas Shepherd and was ejected from his living in 1662 for Nonconformity. However, one can hardly describe the education he provided for his charges as doctrinarian. Even those Puritan families who entrusted

Holbeach with the education of their children – and the schoolmaster helped raise all four sons of Oliver Cromwell – had no desire for a solely religious education. Not that they did not seek assurances of Holbeach's personal piety and Calvinist convictions, which indeed may have been translated into the catechism of the school. However, having ascertained Holbeach's religious proclivities, what parents sought – and found – in Felsted were stern discipline and solid grounding in classical learning.[7] From all accounts Holbeach was indeed an exceptional teacher and firm disciplinarian. John Wallis, who had been a student of Holbeach a decade before Barrow, stated that in his time there were "above a hundred or six score Scholars; most of them Strangers, sent thither from other places, upon reputation of the School; from whence many good Scholars were sent yearly to the University."[8] So it is not surprising that Thomas Barrow, in despair over his son's poor scholastic achievement and unruly character, sent Isaac to Felsted.

The brief description provided by Wallis of the curriculum at Felsted indicates that under Holbeach the school modeled its curriculum – in terms of both structure and depth of study – after such great grammar schools as St. Paul's and Westminster. Proficiency in Latin and Greek was the main objective, and to such an end the students were intensively drilled both in the rules of grammar and syntax and in the habits of language, and fed a strenuous diet of classical texts adjoined with contemporary helps for study. The Latin and Greek grammars of Lily, Camden, and Farnaby and the dictionaries and lexicons of Stephanus, Pasor, and Schrevell were intended to aid the reading of the classics. Certain phrase books and colloquial exercises were also assigned. Important, and effective, tools employed in the education process were written and oral exercises of double translation, not only from Latin or Greek into English (and vice versa) but also from prose into verse, as well as the opposite. To reinforce such skills, the students were instructed to speak only Latin or Greek, even outside the classroom. Written and oral exercises consisted of orations and epistolary writing imitative of a particular style or purpose. Also, ex tempore declamation that was intended to enhance facility in the spoken tongue – as well ingenuity of improvisation – was widely practiced. It was firmly believed that a rigorous combination of written and oral exercises would not only develop the student's memory – an important underlying concept in contemporary education – but inculcate in him a refined taste for the best classical masters, whose style could then be imitated. The effectiveness of such techniques was verified by Wallis, who recalled, "I had been used . . . to speak *Latin;* which made the Language pretty familiar to me; which I found to be of great advantage afterwards." The upper forms were also taught Hebrew – using the Hebrew Bible and Psalter as textbook, alongside a grammar and a dictionary – as well as elementary French. In order to round off their preparation for university, they were also introduced to logic.[9]

The chronology of Barrow's younger years is confused, but it seems that he came to Felsted c. 1640, and stayed for some four years. By 1642, the elder Barrow was facing financial ruin owing to the Irish rebellion, which evidently had destroyed his extensive trade with Ireland. (Thomas told Aubrey his losses amounted to "neer 1000 pounds.") When Holbeach was informed of the predicament of the elder Barrow, he responded by removing Isaac from the boarding home where he had been residing and taking him into his own house; later he appointed him "little Tutour" to Thomas Fairfax, who in 1641 succeeded as fourth viscount Fairfax of Emely in Ireland.[10] The appointment, besides demonstrating Holbeach's confidence in the talents of the young man, must have also carried with it a small, but much needed, remuneration. Then, on 15 December 1643, Isaac was admitted a foundation scholar at Peterhouse, Cambridge. It is possible, as Gascoigne suggests,[11] that the choice of Peterhouse reflected "the strong Royalist and episcopal sympathies of his father." Perhaps more relevant, however, is the fact that Barrow's namesake uncle – and future bishop – was a fellow of Peterhouse at the time and probably procured the scholarship for his nephew.[12] Be this as it may, Barrow never enjoyed his scholarship, for the senior Isaac was ejected the following month by the parliamentary visitors – and moved to Oxford, where he joined Thomas Barrow – while the scholarship was given to another. So Barrow remained with Fairfax at Felsted.

The association with Fairfax, however, nearly sealed Isaac's scholarly career before it began, for the young nobleman fell in love with a local girl – with a dowry of "only" £1,000 – and married her, despite the fierce opposition of his guardian, Lord Saye and Sele. The latter, faced with the fait accompli, mercilessly cut off all his ward's sources of income. Sometime in 1644 Fairfax left for London, taking Barrow with him, but as his wife's portion could not last long, Barrow quickly found himself destitute of both patron and income. Again Holbeach came to the rescue. Having searched out his former student in London, the schoolmaster sought to lure him back to Felsted – even promising to make Barrow his heir. Isaac, however, declined the kind offer and instead went off to visit a former school fellow, a member of the Harley family in Norfolk, who invited Barrow to accompany him to Trinity College, Cambridge, where he promised to support him. If Aubrey's information is correct, the person must have been Thomas Harley, who was admitted pensioner on 10 March 1645/6, two weeks after Barrow's own admittance (on 25 February) as subsizar under the tutorship of James Duport.[13]

Trinity College, Cambridge

In 1646, when Barrow entered Trinity, the college had just celebrated its first centenary. Yet the event passed almost unnoticed owing to the Civil War and the Puritan rise to power. The immediate, and devastating, result for the college was the ejection of most of the fellows (forty-nine of sixty) during 1644–5 by the parliamentary visitors – including the master Thomas Comber, the poet Abraham Cowley, and the scholars Herbert Thorndike and Robert Boreman. The purge was followed by the intrusion of many "loyal" scholars and fellows of other colleges, chief among them the new master, Thomas Hill, who was destined to play a significant role in Barrow's career at Cambridge.[14]

On the face of it, Hill was the most unlikely candidate to serve as Barrow's patron. Reputed to be the most dogmatic and rigid Presbyterian of the eight intruded masters who were members of the Westminster Assembly of Divines, Hill quickly acquired a reputation as a seeker of sweeping religious reforms at Cambridge. As early as 1643 he urged Parliament to effect a "purging and pruning" of all Laudian innovations, a task he took quite seriously during the first years of his mastership. Certainly, in matters of religion he was uncompromising. Oliver Heywood, an exact contemporary of Barrow, recorded that Hill "would sometimes lay his hand upon his breast, and say with emphasis, 'Every Christian has something here that will frame an argument against Arminianism.'" In his funeral oration of Hill, Anthony Tuckney recorded that Hill "had made a fair progress in a learned confutation of the great daring champion of the Arminian errors [John Goodwin], whom the abusive wits of the university, with an impudent boldness, would say none there durst adventure upon." At Trinity, not only was Hill zealous for religious reforms, but he also severely punished a scholar in 1646 who, to the master's chagrin, maintained Parliament to be more rebellious than the Irish.[15]

From about 1647, however, Hill's reforming zeal appears to have slackened. Though his distaste for Arminianism remained strong, he nonetheless displayed more tolerance toward certain people thus affected. Such conspicuous forbearance may have been a direct result of Hill's realization that the religious reformation he had so vigorously sought had been achieved, thus making his continued vigilance no longer necessary. More likely, however, it had its roots in Hill's reaction to the swift rise to power of the Independents, who not only were encroaching rapidly upon the political power of the Presbyterians in church and state, but whose attack on the structure and curriculum of Oxford and Cambridge dangerously imperiled both universities. Thus, Hill, who respected learning (even though he himself was not an exceptional scholar), found himself, no doubt with some surprise, in the position of a staunch defender of the universities. Indeed, fear of the religious and political extremism that many Presbyterians associated with the Independents resulted in a silent coalition of interest between Presbyterians and Royalists – much more pronounced at the universities than in the country at large – which, in turn, resulted in a unified front to defend the universities against any attempts at further reformation or ejection of a large number of fellows, who would then be replaced by supporters of the Independents. Such an alliance, forged in the face of a common enemy, may explain the religious and political pluralism that existed in Cambridge during the 1650s and that pervaded life at Trinity College, as we shall see shortly.

Be this as it may, by the late 1640s Hill had sufficiently mellowed to tolerate even Arminians if exceptionally learned. An example of this new attitude may be found in a sermon he delivered in 1648, in which he un- equivocally praised the character and learning of the very Royalist head- master of Westminster School, Richard Busby, whom he designated as "very able and industrious" and who made his school great, even though Hill desired that the school's "Moralls, and Spiritualls" be "answerable to the *Intellectualls*." That he also held such a view as regards Trinity College may be seen from a sermon delivered later that year: "I earnestly desire wee may have a learned, as well as a religious Reformation."[16] When translated into the management of the college, such respect for learning caused Hill to come close to the practice of his friend and intruded master of St. John's College, Anthony Tuckney, who, when once criti- cized for not taking sufficient regard of the "godly" in elections for fellow- ships, replied: "No one should have a greater regard to the truly godly than himself, but he was determined to choose none but scholars," adding that "they may deceive me in their godliness, they cannot in their scholar- ship."[17] In fact, by 1649 Hill was even taken to task for allowing two fellows to retain their fellowships despite their failure to sign the Engage- ment.[18] And as we shall later see, Barrow was perhaps the chief bene- ficiary of Hill's protection at Trinity.

Indeed, it was precisely during 1648–9, when the Independents were gathering forces and the university was under attack, that Hill did his best to support teaching and research at Cambridge. Thus, he convinced Sir John Wollaston, lord mayor of London, to consider endowing a math- ematical professorship at Cambridge, the first stage of which was Wol- laston's gift of a yearly stipend of £20 for a reader, a sum he promised to augment to £60 and make into an endowed lectureship if the university approved.[19] In addition, Hill was instrumental in conspiring with John Selden to secure for Cambridge the magnificent Lambeth Palace Library, a gift that soothed some of the envy Cambridge men had felt for half a century for Oxford and its Bodleian Library.[20]

Although the university that Barrow entered may have been reminiscent of the medieval university, in some aspects it had become a new institu- tion. The most important of these involved structural transformations that had occurred during the preceding century, specifically the shift in university government from a collective body of regent and nonregent masters to a collective body of heads of house. This shift was accompanied by the momentous rise of the college as a major center of undergraduate teaching, which, while not making university lectures obsolete (as is often assumed by historians), nonetheless had far-reaching consequences for the shape and nature of university education. Most important was the gradual disappearance of the medieval division of the disciplines between the B.A. and M.A. curriculum. By the turn of the seventeenth century, the entire arts and sciences curriculum had collapsed into the four years of undergraduate study, thereby transforming the M.A. sequel into a course of independent study aimed at expanding and elaborating the foundations of knowledge previously acquired.

This new trend is already evinced in the 1570 Elizabethan statutes. Having stipulated that bachelors of art should attend the lectures in philosophy, astronomy, optics (perspective), and Greek, the statutes continued, "and that, which had been before begun, they shall complete by their own industry." Seven years later, William Harrison was able to describe the actual practice of the new incoming system. The undergraduates, he wrote, receive their bachelorship of arts after they have grounded themselves in both trivium and quadrivium, and "from thence... giving their minds to more perfect knowledge in some or all the other liberal sciences and the tongues, they rise at the last to be called masters of art." By 1608 the official statutes were duly amended to indicate the completion of the process of incorporating the seven liberal arts and the three philosophies into the undergraduate curriculum; the vice-chancellor and heads of house provided an interpretation of the statutes that dropped residency requirements on the grounds that "a man once grounded so far in learning as to deserve a Bachelorship in Arts is sufficiently furnished to proceed in study himself."[21] This did not mean, however, that most students now departed from the university after obtaining the B.A. degree, only that the university authorities had acquiesced to the new reality without having resorted to the sensitive task of revising the existing statutes.

The reasons for the restructuring of the curriculum were twofold. First, it accommodated the new type of students who were entering the university in growing numbers: Desirous of a well-rounded education yet not interested in acquiring a degree, they never intended to remain for the required four years stipulated for the bachelorship, let alone the additional three years stipulated for the M.A. Thus, the increasing transfer of much of the education to the colleges facilitated the regulation of the undergraduate course on a more individual basis; the respective tutors could and would determine the course of study, its breadth and depth, in accordance with the background, requirements, and desired length of residence of their students. Second, the printing revolution had made obsolete the medieval system of instruction whereby the master dictated from a text he alone possessed. The fact that now all students owned the necessary textbooks forced teachers to devise a new system of instruction. As

John North put it, "Since books are so frequent as now they are, public lectures are not so necessary, or (perhaps) useful, as in elder times when first instituted, because the intent of them was to supply the want of books, and now books are plenty, lectures might better be spared, and the promiscuous use of books come in the place of them."[22]

Nevertheless, there was nothing in these major changes to alter a cardinal facet of the inherited medieval world view, namely, the ideal that a unified body of knowledge be shared by all recipients of higher education. Contrary to the prevailing assumption among contemporary scholars, the influx of "gentlemen" into Oxford and Cambridge did not result in the creation of two parallel educational systems: one for "clerics," who intended to graduate and proceed into the church and were therefore instilled with traditional "scholastic" education, that is, logic, antiquated natural philosophy, and metaphysics; the other for members of the upper class, who sought no degrees but were destined for the Inns of Court and the management of estates and were therefore treated to the more liberal and "pleasing" subjects, such as cosmography, history, astronomy, and modern languages.[23] In reality, the system remained largely homogeneous in terms of offerings to *all* students. As we shall see shortly, the prevailing ideal of a shared culture and a unified body of knowledge caused an almost identical course of study to be offered to both groups, the significant difference being that members of the upper class obtained a more "condensed" – that is, shorter and, at times, more superficial – version and were more personally supervised by the tutor. It must be stressed, however, that neither did they dispense with the "scholastic" requirements of logic, ethics, and metaphysics nor were the "clerical" students deprived of "modern" subjects.

The rise of the colleges as centers of undergraduate teaching during the sixteenth century introduced two major features into English higher education: the crucial role of the tutor in guiding a student's course of study and the institution of lectureships within the colleges in order to supplement tutorial instruction. Yet the tutorial system itself was rapidly evolving and by the mid-seventeenth century had undergone an important modification. If during the Elizabethan period the students had been more or less evenly distributed among the fellows of a college, by 1650 most undergraduates were assigned to a handful of the more junior fellows. This change was prompted by the recognition that the multiple employments of the seventeenth-century college necessitated a more sensible division of labor among college fellows. A college such as Trinity, for example, employed nearly a third of its fellows as either lecturers or examiners; the senior fellows, for their part, were burdened with administration or ministerial duties. Consequently, there emerged a system in which for some time the junior fellows were in charge of teaching and the tutorials of the majority of students; thereafter, they would be "promoted" to more weighty offices and allowed more time to pursue research. As Thomas Hill, the master of Trinity during Barrow's student days stated, "The *Seniors* need not trouble themselves with taking Pupills, their Colledge businesses being so many, and their pittance of time (all things considered) but small for the improvement of themselves."[24]

This system invited abuse, and the annals of Cambridge and Oxford indeed indicate widespread abuse during the eighteenth century. Yet, at least during the seventeenth century, the new structure was quite successful. Following his arrival at university, the incoming student was given a sustained period of attention. His tutor would acquaint him with the course and order of study, provide him with introductory manuals delineating the basics of the various disciplines, and then set down the course of study, which the student was expected to pursue independently, meeting daily with the tutor to report his progress or enlist help. Joseph Mede's biographer provides a good illustration of the system:

> After he had by daily Lectures well grounded his Pupills in
> Humanity, Logick and Philosophy, and by frequent converse
> understood to what particular Studies their Parts might be
> most profitably applied, he gave them his Advice accordingly.
> And when they were able to go alone, he chose rather to set
> every one his daily Task, then constantly to confine himself and
> them to precise hours for Lectures.[25]

This type of instruction may help explain the sudden proliferation of manuals and "advices to students" by the middle of the seventeenth century. These guides, prepared by tutors for the benefit of their charges and designed to provide rules of conduct as well as general instruction on books and methods of study, were intended to substitute for the lack of continuous lecturing and sustained personal tutorials. It was thus possible for a tutor to take charge of a large group of pupils. Not solely responsible for their scholars' education, the tutor's task was rather to ensure that they attended the daily college – and, when appropriate, university – lectures and to supervise their readings. What was equally important and time-consuming, he was to guide their religious instruction, supervise their behavior, and be responsible for their finances.

The students, for their part, enjoyed a certain amount of intellectual freedom that prepared them for the independent course of study they were expected to follow after graduation. No less important, the system allowed students to benefit from the instruction of fellows other than their official tutors. This was especially significant in specialized topics such as Semitic languages and the sciences, for now the tutors could "farm out" students to other fellows (at times for a fee) who were better qualified to teach them. As we shall see, such was the practice of James Duport, who directed his students to Isaac Barrow for part or all of their instruction in mathematics. Another example is Henry More, who, though reluctant to take on the tedious responsibilities of tutor, was nonetheless prepared to provide scientific instruction to certain students.[26]

If Barrow was fortunate to find a protector in Thomas Hill as master of the college, he was blessed in having James Duport as his tutor. Duport came to Trinity from Westminster in 1622 and was elected fellow five years later, graduating M.A. in 1630. From his student days he was renowned for his linguistic skills - particularly Greek - as well as for his wit. The latter talent earned him the appointment of *praevaricator* (or commencement jester) for the year 1631, when he expounded with elegance and humor on the topic "Aurum potest produce per artem Chymicam." In performing the task, Duport appears to have achieved the rare feat of amusing his auditors without provoking the wrath of the university authorities, as was often the case. Following the untimely death of Ralph Winterton in 1639, Duport was elected his successor as Regius Professor of Greek. He proved to be both a popular and a skillful lecturer. His lectures on the characters of Theophrastus and the orations of Demosthenes still survive, the latter seeing publication during the nineteenth century. His rendition of the Book of Job into Greek meter became an immediate success, and his other attempts at producing religious literature in Greek translation - particularly the Book of Common Prayer and the Psalms - were also popular. Duport's main scholarly publication, however, was the erudite *Homeri gnomologia,* which, though expounding the popular seventeenth-century topic of "proving" the affinity of Greek to Hebrew, demonstrated more than a common knowledge of classical and biblical studies.[27]

Upon his election as Regius Professor, Duport also obtained a mandate from Charles I that allowed him to continue to serve as college tutor and retain his seniority and the college dividends, notwithstanding the contrary stipulation of the statutes governing the professorship.[28] Duport, however, never exercised this dispensation before 1645, at which time he began assuming the tutorial responsibilities for a significant share of the yearly intake of undergraduates at Trinity College. In fact, during the fifteen years of Puritan hegemony (1645–59), Duport served as tutor to some 180 students, or more than 20 percent of *all* undergraduates admitted to Trinity during that period! Undoubtedly, Duport assumed such a tiresome and taxing new role as a consequence of the upheaval in both state and university that had resulted in the mass expulsion of Royalist and Anglican fellows, leaving Duport virtually the only tutor at Trinity who publicly espoused Royalist and Anglican sympathies. Yet Duport's continuance to hold his professorship - as well as the adjoining Trinity fellowship, albeit on new terms - was tolerated in part because of his popularity and in part because, as Monk put it, "there probably would have been a considerable difficulty in supplying Duport's place with a Presbyterian Greek Professor." In any event, Duport used his elevated position to take charge of – and to protect – the sons of many Royalists who had been sent expressly to Trinity to be under Duport's tutelage. And it is significant that none of the sixty or so Trinity men who were educated at Cambridge during the Civil War and Interregnum, and who were ejected in 1662, were students of Duport (with the notable exception of John Ray). Duport became, indeed, the "official" Royalist tutor, and it was to his secure care that the young Isaac Barrow was also entrusted.[29] The tutor-professor became so fond of his new pupil that, like Holbeach before him – having been informed of the father's financial situation – he forwent all tuition fees from Barrow and even took the youth into his own rooms gratis.[30]

Today, however, Duport is remembered – if at all – for his alleged die-hard conservativism. His former renown as a classicist and a wit has given way to a notoriety of sorts as a tenacious Aristotelian, antagonistic to the new science; scornful of the ideas of Galileo and Descartes; and hostile toward the Royal Society.[31] The evidence for such a historical judgment derives from two sources: some verse extracted from Duport's massive *Musae subsecivae* and two or three comments included in his "Rules to Be Observed," a manual of university conduct he composed expressly for his students. These passages, which purport to exhibit a decisively conservative, Aristotelian cast of mind, have been categorically adduced by some scholars to reflect Duport's innermost, unqualified "prejudices" and to have been translated into his teaching. I should like to scrutinize Duport's views and the validity of such a historical judgment at some length, both because Duport was Barrow's tutor – and exerted great influence on him – and also because a better understanding of Duport's alleged "conservatism" may serve as a key for unlocking the complex cluster of perspectives that he shared with many other university men who also have been unequivocally dismissed as inimical to the new philosophical and scientific spirit of the seventeenth century.

It cannot be denied that the Regius Professor displayed noticeable irritation with the anti-Aristotelian, pro-Cartesian temper that abounded in post-1640s Cambridge. Yet Duport's pique was triggered less by his intolerance of new cosmologies than by his perception of what he feared would be the consequence for religion and belles-lettres if such world views were allowed to take root. (In retrospect, of course, Duport was proved correct on both counts.) Thus, his "Apologia *pro* Aristotle, *contra Novos Philosophos. Seu Pro Veteribus contra Nuperos Novatores,*" a piece often quoted as illustrative of Duport's quintessential conservativism, should be understood as a "manifesto," in which Duport voiced his belief that those who worshiped Ramus and Descartes sanctioned Philistinism. The new "sect" of English "innovators" who thoughtlessly championed everything novel and fashionable had already created a climate of opinion, he felt, in which the Greek and Latin heritage was despised, Aristotle attacked by parasites, and Seneca and Cicero impetuously condemned. Duport attributed the slander of contemporary Englishmen to their basic ignorance of such authors as Homer and Aristotle, whom they were unable to read in the original (knowing little Latin and no Greek) and hence styled "displeasing," as if to excuse their want. Duport's critique, therefore, was aimed only incidentally at the philosophers referred to in the title; his main concern was language and the *literare humaniores,* which he was determined to vindicate from the calumnies of its "ignorant" detractors. Most important, in voicing such a position, Duport was joining the growing chorus of scholars and other educated men who were genuinely concerned for the future relations between literature and science.[32]

Other examples of misunderstanding of Duport's "conservatism" abound. When, for example, Duport chided his former pupil Francis Willughby in the early 1660s for the latter's infatuation with Cartesianism, astronomy, botany, and the Royal Society, his good-natured reproach was based on his concern for sensibilities, not hostility toward the new science. Duport feared that Willughby's scientific interests, carried to excess, would lead inevitably to his neglect of the more pleasing polite arts. Such paternal concern was exacerbated by Willughby's delicate constitution and poor health, which the former tutor feared would deteriorate in the wake of rigorous scientific study – which was indeed the case. Similarly, in another piece, Duport made use of the Trinity College fountain as a metaphor for his lifelong devotion to the classics, particularly his passion for Greek poetry. Although he extolled the pure fountain of Aristotle, in the depth of which "truth" shines, his real subject was not Aristotelian philosophy, but rhetoric and poetry, Cicero and Livy, and the greatness of the Athenian and Roman heritage. To drink from such a fountain would last a student a lifetime, long after the fame of the Cartesians had vanished. Duport's verses, then, were an expression of profound concern for the fate of the muses and his battle against what he perceived to be the pernicious effects of the new science upon polite learning and revealed religion; they were not, as has often been assumed, intended to be a defense of Aristotelian philosophy against the new science. Even his "In hypothesin Copernicanam, de motu terrae, & quite coeli," which certainly lends itself to an anti-Copernican reading, may, in fact, be read instead as a classical "paradox": the defense of contraries as a true test of wit. Thus, when Duport contrasts Copernicanism and common sense, claiming he dislikes contradictions, or when he complains of "the new class of Philosophers [who] would bring forth a new kind of Theology," he may very well be expressing his religious conservativism. Yet in view of his predilection for puns and hostility toward Catholicism, he could just as easily be mocking the placing of Copernicus on the

Roman Index and the subsequent condemnation of Galileo. After all, the piece is "addressed" to Galileo, who is "admonished" not to displease the cardinals of the "stationary" earth, lest he should lose his dogma in prison.[33]

The other document brought as evidence for Duport's alleged deep-seated Aristotelianism is his manual for his students. Here too, however, excerpts have been extrapolated with little concern for either their precise meaning or their context. The manual was intended mainly to instruct the student in matters of religion, morals, and academic conduct (Curtis styled it "full and even repetitious"), not to stipulate rigidly which philosophical systems to follow or authorities to obey. The philosophy of education that informs the manual correlates with Duport's own predilection for the liberal arts as the cornerstone of a well-rounded education and a full life, regardless of one's future vocation or preference of study. Hence, his primary concern was to inculcate into his students what he believed to be of prime importance for every educated man: a broad range of humanistic learning based upon the mastery of languages and literature. His objective was to cultivate healthy reading habits, a firm capacity for reasoning and eloquence, and solid compositional skills. He made no mention of the mathematical sciences and of natural philosophy, because *for him* they were secondary in merit (not meritless) and, more important, because he usually assigned the *advanced* teaching of such subjects to others (such as Barrow).

However, even within the framework of humanistic studies charted by Duport, Aristotle was commended for specific purposes, not slavishly adopted en bloc, a point usually glossed over by historians. On the three or four occasions that Duport cites Aristotle, his intent is the following: to assure the student that despite the serious challenges undermining Aristotle's authority in matters of natural philosophy, an important portion of the Aristotelian corpus is still worth preserving. The message is one of moderation: "*If* at any time in your disputation you use the Authority of Aristotle, be sure you bring *his own words & in his owne language.*" Similarly, "In your answering [at a disputation] reject *not lightly* the authority of Aristotle, *if* his owne words *will permitt a favorable, and a sure interpretation.*" Even when Duport advises the student not to follow "Ramus in Logick nor Lipsius in Latine, but Aristotle in one and Tully in other," his stance is not dictated by a "reactionary" dismissal of the "modern" Ramus. Instead, his articulated preference for a particular style echoes a truism shared by most informed seventeenth-century observers: The sweeping "reform" promised by Petrus Ramus a century earlier – which was enthusiastically embraced by many Protestants following Ramus's martyrdom at the St. Bartholomew Day Massacre – was little more than a reckless iconoclastic critique of Aristotle and an irresponsible attempt to reorder the disciplines.[34]

Such a corrective to the traditional view of Duport hardly qualifies him to be ranked among the "moderns" of science and philosophy. But neither should he be indiscriminately considered a reactionary. Duport's scholarly concerns and persuasions must be characterized as reflective of *taste,* not *opinion*. In this sense he was not unlike Sir William Temple, for whom, as Garrod pointed out, "'Old Books to Read' still meant something" in a "scientific" age that placed increasingly less value on literature. And Duport, like Temple, genuinely preferred the belles-lettres to science; his poems and epigrams (like Temple's essays) did no more than "express the character of one governed by such genteel prejudices."[35] Perhaps a good analogy to Duport's attitude toward contemporary philosophy is that of his great predecessor Isaac Casaubon, who was also hostile toward Petrus Ramus and his disciples. As Glucker stated, "One can perhaps say that Casaubon's attitude to what was the philosophy of his day was mainly a humanist's attitude to those who were incapable of appreciating the 'bonae literae' he stood for. It was his natural standpoint as a 'grammaticus' in the Battle of the Ancients and Moderns."[36]

Finally, it is remarkable that an examination of Duport's library catalogue indicates a person who possessed virtually no scholastic texts. Of more than two thousand volumes, only a dozen or so can be designated as "scholastic": in logic, the treatises of Marcin Smigleckius and Robert Sanderson; in natural philosophy, the works of Adrian Heereboord and Libertus Fromondus; in metaphysics, Fromondus again as well as Christoph Scheibler; and in moral philosophy, Eustachius of St. Paul's *Ethica*. Such a paucity of scholastic-Aristotelian texts confirms the priorities of a man who was, body and soul, a devotee and apostle of the *literae humaniores*. The classics, the learned languages and literature – including contemporary English literature – and noncontroversial theology were his passion, and it was these tastes that he tried to inculcate into his students. Certainly, the fact that so many of his students would either distinguish themselves as men of science, or be consumers of such literature, demonstrates the openness of mind displayed by Duport, who obviously made no attempt to stifle his students with philosophical quibbles, but rather allowed them to follow their own inclinations so long as they grounded themselves in the humanities. Equally telling, despite his passion for literature and the exclusion of books of scholastic philosophy from his shelves, Duport's library included quite a few "modern" scientific books. It would not be far-fetched to muse that Duport allowed his brilliant students – Barrow, Ray, and Willughby – to lure him deeper into the new science, since most of his scientific books were published (or purchased) after 1660.[37]

The purpose of this lengthy reappraisal of John Duport has been to demonstrate the inadequacy of applying simplistic categories to an evaluation of the mind and worth of a midcentury scholar. Duport was representative of an important segment of the Cambridge intellectual community who have been traditionally dismissed as conservatives inimical to change and opponents of the new science. Yet as our study of Duport illustrates, when – and if – these scholars criticized the new science, they did so in accordance with reasons and priorities different from those usually ascribed to them. In addition, Duport's preoccupation with the *literae humaniores* played an important role in the intellectual development of Barrow, whose love of languages and Greek and Latin poetry set him apart from the accepted modern image of the scientist, but was still very much a part of the seventeenth-century ideal of the general scholar; and it must be stressed that Barrow, like such scholars as John Bainbridge, John Greaves, and Edward Bernard, never embraced science as a vocation. Consequently, to understand the vision of science that animated Barrow and many of his contemporaries, the historian must first appreciate its place in a larger cultural framework.

Unfortunately, direct information concerning the course of studies pursued by Barrow is lacking. However, the survival of Duport's "Rules," in conjunction with Barrow's scholastic exercises and certain commonplace books and library catalogues of his contemporaries at Trinity – and Cambridge at large – shed much light on his probable course of study.[38]

The first half of the seventeenth century witnessed a decisive redirection of intellectual taste back to a more invigorated version of humanism than had existed after the English Reformation. This shift was characterized by a strong preoccupation with grammar and language. Not only was emphasis placed on the need to acquire a correct and stylized Latin in both discourse and writing, but the study of Greek was elevated to unprecedented heights. In addition, the period can also be seen as the golden age of the earnest pursuit of the Semitic languages – Hebrew and, to a lesser extent, Arabic – not to mention of the cultivation of the modern European languages. Such attention to language and grammar was accompanied by a revival of the other preoccupation of the Italian humanists – the pursuit of literary classics, rhetoric, poetry, history, and moral philosophy. These seventeenth-century English humanists, however, retained their interest and faith in logic, although it had now assumed the shape of a diluted Aristotelian logic, one modified by Renaissance dialectics.

The incoming student, who arrived at university from grammar school solidly grounded in grammar and language, was expected to devote much of the first two years of his university stay to strengthening the foundation hitherto laid in the ancient tongues. The ideal was to bestow upon him the capacity for reasoning and engaging discourse: to think clearly and orderly, and to express such thoughts, both orally and in writing, with precision and persuasiveness.

The actual exercises of the students reflected the common belief that style and language were the fruits of practice and that inventiveness and memory could also be thus cultivated and acquired. Duport repeatedly imparted to his students the desirability of selecting "the best Authors in every faculty," as well as the necessity of "double" translation of Greek and Latin as the tool with which to "better come to learn the Genius and Idiome of both Tongues." He specifically commended Virgil, Cicero, and Seneca among the Latin authors, and Homer, Plutarch, and Aristotle among the Greeks, as the best authors for both language and content. Other authors universally espoused and diligently studied included Horace, Ovid, Terence, and Quintilian, as well as Martial, Lucan, Plautus, and Juvenal. No less important was the prescription of modern authors who could supplement and enrich the classics either in style or in form and content. Thus, the older works of Laurentio Valla and Erasmus retained much of their original appeal, and among the admired recent Latin authors Lipsius, Barclay, Scaliger, and Heinsius were recommended. The most frequently assigned modern rhetoricians were the Continental Vossius and Caussinus and the English Farnaby.

As previously mentioned, during the seventeenth century heavy emphasis was placed on the study of Greek, without which, Duport informed his students, "never think you can be a Scholler indeed or well skilled in humane learning."[39] For most students, even those who had acquired a certain command of the language in grammar school, the assigned introductory texts included the Septuagint and the New Testament in Greek or Aesop's fables. Subsequently, students pursued "approved" authors, such as Homer and Demosthenes in poetry and oratory, respectively. Additional texts included Hesiod, Aristophanes, Theophrastus, Lucian, Plato, Isocrates, and the minor Greek poets. Increasingly, other disciplines, such as logic and moral philosophy, were made part of the study of Greek. Thus, students (such as Newton in 1661) were assigned such appropriate Aristotelian texts as the *Organon* or the *Ethics* in the original, often in a manner indicating that mastery over the language was deemed as important as acquisition of the content – an attitude that, we shall see below, was already successfully employed in the study of history. On a few occasions Barrow commented on Duport's Greek instruction. In 1654, for example, he praised highly the benefit reaped by students from Duport's reading of Plato and Aristotle, as well as of the poets, historians, and philosophers in Greek – praise he was to repeat a few years later upon his appointment as Regius Professor of Greek. Of his own studies, two commonplace books he compiled c. 1648 have survived. One contains excerpts from various Greek tragedies and comedies; the other, a similar collection from Greek historians.[40] One of Barrow's colleagues at Trinity, Walter Needham, also provides testimony of his reading of Aristotle, Ptolemy, and Sophocles in Greek.[41]

Nothing approaching the proficiency he acquired in Greek did Barrow attain in the Semitic languages. Duport, indeed, demanded that his students read the scriptures in their original tongues, claiming that they could "not see clearly into Gods word without the two eyes of Greeke and Hebrew." Yet, unlike the study of Greek, which was sanctioned both for the purpose of reading the scriptures in the original and for the joy and profit to be had from profane authors, the study of Hebrew was strictly confined to a single purpose: the ability to read the Old Testament in Hebrew. This motive was strong enough to recommend the language to most students – or their parents – and we find a Hebrew Bible, the Psalms, as well as a Hebrew grammar and dictionary in the catalogues of most Cambridge undergraduates regardless of social class. (Not a few owned considerably more.) Nevertheless, the majority of these students never acquired more than a superficial familiarity with this difficult language. Barrow himself only rarely referred to Hebrew. In an oration he delivered in 1654 he exclaimed that proficiency in the language, which "once seemed beyond human skill without diabolic assistance," was now attained even by undergraduates. His own library catalogue contained half a dozen Hebrew books, including a Bible, the standard works of Johann Buxtorf, and a Talmud.

Arabic, we are informed by Barrow's biographer, was the only discipline to have defeated him. Calamy supplied the anecdote that as an undergraduate, Barrow and a fellow student of Trinity, Samuel Sprint, went to Abraham Wheelock, the Adams Professor, "to discourse with him about the Arabicke Language, which they were desirous to learn. But upon hearing how great Difficulties they were to encounter, and how few Books there were in that Language, and the little Advantage that could be got by it, they laid aside their Design."[42] Possibly the ailing and impoverished Wheelock did not wish to be bothered; and Barrow may not have acquired even the rudiments of the language, not even later during his stay in Turkey.

Finally, the study of modern European languages had reached new heights by the middle of the seventeenth century. In his 1654 oration, Barrow enthusiastically commented on the great increase in the cultivation of French, Italian, and Spanish among Cambridge students, and we know that he himself was well versed in at least the first two. Many of his friends and colleagues at Trinity left some evidence of their active pursuit of such studies. Thus, one of Barrow's closest friends both at Trinity and later, John Mapletoft, was proficient in all three of the languages, while another, Paul Rycaut, was not only versed in French and Italian, but capable of assisting his fellow students in the study of Spanish, which he knew from childhood. Similarly, John Nidd owned some fifty volumes in French and Italian, attesting to his proficiency in those languages, and other students, such as John Dryden and Gilbert Havres, left Trinity College endowed with similar attainments.

The study of history was as integral to the humanistic undergraduate education as poetry and, at least initially, was part of Greek and Latin studies. Students were often advised to begin by reading the epitome by Florus or the equally popular history by Justinus. Thereafter, they proceeded, more or less systematically, to the study of Thucydides, Polybius, Herodotus, Plutarch, and Xenophon - and to a lesser extent Diodorus Siculus and Dion Cassius[43] - among Greek authors, and Livy, Tacitus, Suetonius, and Sallust among Roman historians. Nor were modern authorities neglected. Some of the sixteenth-century authorities, such as Bodin and Sleidan, retained their popularity, alongside more contemporary commentators such as Vossius and Cluver. Among English commentators of universal and Roman history, Cambridge students usually read Thomas Godwin, Sir Walter Raleigh, and Degory Wheare. Increasingly popular were the "modern" histories, whether the histories and antiquities of England, the likes of William Camden, Sir Francis Bacon, and Lord Herbert of Cherbury, or the variety of French, Italian, Dutch, and German histories that became widely available and were to be found on the shelves of all students.

"Chronology and Geography, were two inseparable Sisters, and the two Eyes of History, without which she must inevitably be either Blind or very Obscure."[44] This piece of common wisdom quoted by Thomas Hearne from one of Gerardus Vossius's popular works was widely appreciated and practiced by early modern authors of geographical and chronological textbooks, college tutors, and their students,[45] and Duport imparted a similar lesson: "In reading an History know when, & where you are, I mean carry the Chronology & Geography along with you, or else you will miserably loose your self."[46] With this objective in mind, the students were advised to purchase maps, a good atlas, and a variety of ancient and modern treatises of geography. Chief among these aids were the ever popular works by Mercator and Ortelius, joined by more recent commentaries by Fournier, Varenius, Heylin, and Cluver, while Scaliger, Petavius, and Calvisius provided the necessary groundwork for the study of chronology.

Obviously, the universality aimed at, and the mastery attained, in the domain of the arts curriculum could not be matched in the realm of mathematics. Then, as now, it was wishful thinking to expect more than a minority of scholars to seek - let alone obtain - more than a tolerable proficiency in mathematics. To be on the cutting edge was rarer still.

Many would have concurred with the common wisdom of Thomas Fuller, who believed that the general scholar should study mathematics but moderately, and certainly not allow "it to be so unmannerly as to justle out other arts." Others would have favored the beliefs of Henry Hammond, who viewed the study of mathematics as a pleasant divertisement: "History and Geometry, and the like, go down pleasantliest with those which have no design upon Books, but only to rid them of some hours." Still others would have found themselves in agreement with Richard Baxter, who was totally disinclined toward mathematics and "never could find in [his] heart to divert any studies that way."[47]

Nevertheless, the study of mathematics was an integral part of the Cambridge undergraduate curriculum. Most students would have encountered mathematics in their second or third year at university, the course consisting of arithmetic, geometry, and optics; astronomy would be taught either independently or as part of natural philosophy. Again, as was the case with other disciplines, the students were neither expected nor encouraged to specialize in mathematics; such a stage could come only after the student had completed the cycle of studies and graduated B.A. It is not surprising, therefore, that evidence of sustained study of mathematics almost invariably came simultaneously with graduation, when those students who had the inclination could turn to it their undivided attention.

However, even for those who had delighted in mathematics when young, their later mandatory application to theology usually meant the cessation of mathematical studies. For example, late in life Joseph Mede told one of his correspondents that in his younger years he had "studied the Grounds of those Sciences, but ever since neglected them." His disciple, Henry More, who had consolidated a reputation as a mathematician during the 1640s, expressed a similar sentiment. Writing to Ann Conway in October 1664, More said, "I have renewed my acquaintance with such Mathematicall Theorems as I was in some measure conversant in before I fell a Theologizing, which was so long, that I had almost forgott all that little I knew in Geometry."[48]

Barrow probably began studying mathematics in earnest in 1648 or 1649. This was a particularly opportune moment, for in November 1648 John Smith of Queens' College embarked upon the teaching of mathematics in the lectureship just founded by Sir John Wollaston. Although there is no concrete evidence that Barrow attended Smith's lectures, he referred affectionately to the memory of the Cambridge Platonist in a later oration. Significantly, Smith almost certainly chose to lecture on Descartes's *Geometry*. This conclusion may be drawn from John Wallis's later use of his correspondence with Smith during November 1648 on the *Geometry* to claim his independent discovery of a method of resolving cubic equations. Unfortunately, apart from telling us that the "professor of mathematics" at Cambridge desired his assistance in clarifying certain obscurities in Descartes's text, Wallis reveals nothing of the content of Smith's lectures. However, judging from the catalogue of Smith's library – which was bequeathed to Queens' College upon his death in 1652 – his interest in mathematics in particular, and science in general, was considerable. Among the authors most relevant to the teaching of the new mathematics were Descartes, Galileo, Oughtred, Archimedes, de Billy, Cavalieri, Mersenne, and Van Schooten.[49]

Smith, in fact, may not have been the first to expound the new analysis at Cambridge. In 1642 another of the Cambridge Platonists, Ralph Cudworth, had been a university lecturer in mathematics. Again, nothing is known of the content of his lectures, but some evidence suggests that he was a highly regarded mathematician, so much so that he was recommended to John Selden as the person most qualified to assist the venerable scholar in the complicated astronomical and chronological studies he was engaged in. A copy of Cudworth's library catalogue confirms his deep interest in the mathematical sciences, which lasted at least throughout the 1640s and 1650s. Among his books were the 1637 edition of the *Geometry,* as well as Schooten's Latin edition. Cudworth also possessed the works of Briggs, Gellibrand, Harriot, Viète, Cavalieri, Torricelli, de Billy, Wallis, and Galileo. Certainly, both in the early 1640s as well as later, Cudworth would have been in a good position to encourage the studies of interested students.[50]

Insight into the impact made by Descartes's mathematics at Cambridge in the 1640s is gained from the reaction of an amateur like John Hall, who was pupil of the mathematically inclined John Pawson at St. John's as well as an acquaintance of Henry More. Hall hoped that a new philosophy could be erected on the foundations of the new mathematics:

> But yet methought that there was somewhat in them which was yet hid from us, and that the ancient founders of these sciences had been content to retaine somewhat not fully discovered. For I found most men imployed onely in learning those immense heaps of Demonstrations *they* [the ancients] had left us, but seldome enlarging them or going forward, which made me fear the key of these Sciences were hid, and that without such a key, or engine it had been impossible to reare up such a huge superstructure of vast consequences. But this I found two or three great spirits had already light on, and had directed a way which if well followed, will make our *Mathematicall* reason nimble and apt to finde the fountain head of every *Theoreme,* and by degrees, as we may hope, inable us to the solution of any *Probleme* without any more assistance then pen and inke... and direct us to more exact and easie instruments then any have been yet knowne, and recall mens minds by delicate ravishing contemplations, from the sordid jugling use of those instruments on which they now so perversly and unanimously doat.[51]

Apart from Smith's lectures, Barrow could also have benefited from the in-house mathematical lectures at Trinity College. In particular, two fellows intruded in 1645, Nathaniel Rowles and Charles Robotham – university lecturers in mathematics in 1648-9 and 1652-3, respectively - could have encouraged him. A connection with Robotham is documented, for the latter contributed verses to Barrow's edition of Euclid. Moreover, as we shall see shortly, the two fellows, and especially many of their students, were part of the scientific community that flourished at Trinity during the late 1640s and 1650s.

Unfortunately, only fragmentary evidence of the nature and content of the study of the mathematics at Cambridge during the Interregnum has survived. What little evidence we possess on Barrow's own studies will be discussed in the context of his first publication. At this point, however, something should be said of Barrow's major efforts to encourage and strengthen the teaching of mathematics at Cambridge.

Two main problems faced all those who sought to cultivate the mathematical sciences: to find young and eager students who possessed both aptitude and inclination for such studies, on the one hand, and to find prospective patrons willing to bestow the necessary funds to establish these sciences on a solid footing, on the other. Barrow himself commented on the first difficulty in his 1651 oration when he lamented how the mathematical sciences linger in obscurity, since so few students venture "to get below the outer surface of divine Geometry and none descend into its solid and profound secrets...[and] two at the most may be engaged in following the clews to the dark and involved labyrinths of Numbers." A few years later Barrow's friend and colaborer in the study of mathematics, John Ray, repeated the point in a lecture he delivered at Trinity: "I am sorry to see...that those ingenious Sciences of the Mathematicks are so much neglected by us: and therefore do earnestly exhort those that are young, especially Gentlemen, to set upon these Studies, and take some paines in them."[52]

Barrow, however, did not allow these factors to deter him from relentlessly pursuing his own mathematical studies as well as encouraging those of similar inclinations to follow suit. Certainly, his own considerable knowledge of, and enthusiasm for, ancient and modern mathematics, combined with his willingness to assist those eager to learn, made it possible for him to draw not a few people to the study of mathematics. Gilbert Clerk was not exaggerating, therefore, when he told Newton in 1687 that he and Barrow "contributed neare 40 yeares since, as much or more than any two others, (to speake modestly) in *diebus illis,* to bring these things into place in ye university."[53]

Barrow, who never boasted of his personal accomplishments in mathematics, made an exception in the case of teaching and was particularly proud of his contribution to the cultivation of mathematics among Cambridge students. As early as 1654 he was pleased to see that the "names of Euclid, for instance, Archimedes, Ptolemy, Diophantus, once so ghastly, many of you can now hear without tingling ears"; even more, that "many

brave men have triumphed over that horrific monster named Algebra."
A decade later, in his inaugural oration as Lucasian Professor, Barrow
spelled out more clearly his contribution to Cambridge mathematics:

> While I was a private Person nor otherwise obliged, being
> enamoured only with the Loveliness of the Thing, I showed
> such hearty Desires and Endeavours to have these Sciences in
> the highest Degree recommended to you; it cannot now be
> doubted but by Reason of my publick Office, and more solemn
> Engagement, I will more diligently apply myself to their
> Promotion according to my slender Ability, since what was
> then an Inclination becomes now a Duty. But what the Laws
> do strictly require of me . . . I have always shewed the greatest
> Readiness to perform.[54]

Cambridge intellectual life

Barrow was elected fellow of Trinity in 1649 on merit alone, as
"nothing else could recommend him who was accounted of the contrary
party." Certainly, Barrow never disguised his convictions. He may even
have been a ringleader among the young Royalists in the college, for on
27 March 1648, Barrow, Paul Rycaut, and Thomas Jolley received an
"Admonition, tending expulsion, for the rude behaviour, upon the 24 of
the same Month after Supper." Since 24 and 27 March were the ascension
days of James I and Charles I, respectively, the nature of the offense seems
obvious.[55] A few months later, in January 1649, Barrow actually subscribed to the Engagement, but immediately afterward rushed back and
had his name struck from the book. Nevertheless, Barrow was spared
expulsion. Thomas Hill, the master, obviously protected him. Perhaps
this was the occasion when Hill laid his hand over Barrow's head, saying,
"Thou art a good Lad . . . 'tis pity thou art a Cavalier."[56] A couple of years
later, on 5 November 1651, Barrow managed again to infuriate the Puritans. This anniversary day of the Gunpowder Plot was effectively used
by Barrow to pronounce his deep-rooted sympathy with the house of
Stuart, under the disguise of delivering an anti-Catholic speech. Again
the master came to Barrow's defense; he responded to those fellows who
demanded Barrow's expulsion by insisting, "Barrow is a better Man than
any of us."[57]

The zeal Barrow displayed in defending the cause of the monarchy and
the Anglican church soon brought him into the orbit of Henry Hammond, who did more than anyone to keep the church alive during its
period of crisis. By 1651, if not earlier, Barrow had become a beneficiary
of a fund that had been established by Henry Hammond to assist deprived Anglican ministers and that was evidently expanded to include

young, promising, and loyal scholars such as Barrow, Peter Gunning, and Richard Allestree. In return, Barrow appears to have been active during 1651–2 in the cause of the "underground." The nature of his activities was twofold. First, he was employed in some teaching capacity for the sons of loyalists. In January 1651/2 Hammond expressed the hope that Barrow would "continue [in] his employments till he had been forbidd[en] particularly," and as late as the following August Hammond still discussed "the agreement [that was] made for Mr. Barrow to undertake the charge of teaching for this winter at least." However, no more is heard of this subject. Second, he was to serve as go-between for the more senior Anglicans – Hammond, Matthew Wren, Gilbert Sheldon, and Robert Sanderson – delivering letters, news, and books, as well as manuscripts for the press.[58]

Barrow's employment by Hammond seems to have ended by 1652, and Barrow returned to the life of a scholar. In the meantime the political atmosphere at Trinity had been changing rapidly. As we had occasion to speculate earlier, by the late 1640s Thomas Hill had relented in his reforming zeal. The execution of Charles I and the growing alarm at the rise of the Independents may have further contributed to Hill's patronage of Barrow, in particular, and to his toleration of a subtle shift toward Royalism in the college in general. By the early 1650s, some of the fellows who had been intruded in 1645 – such as John Templar, Alexander Akehurst, and perhaps Charles Robotham – as well as some of the more recent fellows had become more amenable to Royalism. An indication of this slow shift may be discerned in some of the events of 1653–4. Akehurst, who was vice-master in 1653, "allowed one of his students, Nicholas Hookes, to dedicate to him an openly Royalist volume of poems, the *Miscellanea Poetica*." A few months later, on 4 July 1654, Akehurst was apprehended, and a committee – which included Lazarus Seaman, master of Peterhouse and the present vice-chancellor; John Arrowsmith, the new master of Trinity; and Anthony Tuckney, master of St. John's – was appointed to investigate the charges that Akehurst had uttered "some horrid blasphemous expressions... to the dishonour of God and Scripture." The exact nature of his crime is not clear. Oliver Heywood, who had been Akehurst's pupil, accused him of "having grievously apostasized, becoming a common Quaker"; but it is almost certain that it was the vice-master's Royalism that precipitated the action, precisely during the period during which the university was under a new attack by religious enthusiasts. James Jolley, Thomas's brother, wrote a letter in defense of Akehurst, who was eventually released without punishment on 1 March 1654/5, since no charges had been brought against him during the eight months of his incarceration. Akehurst left Cambridge, but indicative of the new political mood of the college was the almost immediate election of an even more pronounced Royalist, James Duport, as vice-master by

the senior fellows only a short while after Duport was forced to relinquish his professorship of Greek![59]

During this period Barrow was slowly consolidating his reputation as
a scholar. A few months before his Gunpowder speech, he was elected
moderator in the schools, a clear indication of the wide acknowledgment
of his intellectual prowess while still a bachelor of arts. Barrow's speech,
delivered 30 April 1651, is important as an early example not only of his
accomplished style and wit, but of the views and beliefs that would guide
him for the next two decades. Already we find a covert reference to "tri-
umphant barbarism," the unfortunate effects of the Puritans coming to
power on the free and serious cultivation of the arts and sciences. Barrow
laments the paucity of students willing to cultivate the most demanding,
and least rewarded, disciplines – citing specifically the mathematical sciences and oriental languages. Equally disenchanted with the predilection
of the godly to elevate "true" divine knowledge over scholarship, he snipes
at those who cultivate Hebrew "with blind devotion rather than known."

More important still was Barrow's first articulation of his faith in the
search for truth. Truth, he asserted – and would repeatedly assert in future years – was the object and goal of all studies and inquiries, wherever
one might find it. Though he hoped to see Aristotle's authority endure,
yet "the yoke of truth alone" should reign supreme. In fact, his praise for
Aristotle was intended to be a preparation for the warning he would voice
against excessive disparagement of the past in the search for truth. He
cautioned his listeners to allow "no sectarian zeal, no wilfulness of opinion, no love of novelty, no prejudice of antiquity hinder the free course
of Truth." In such opinions, Barrow revealed himself to be a true disciple of Duport. Like his mentor, he insisted that the ancients deserved
honor and credit, for they "led the way, borne on the torch, discovered
the arts and wisdom, and pointed out the paths of Science." Similarly, he
condemned novelty for novelty's sake. However, he went beyond Duport
in articulating the possibility, even desirability, of discriminating between
past contribution to the humanities and to the sciences. As he put it,
even those

> unpleasant Scholastics do not deserve to be laid aside alto-
> gether: they are not fruitless, nor without utility, though neither
> ornamental nor spruce.... None of them may have erected
> much in the way of experimental science but rather elaborate
> structures on flimsy foundations. They may, in short, have
> contrived little out of Nature's bowels, though many subtilities
> out of their own. Yet, however seldom they may have tracked
> down Truth, let them serve you as a whetstone.

To expose error, Barrow seemed to be saying, was secondary only to discovering truth. And the scholastics were capable of instructing contem-

poraries in the art of reasoning and disputing, of resolving doubts and refuting fallacies.[60]

If in his capacity as moderator Barrow was careful to emphasize the dual function of a true university – the transmission of past knowledge and the advancement of its boundaries – when the time came for him to deliver his M.A. oration, he made the nature of his scientific beliefs clear. However, in order to appreciate Barrow's views on Descartes and the new philosophy both in 1652 and later, it is necessary to survey the fortunes of Cartesianism in Cambridge from the 1640s to the 1660s.

The significant, though not conclusive, role played by Henry More in spreading Cartesian ideas in Cambridge is well documented.[61] He himself described his initial enthusiasm for the writings of Descartes in words that must have echoed an enthusiasm experienced by many contemporaries in the 1640s: "At the first sight of Cartesius, I confesse that I was mightily taken with the design; as having so mechanicall a head of my own, that I alwais took more delight in Mechanicall experiments and deductions from them, than in Metaphysicall speculations."[62] Thanks to More, other young students at Christ's College were converted. Henry Power, for example, told Sir Thomas Browne in 1648 that "the two Incomparable Authors, DesCartes & Regius . . . were the only two that answer'd [his] doubts & Quaeres in that Art [anatomy]."[63] Other students, such as Thomas Baines, Thomas Plume, Thomas Leigh, and Thomas Burnet, studied (or taught) Cartesian philosophy during the following years.[64] In other colleges, too, the ideas infiltrated rapidly. At St. John's, for example, John Hall became acquainted with, and quickly a convert to, Cartesian ideas shortly after his arrival at Cambridge in 1646. At Trinity, John Ray was studying Descartes by 1650 at the latest, and John Felton of Gonville and Caius College became the author of a very popular manuscript epitome of Descartes. More certainly did not exaggerate, therefore, when he informed the French philosopher that "there are those among the English who hold you and your works in great esteem."

By the 1660s Cambridge was seen, by some people at least, as a Cartesian stronghold. In 1663 Gilbert Burnet visited Cambridge, and his impression was that the "new philosophy was . . . much in all people's discourse, and the Royal Society much talked of." John Beale was informed in 1669 that his alma mater "is entirely Anti-Calvinian, & Anti-Aristotelian, generally Cartesian, a great growth there in Mathematics, & not without a Club of Phylosophicall Chemists."[65]

However, hand in hand with the rapid spread of Cartesian ideas in Cambridge there evolved some serious concerns over the implications of the Frenchman's ideas for both religion and morals, partly because of Descartes's metaphysics and excessive rationalism, but more seriously because of the alarm raised by many Englishmen following the publication of Hobbes's major works during the 1650s. Within a short time everything involving matter and motion would be viewed as Hobbist. And though many men of science sincerely attempted to differentiate between the two philosophers – and even between Descartes's natural philosophy and his metaphysics – the critics were not at all convinced. Cambridge Puritans of the 1650s, like their Anglican successors in the following decades, genuinely feared the pernicious effects that the general attraction to Descartes's ideas would have on the minds and souls of students.

Already in 1656 the very puritan master of Emmanuel, William Dillingham, "publicly asserted the need to combat the spread of Cartesian ideas and urged reverence for the Word as against a preoccupation with 'naturall reason.'"[66] A younger Puritan holding similar views was Thomas Arrowsmith, a kinsman of the master of Trinity College, who was mocked in 1656 for espousing similar views in a disputation, although, insinuated the anonymous author, such zeal went a long way in obtaining Arrowsmith a fellowship:

> Sr Arrowsmith prooved that the earth stood still;
> Confuting Des Cartes by Zabarell.
> But by his good will he disputed but ill
> For him we know that the World goes well.[67]

A year later, George Lawson, a former member of Emmanuel, published a confutation of Hobbes, his *Examination of the Political Part of Mr Hobbes, His Leviathan,* the writing of which was provoked, he explained, because "many gentlemen and young students in the universities were revealing a distressing tendency to take the Leviathan as a 'rational piece.'"[68]

After 1660 criticism of Cartesianism intensified mainly because of the belief that it underpinned Hobbism. Again, the main concern was the attraction of fashionable Cartesian ideas for undergraduates, who lacked "proper judgment" to the detriment of their character and religion. Thus, when John Covel of Christ's delivered his M.A. oration in 1663, he complained about how the incoming undergraduates absorbed Descartes and Gassendi with the alphabet, the upshot being a blind criticism of Aristotle. Undoubtedly, Covel's concern was Hobbes, not so much Descartes; like Duport, he lamented the uninformed scorn for Aristotle, who would have provided an effective defense against Hobbes. Similarly, Roger North recalled that when he had studied the "heretic" Descartes in 1667, most of the young students were invariably inclined that way.[69]

The combined effect of the serious concern for religion and moral philosophy, and the alarming attraction of the "younger sort" to Cartesian ideas, prompted university authorities to strike out against such ideas. In November 1668, shortly after his appointment as vice-chancellor, Edmund Boldero issued a decree prohibiting sophists and bachelors of art from basing their disputations in the schools on Descartes, stipulating the use of Aristotle instead.[70] This injunction, it must be made clear, came in the wake of the Scargill Affair, in which a young fellow of Corpus Christi College, Daniel Scargill, was accused of publicly espousing Hobbist ideas. The reaction was particularly strong since Scargill was reported to have always been surrounded by undergraduates. As a consequence, Scargill was expelled and allowed to return only after the intervention of Archbishop Sheldon and a humiliating act of repentance.[71]

Timothy Puller, a fellow of Jesus College – of whom Boldero was master – provides us with a clue to the mind of the vice-chancellor. Writing to Worthington in late November 1668, Puller said: "Dr. Boldero is mightily busied in reformations, putting down Ale Houses, and the like. This week, he did an exemplary piece of justice in suspending [Daniel Scargill] . . . for holding Hobb's opinions in publick disputations,"[72] and Worthington himself was concerned over the same issues. In fact, only the previous year he had written a letter to Henry More, whom he credited with introducing Descartes's ideas to England, requesting the Cambridge Platonist to compose a sound textbook of natural philosophy that would eliminate the attraction of students to the exciting, yet pernicious Cartesian texts:

> I wish you would refresh and joy me with the news of your being about a Body of Natural Philosophy. . . . [Y]ou have as highly commended Des Cartes, as is possible, and as knowing no better method of Philosophy, you recommend it effectually in some parts of your books, whereby you had so fired some to study it, that your Letter to V.C. (which came long after) could not coole them, nor doth it yet: but they are enravisht with it, and derive from thence notions of ill consequence to religion, you being sensible of this great evil, as also of the uncertainty of his Philosophy throughout (as you intimated to me) you cannot but think, how much it concerns you to remedy this evil. And seeing they will never return to the old Philosophy, in fashion when we were young scholars, there will be no way to take them of from idolizing the French Philosophy, and hurting themselves and others by some principles there, but by putting into their hands another Body of Natural Philosophy, which is like to be the most effectual antidote.[73]

More never attempted such a task. However, his own growing disenchantment with Cartesianism began to take root by the early 1650s, and it flowered into open hostility during the following decades. Yet More never ceased to teach Descartes's natural philosophy, partly because he was able to distinguish between Cartesian science and metaphysics and partly, as Gabbey pointed out, because he realized the wisdom of he himself exposing the "impious implications of universal mechanism," rather than allowing students to encounter Descartes's writings unprepared and unaided.[74]

This background is necessary for our understanding of Barrow's own progress from an enthusiastic, if discerning, advocate of Cartesianism while a student to an opponent after the Restoration. By the 1660s for him, too, the advent of Hobbism signified the need to counter the rationalist and materialist implications for religion, and subsequently his efforts to confute Hobbes spilled over to Descartes as well. Still, in the end, he would retain more respect for Descartes as a mathematician and a philosopher of nature than would More.

While an undergraduate, however, Barrow had been among those who quickly converted. His biographer tells us that his "judgment was too great to rest satisfied with the shallow and superficial Physiology then commonly taught and received in the universities" and that "he applied himself to the reading and considering the Writings of the Lord Verulam, Monsieur Descartes, Galileo and the other great wits of the last age, who seemed to offer something more solid and substantial."[75] The fruits of his careful reading of a wide range of modern sources are indeed displayed in his exercises toward the M.A. degree. The sterility of natural philosophy after Aristotle, he begins his 1652 oration, has changed of late owing to the proliferation of alternative philosophies of nature, whether ancient or modern. He specifically mentions Gassendi, Magnenus, Digby, Boulliau, and Roberval among those who revived ancient theories, while Telesio, Campanella, Gilbert, and various chemical philosophers are "amongst those who struck out something new." However, it is Descartes "who seems, not undeservedly, to shine with special pre-eminence among these philosophers." Descartes, says Barrow, has given us the best philosophy to date by virtue of his

> extraordinary skill as a mathematician, his unwearied thought concerning the nature and use of meditation, his judgement stripped of all prejudice and disengaged from the snares of popular error, equipped moveover with a great deal of well devised empirical work...not to mention his incomparably acute apprehension and his excellent ability, not only to think clearly and simply but also to explain his own mind fully and lucidly in the fewest possible words.

Such praise, however, is mitigated by a sensitive criticism of some important Cartesian features. He faults Descartes for his extreme rationalism and defiant dogmatism. The source of Descartes's error, argues Barrow, is that

> it seemed good to him, not to learn from things, but to impose his own laws on things. ... first, he collected and set up metaphysical truths which he considered suitable to his theory from notions implanted in his own mind...next, from these he descended to general principles of Nature; and then generally

advanced to particulars explicable from principles which, forsooth, he had framed without consulting Nature.

But Barrow criticizes Descartes for more than the Frenchman's neglect of an experimental study of nature, pointedly contrasting him with Bacon and Gilbert. He is equally concerned with the grave religious implications of Cartesianism rooted in the banishment of the spiritual and immaterial. Anticipating the criticism of Henry More and Ralph Cudworth, Barrow expressly states that in this respect the ancient writers, including Aristotle, are more satisfying, for they believed in spirit and soul:

> He thinks unworthily of the Supreme Maker of things who supposes that he created just one homogeneous Matter, and extended it, blockish and inanimate, through the countless acres of immense space, and, moreover, by the sole means of Motion directs these solemn games and the whole mundane comedy, like some carpenter or mechanic repeating and displaying *ad nauseam* his one marionnettish feat.[76]

Barrow's university orations serve as a barometer of both the rapid spread of Cartesian ideas in Cambridge during the late 1640s and the 1650s and the subsequent alarm at the implications they raised. They also offer an accurate indication of the dramatic increase in the number of university men actively engaged in scientific investigation during the same period. Historians have studied the growth of science at Cambridge with less rigor than they have the parallel growth at Oxford, in part because it was the Oxford group of scientists that established the model on which the Royal Society was founded in 1660 – and Oxford men constituted almost exclusively the active core of the new society – and in part because nothing like the spectacular arsenal of scientists appears to have existed at Cambridge. As Shapiro rightly points out, however, much of the reason for this difference was the fact that Cambridge "seems to have been put at a disadvantage largely by the absence of good chairs comparable to the Savilian professorships, and of good recruiters comparable to John Wilkins."[77] Nevertheless, the political and religious background underlying the intellectual developments at both places was very much the same, and before an attempt is made to present some of the evidence for the existence of a flourishing scientific community at Trinity College during Barrow's student days, it is important to comment on the effect exerted by Puritanism on the intellectual life at Cambridge.

This is not the place to narrate in detail the attitude of Puritanism toward "secular" knowledge, scientific knowledge included. Suffice it to say that even before the Civil War the predominant interest of Puritans in and out of the universities was the creation of a preaching ministry; a disinterested love of scholarship was rarely relevant, let alone encouraged. This predilection was commented upon both by Puritans who appreciated learning and by their opponents. In 1628, for example, John Milton wrote a letter to his former teacher at St. Paul, Alexander Gill, bitterly complaining of the intellectual attainments of some of his fellow students at the very Puritan Christ's College:

> There is really hardly anyone among use...who, almost completely unskilled and unlearned in Philology and Philosophy alike, does not flutter off to Theology unfledged, quite content to touch that also most lightly, learning barely enough for sticking together a short harangue by any method whatever and patching it with worn-out pieces from various sources – a practice carried far enough to make one fear that the priestly Ignorance of a former age may gradually attack our Clergy.[78]

Similar sentiments were shared by William Laud, archbishop of Canterbury who, a few years later, blasted what appeared to him to have been an exclusive preoccupation of the godly with Calvinist theology even while undergraduates: "A man would think those two [probationary] years" before scholars of the Winchester Foundation were elected fellows at New College, Oxford,

> and some years after, should be allowed to logic, philosophy, mathematics, and the like grounds of learning, the better to enable them to study divinity with judgement: but I am of late accidentally come to know, that when the probationers stand for their fellowships, and are to be examined how they have profited, one chief thing in which they are examined is, how diligently they have read Calvin's Institutions. . . . I do not deny but that Calvins Institution may profitably be read...when they are well grounded in other learning; but to begin with it so soon, I am afraid doth only hinder them from all grounds of judicious learning, but also too much possess their judgments before they are able to judge.[79]

The criticism leveled by both Milton and Laud was aimed at the prevailing attitude among many Puritans that sought, if not to disparage humane learning outright, then to subdue it to religious ends. Puritan scholars were often reminded that their duty was to spread the Gospel by preaching and that indulgence in "pure" learning bordered on the sin of pride. Such was the message of William Perkins, who, articulating the official position among his peers, reminded his audience that "the end we aime at, is not humane nor carnal; our purpose is to save souls." Students, he continued, should not linger on "too long in their *speculative* courses," but rather "yield their service to the Church" as soon as they were but "competently furnished with learning, & other qualities befitting" the calling of the ministry. The latter aspect referred to the edification meetings so often recommended for the godly. Thus, from the time the celebrated "prophesying" sessions were instituted during the 1580s, through the formation of the devotional and biblical study groups for young students, moderate learning and extreme piety remained the predominant combination among Puritans.[80]

Thomas Wadsworth, who matriculated from Christ's College in 1647, provides us with a good illustration of the system in full gear during the Interregnum. While at Cambridge, he associated "with an honest Club of Scholars of his own and other *Colledges,* as were not only daily conversant in *Philosophical* Excercises, but did frequently meet to promote the *great business of real godliness and growth in grace.*"[81] The dominant element in such groupings was not only their solidarity and commonality of purpose, but their implied exclusionist attitude. After 1640, the failure to be counted among the godly at Cambridge could be detrimental to one's career. A story is often told of how John Wilkins, shortly after his appointment as master of Trinity, was confronted with a delicate situation in which some of the "godly" in the college attempted to obstruct the election to a fellowship of Robert Creighton, one of Duport's students and a future Regius Professor of Greek; he was accused of profanity, irreligion and "'twas added that he never came to their private (prayer) meetings."[82] By the late 1650s, as we saw previously, the religious atmosphere at Trinity had shifted away from sectarian Puritanism, and Wilkins was able to foil such partisanship easily; yet in previous years matters at Trinity (and Cambridge at large) had been different, as Barrow himself was witness on those occasions when he himself faced the prospects of expulsion.

The coming of the Puritans into power and the ejection of a large number of Cambridge fellows aroused much concern over the possible effects of the new insistence on godliness in learning. Not only was it argued that by and large the intruded fellows – even if well qualified – were chosen for reasons other than learning, but many feared the implications of the suspicion inherent in Puritanism for secular knowledge and its autonomy, as well as for standards of worth and promotion at the university.

A Royalist propagandist may indeed have exaggerated the case when he argued that "the garland had been torn from the Head of Learning and placed on the dull brows of Disloyal Ignorance," but his premise held true. Thomas Fuller certainly concurred when he argued that the intruded fellows were "short of the former in learning and abilities [though] they went beyond them in good affection to the parliament." Even John Hall, a self-appointed propagandist for educational reform, concluded that the ejections were largely the result of an intent to satisfy political goals and thereby "removed many persons of a more thriving and consistent growth in learning, then it either left there, or planted in their steads."[83] A contemporary of Barrow at Trinity further substantiated this view.

Oliver Heywood, who came up shortly after Barrow, on 12 June 1647
(graduating B.A. in 1650), censured himself for adhering to the same
practice of theology instead of philosophy:

> All the time I was in the university my heart was much deadned
> to and in philosophical studys, nor could I as I desired apply
> my mind so close to humane literature, tho I blame my selfe
> for it . . . but my time and thoughts were most imployed in
> practical divinity, and experimental truths were vital and vivi-
> fical to my soul. I preferred Perkins, Bolton, Preston, Sibs far
> above Aristotle, Plato, Magyrue and Wedreton tho I despise no
> laborious authors in these inferiour subservient studys.[84]

Small wonder, then, that Thomas Smith, a colleague of Henry More at
Christ's College, could write Samuel Hartlib:

> Truly Sr, it grieves me to see (wt I should not have beleeved
> had I not been ye witnes) yt this University is so far degen-
> erated. . . . A generall opinion hath bespread or young scholars
> (I wish I might not say or Felowes too) yt humane learning
> is no way profitable, much lesse necessary to Divinity, wch
> they account ye only knowledge. For great Scholars, they tell
> me, scarce one of 10 goe to heaven, & therefore we should
> rather avoide than strive for that wch will probably encrease
> our condemnation.[85]

More research is needed to gauge fully the impact of the religious zeal,
which was never sufficiently harnessed in view of the continuing radicali-
zation of a small, but vocal segment of the Puritan camp, always ready
to turn against the universities, as events in 1653–4 and 1659 would demonstrate. Suffice it to say that sufficient evidence exists to demonstrate
that various members of the university could be – and were – deterred
from pursuing "prophane learning." Barrow himself, we noted before,
censured the impediment to the pursuit of true and free knowledge. One
of his colleagues at Trinity, Walter Needham, actually encountered such
an impediment, when his devotion to experiments was judged excessive
by Thomas Senior, a fellow of Trinity who would be ejected after the
Restoration and who not only reproached Needham personally, but apparently complained of him to Richard Busby, the youth's former headmaster at Westminster.[86]

Be this as it may, many Royalists abandoned the choice of a career they
otherwise would have followed – be it one of church or state – and consequently found themselves with "free" time to pursue "secular" studies.
Even those who opted for the only viable profession open to them, medicine, found themselves engaged in the pursuit of scientific studies. Many
contemporaries commented on the situation. In 1657, for example, Walter
Charleton wrote:

Our late Warrs and Schisms, having almost wholly discouraged
men from the study of Theologie; and brought the Civil Law
into contempt: The major part of young Schollers in our Universities addict themselves to Physick; and how much that
conduceth to real and solid knowledge, and what singular
advantages it hath above other studies, in making men true
Philosophers; I need not intimate to you, who have so long
tasted of that benefit.[87]

Robert Sparkling concurred: "For *Cambridge* can never forget, that when
her Theology and Law lay bleeding and expiring by the Swords of Rebels
and Usurpers, Physic alone praeserved her perishing fame alive."[88] The
study of medicine, subsequently, was pursued by a large number of Royalist students, such as Barrow, Sancroft, and Tenison, not to mention
many members of the Oxford group.[89] In fact, even those who were not
necessarily rigid Royalists came to follow the path for similar reasons.
Samuel Rolle, for example, who was elected scholar at Trinity in 1646
and graduated M.A. in 1650, wrote to Sancroft in 1676 that "he had delighted in the study of medicine for almost 30 years 'though first driven
to it by the Great Suspition which [he] had of the Approaching ruine of
Schollars in the late time.'"[90]

The years under Puritan rule, therefore, forced a large number of individuals not in favor with the powers that be to divert their attention to
medicine and related areas. Another important "contribution" of Puritanism vis-à-vis the Royalists was the tendency of like-minded "outsiders"
to bond together into a community of purpose. Such a response to Puritanism enabled the Anglican church to survive the Interregnum and, at
least in part, explains the background of the formation of scientific com-
munities at the universities. As Thomas Sprat said:

To have been always tossing about some *Theological question,*
would have been, to have made that their private diversion, the
excess of which they themselves dislik'd in the publick: To have
been eternally musing on *Civil business,* and the distresses of
their Country, was too melancholy a reflexion: It was *Nature*
alone. . . . The contemplation of that, draws our minds off from
past, or present misfortunes, and makes them conquerers over
things, in the greatest publick unhappiness.[91]

Recently Paul Hammond has denied that there existed a "community of
purpose" at Trinity College in the late 1640s and early 1650s whose members were interested in the new science and who gathered together for the
purpose of conducting experiments. For him, Barrow, Ray, and Needham were men of different tastes; as for the other individuals often mentioned as part of this so-called fictitious group – usually "friends" of John
Ray – they were merely physicians who should not be dignified by the

appellation scientist.[92] These claims, however, are widely exaggerated. Barrow, Ray, Needham, and Nidd did, in fact, form the nucleus of a large and highly active community at Trinity whose members – despite individual tastes – shared many of their colleagues' interests and who met regularly for the purpose of carrying out experiments and scientific discussions. That many of them were physicians or students of medicine was, we noted earlier, a byproduct of their religious and political views. As for the emphasis on chemistry, anatomy, and botany, it reflected not merely the preoccupation with disciplines related to medicine, but the genuine belief that new and exciting developments were taking place in precisely such disciplines.

What is significant about the Trinity group is its homogeneity. It appears that, with few exceptions, membership was confined to a coterie of four tutors and their pupils – Alexander Akehurst, John Templer, Charles Robotham, and Nathaniel Rowles – and many of the students of a fifth tutor who apparently did not take part in the activities, James Duport. Two other members of the group were John Nidd and William Lynnet, distinguished from the other tutors by the fact that they tutored only a few students, none of whom are known to have participated in the meetings.

Alexander Akehurst, who was arrested in 1654 for his antigovernment statements, was apparently a superb teacher; even Heywood, who differed greatly in his religious beliefs from Akehurst, regretted that he had not studied natural philosophy with his tutor. Our knowledge of Akehurst's interests is limited to chemistry. Hartlib, to whom he was known as "Chymically given," records in his diary for February 1653 that Johann Fortitudo Hartprecht was about to install a furnace in Akehurst's chambers at Trinity College; quite possibly, he was also to serve as Akehurst's operator.[93] Among his students we find Thomas Jolley (admonished with Barrow in 1648), who arrived at Trinity in January 1646 but never graduated. Hartlib described him as "a kind of Rosecrucian or Adeptus making some lucriferous experiments whereby hee lives and hath repaid the mony which he got at Cambridge."[94] Another student of Akehurst was John Holland, who arrived in April 1646, taking his M.A. in 1653. Holland's interests gravitated more toward astronomy and practical mathematics, for he constructed, and presented to the college in 1650, a double horizontal dial after Oughtred's design and that same year purchased a copy of Kepler's *Epitome astronomiae Copernicanae*.[95]

John Nidd was perhaps closest in interests to Akehurst. He "kept a vivarium in his room and studied the breeding of frogs"; Ray recorded an occasion in which four birds were dissected in Nidd's rooms; and he assisted Ray in gathering the material for the latter's *Cambridge Catalogue*. But Nidd was also an avid chemist. His library included a large number of books on the subject, and a cryptic reference in one of Ray's letters indicates an interest in experimentation as well. Close friends and scientific colleagues of Nidd included Thomas Millington, Thomas Pockley, and William Lynnett – all of whom appear to have had free access to Nidd's chemical library and received bequests of books following Nidd's death in 1659 – as well as Daniel Foote, of whom we shall hear more later.[96]

John Templer graduated B.A. from Emmanuel College in 1645, the year he was intruded fellow at Trinity. Remembered mainly as John Dryden's tutor, Templer appears to have had at least a modest interest in science. His post-Restoration publications indicate a respectable reading in contemporary natural philosophy, and his correspondence with the Royal Society in the early 1670s, conducted via his former pupil Walter Needham, demonstrates an interest in natural history.[97] Templer's most famous scientific pupil was Walter Needham, who graduated B.A. in early 1654. Needham's apparent chastisement by Senior for his addiction to experiments was noted previously; in the present context the long letter that Needham sent to Busby in defense of his scientific interests is most revealing. Needham proclaims his faith in Copernicanism, continuing with a tirade against Aristotelian natural philosophy; extols Bacon – whom he praises as the greatest of all opposers of Aristotle – as well as the experimental philosophy in a manner reminiscent of Barrow's 1652 oration; and concludes with his own ambition to write a commentary on Aristotle comparable to Gassendi's commentary on Epicurus. Indeed, many of Needham's anatomical and embryological dissections upon which his *De formato foetu* (1667) was based (and in some of which Ray was involved) were conducted at Trinity from 1654 onward.[98]

Less well known is another of Templer's pupils, Francis Jessop, who came to Trinity in 1654, becoming a close friend of Ray, Willughby, and perhaps Barrow, too. Years later, Barrow included an elegant solution by Jessop in his geometrical lectures, and certainly the lavish praise he bestowed on him suggests a long-standing friendship. Jessop's interests were wide in the extreme: mathematics, natural philosophy, chemistry, and natural history. During the 1660s, Ray stayed occasionally with Jessop, studying mathematics or conducting chemical experiments. During the 1670s Jessop also circulated a theory of tides he had devised and became involved in a minor dispute with John Wallis, who, having underestimated Jessop's talents, failed to realize that the latter had indeed "found a method for rectifying the arc of an epicycloid."[99]

Charles Robotham, educated at Queens' and Corpus Christi colleges, was a mathematician who also filled the position of university lecturer in mathematics in 1652–3. His fondness for Barrow may be seen in the laudatory verse he contributed to Barrow's edition of Euclid; quite possibly the two also studied together.[100] Another pupil of Robotham was Gilbert Havers, who was admitted in 1649. Havers was interested in chemistry and, like many others of the Trinity group, was a long-standing associate of Ray, with whom he would maintain close contact long after they had both left Trinity. In 1665–6, for example, we find Havers and Ray, together with Martin Lister, experimenting in Paris.[101]

Like Akehurst and Templer, Nathaniel Rowles came from Emmanuel, where he received his M.A. degree in 1644. Mathematically inclined, Rowles was university lecturer in mathematics in 1649–50 and a patron of the resident mathematical practitioner of Trinity, George Atwell, bearing the cost of publishing the latter's *Surveyor* in 1658.[102] Rowles, who became a physician, served as tutor to Thomas Pockley, who was admitted in March 1645/6. In a letter of 1654 to Pockley, Henry Power demonstrates both the extensive range of interests of the Trinity group and their association with like-minded students of other colleges: "W^t Hapynesse I enjoyed at Cambridge I could never properly tell till now that I am removed from it. I now can well understand the greatnesse of the losse I sustaine in the disenjoyment of that worthy Society there, wch used to entertaine mee with such excellent & noble discourses, Physicall mathematicall & Anatomicall." Ray, too, provides us with information on a series of chemical experiments he was planning in conjunction with Pockley in 1658.[103]

An exact contemporary of Pockley was yet another pupil of Rowles, Daniel Foote, whose notebooks were previously utilized in connection with the curriculum at Trinity. The same notebooks, however, also provide information on Foote's avid interest in chemistry. In fact, by comparing the titles of the chemical books that Foote studied against those in Nidd's library catalogue, it becomes clear that unless the two had identical libraries, Foote had free access to Nidd's collection. Foote also provides a link with another fellow of Trinity, John Pratt – yet another Emmanuel man – who was intruded into the medical fellowship at Trinity in 1645. Pratt was not only an admirer of William Harvey but a keen botanist. Foote eventually inherited Pratt's manuscripts, including a catalogue of English plants.[104]

Finally, we come to Duport's students: Barrow, Ray, Willughby, Thomas Burwell, and Millington, before he left for Oxford. All five participated in the myriad chemical, anatomical, and botanical studies shared by so many of the students and tutors thus far singled out. Abraham Hill attested to Barrow's "great progress in the knowledge of Anatomy, Botanics, and Chemistry"; and Barrow himself commended, in his 1654 oration, the propensity of Cambridge men to accompany reason with experiment, praising the wide range of scientific studies practiced at Cambridge and the multitude of scholars who participated in such activities.[105] Clearly, then, far from holding different interests, the members of the Trinity group – like their contemporaries at Oxford – actively participated in, and encouraged, each other's research.

The making of a mathematician

Owing to Barrow's uncompromising Royalist convictions, the growing recognition of his intellectual powers did not lead to an appropriate appointment. Indeed, it is in this political context that the difficulties surrounding Barrow's quest for a suitable position within the university community should be viewed. In 1654, at the height of the period of radical attacks on the universities, James Duport, Regius Professor of Greek, was finally forced to pay for his loyalty to the Stuarts by forfeiting his professorship. Duport, however, had masterminded a scheme according to which Barrow was to replace him. The first indication of this intention is Barrow's delivery of a "Probation exercise," which – Hill informs us – was excellently performed. Duport's support for Barrow was further assisted by two even greater powers within the university. The first was Benjamin Whichcote, provost of King's College, who was a staunch supporter of Barrow on this "as on all Occasions."[106] The second source of support was James Worthington, master of Jesus College – and a former student of Whichcote – who greatly admired Barrow and would assist him in many capacities throughout the 1650s. Worthington, in fact, not only received Episcopalian ordination – and neither took the Covenant nor accepted the Engagement – but "was (and was looked upon in the University) as an Arminian in his opinions."[107] Thus, Barrow's innate talent in and "rational" views on religion were responsible for his ability to win the vote in the university. But this was not enough, for Barrow's opponent was able to enlist even stronger support.

Ralph Widdrington was fifteen years Barrow's senior. Having graduated M.A. from Christ College in 1639, he became a popular tutor. Widdrington's ambitions, however, were greater. He was one of the first to take the Engagement, in 1650, and assist in the ousting of the public orator, the very popular Henry Molle of King's College.[108] In 1654 Widdrington hoped to add the Regius Professorship of Greek to his position of orator. Since Barrow appears to have carried the vote of the university, Widdrington enlisted the support of his elder brother, Sir Thomas Widdrington. Sir Thomas was not only speaker of the House of Commons and commissioner of the Great Seal, but also at this time in favor with the university's chancellor, Oliver Cromwell, who personally intervened on behalf of the young Widdrington. With such formidable backing, and despite the explicit prohibition in the statutes of the Regius Professorships against outside pressure, Widdrington got the appointment.

It is difficult to assess the exact role that Barrow's religious and political views played in his defeat. Obviously, Sir Thomas Widdrington acted more on brotherly than on ideological grounds. Still, in order to alter the popular vote, references must have been made to Barrow's political views. Certainly, his earlier biographers attributed his failure to obtain the chair to his reputed Arminianism. That such "partiality of others against him" indeed existed is evident from an interesting entry in Hartlib's diary. In early March 1655 Hartlib recorded information received from Samuel Morland – mathematician and until recently fellow of Magdalen College, Cambridge – who was in the process of vying for a post in the Cromwellian administration. Inter alia, Morland had the following to say about Barrow: He was "a plodding laborious man, not quick witted, much versed in Euclid and Apollonius but no better then an atheist or for any religion, Socinian, Arminian etc." The deprecatory remarks concerning Barrow's scholarship were certainly the result of Morland's indignation with the former's religious views. A few years later a similar, retrospective view was expressed by a younger contemporary of Barrow at Cambridge: "Mr. Barrow... is lately chosen Gr. Prof. – he should have been chosen about five years ago; but he was hardly dealt with by some, because he seemed not to be of their beloved Dogmata; had he been then chosen he had not travelled."[109] Such views, no doubt, impressed upon Barrow just how precarious his position at Cambridge was, as well as how uncertain any other position he obtained there would be, and he began to entertain ideas of travel abroad.

Still, Barrow's friends did not abandon him, and serious efforts were made to find a position for him in Cambridge, or at least acquire for him a lucrative appointment as a tutor to a traveling nobleman's son. One such opportunity availed itself early in 1655, when it became known that William Cavendish, third earl of Devonshire, was looking for a governor to travel with his son William, the future fourth earl and first duke of Devonshire. What made this position especially attractive was the earl's desire to employ a tutor who was also a competent mathematician. Worthington had high hopes of securing this lucrative appointment – which carried with it the salary of £100 p.a. – for Barrow. But again Barrow was faced with a formidable competitor, this time Thomas Page, the future provost of King's College. Page not only was twenty years Barrow's senior, but had spent much of the two decades since his graduation (M.A. 1636) traveling abroad, which, of course, made him particularly attractive as a tutor. He was also much respected for his learning, and his name headed a list of Cambridge scholars highly recommended by Ralph Cudworth to Secretary of State Thurloe for state employment. Be this as it may, after a short period of uncertainty, Worthington informed Hartlib that Page got the job.[110]

Almost simultaneously, Worthington and Hartlib made different attempts to secure some position for Barrow. Toward the end of February 1655, for example, Hartlib wrote in his diary:

> Col. Drake of Hackney hath a plantation in Barbados which yeelds him yearly 10 or 8 thousand lib. An extreeme lover of

Mathematiques preferring it beyond all other studies whatso-
ever. Mr Peters his factotum with him. To trie what hee will
doe with Mr Barrow or founding a Mathematical lecture.[111]

No more is heard of this effort. Later, when it became evident that Barrow would go abroad, Hartlib hoped to find a spirited man who could
provide for Barrow so that he could travel rather to Joachim Jungius in
Germany. Alternatively, Hartlib suggested Barrow should meet another
correspondent of his, Adam Boreel, in Paris.[112]

The result of all such efforts was frustration, not only for Barrow, but
also for his patron Worthington, who, in a revealing letter to Hartlib
offers valuable insight into both the character of the young Barrow and
the lamentable lack of support for promising scholars:

> For the gentleman [Barrow] is (as many scholars who tumble
> not in the world) not rich; but one of admirable parts, and had
> no body to instruct him in Mathematics, but *proprio Marte* he
> conquered all difficulties in the most crabbed authors. He is but
> a young man. He looks upon his performance upon Euclid as a
> small business to him. He knows Tacquet &c. He is versed in
> Physick, an excellent Grecian. That which he would most direct
> himself to (if he had encouragement to subsist) is Natural Phi-
> losophy, and in that we are most at a loss. He is a free philos-
> opher. He talks of travelling, and but that he is well-principled
> and fixed, the Jesuits would not stick at any thing to give a
> person of such accomplishments. He is a serious and a modest
> person, not vain-glorious and supercilious. It is one of the
> greatest afflictions to my spirit, when I consider that there are
> such accomplished persons here and there in the world, and yet
> because they could not be base or servile, nor scramble for the
> things in the world, are too often without those helps which
> might enable them to do good, as to the advancement of
> knowledge, &c. There are divers rich men, to whom to part
> with £20 or £40 a year would not be more hurtfull than a flea-
> bite, that yet cannot be possessed with any noble thoughts. I
> have sometimes thought, that God would not honour some
> because of their wickedness (though speciously covered) to be
> instrumental to so high and noble ends with their wealth. That
> persons of true worth, and such as might be eminently service-
> able, should want what may in a modest way encourage them,
> is to me an affliction, and make me the more sensible of the
> vanity and slightness of this world.[113]

It was precisely at this critical juncture in his life that Barrow became
engrossed in an ardent study of mathematics. Hill relates an interesting

anecdote about the course by which Barrow embarked upon a "professional" study of the mathematics – beyond the introductory level he had obtained while an undergraduate – and how this study came to supplant his previous study of natural philosophy: "When he read Scaliger on Eusebius, he perceived the dependence of Chronology on Astronomy, which put him on the study of Ptolemy's *Almagest;* and finding that Book and all Astronomy to depend on Geometry, he applied himself to Euclid's *Elements,* not satisfied till he had laid firm foundation." Worthington, for his part, told Hartlib that Barrow "had no body to instruct him in Mathematicks." Contemporaries, unlike some modern historians, were not expected to take such statements literally; they understood them as expressions of praise for individuals whose "genius" transcended the normative bounds of mentors and colleagues rather than as descriptions of the academic neglect of a subject. Barrow, in fact, had both friends and teachers who shared and encouraged his ardor for mathematics. Hill, for example, reports that when Barrow embarked upon his studies he had "the Learned *Mr John Ray* then for his *socius studiorum.*" (Ray had been elected Trinity College Lecturer in Mathematics in October 1653.) Two other scientific collaborators of Barrow were Charles Robotham and Gilbert Clerk, both of whom assisted him in his studies and prefaced laudatory verses to Barrow's edition of Euclid.[114] The outcome of Barrow's intensive studies – and the experience he gained from teaching mathematics at Trinity – was the composition of a series of treatises, or epitomes, on some key texts of Greek mathematics.

Hill's mention of Barrow's study of Euclid recaptures not only Barrow's manner of initiation into the study of Greek mathematics, but also the subject matter of his first publication, the *Euclidis elementorum libri xv breviter demonstrati.* With the possible exception of Barrow's work on Archimedes, Apollonius, and Theodosius, all his mathematical publications were the outcome of his pedagogical work, and his first venture into print clearly demonstrates both his abilities as a teacher and his belief in the importance of teaching. The edition of Euclid appeared early in 1656, when Barrow was already on the Continent, and was dedicated to three young fellow-commoners of Trinity College: Edward Cecil (son of the earl of Salisbury), John Knatchbull (heir of Sir Norton, commentator on the New Testament), and Francis Willughby, the future naturalist, for whose benefit an earlier version of the work had been prepared. All three men came up to Trinity between May and November 1652 as pupils of James Duport, who, appropriately, entrusted their mathematical instruction to Barrow. (As noted previously, such "farming out" of students was a common practice and, no doubt, helped Barrow to augment his meager stipend.)

Barrow's Euclid is interesting for two reasons. First, it provides a clear indication of his dedication to the education of youth – particularly in the mathematical sciences – for which he would become famous and which contributed to the enormous popularity of the book for almost a century. Second, Barrow's introduction to the volume furnishes some important clues to his views of mathematics in his early career. As Barrow himself stated in the introduction, his main purpose was to provide beginners with a clear and concise, yet full and accurate, rendering of Euclid's text. The inspiration for the text came, in part, from Andreas Tacquet's *Elementa geometriae*. Barrow had great respect for the orderly, methodical, and clear text of the Jesuit mathematician, which, he said, would have prevented him from producing his own text were it not for the fact that Tacquet deemed it proper to expound only on the first eight books of Euclid, "having in a manner rejected and undervalued the other Seven, as less appertaining to the Elements of Geometry." Barrow intended, therefore, not the "writing of the Elements of Geometry after what method soever I pleas'd [obviously a pun at the expense of the Jesuit mathematician], but of demonstrating, in as few Words as possible I could, the whole Works of *Euclide*." Evident in this statement is more than high esteem for an ancient text or authority. In his own way, Barrow, by insisting on the necessity of including the subsequent "geometrical" books of Euclid, was advocating "the very near affinity between Arithmetick and Geometry."[115]

As would become clear from his mathematical lectures of the following decade, Barrow's own predilection accorded geometry supremacy over other branches of mathematics. Yet such partiality sought neither to assail nor to deny the importance of the new analysis. Barrow himself was highly knowledgeable of the most recent development in the new branches of mathematics and occasionally employed the new tools in his own work. As for algebra, Barrow not only regarded it with approval – and indeed was instrumental in introducing the subject into the mathematical curriculum of Cambridge undergraduates – but undoubtedly assisted Cambridge scholars in algebraic studies. In the introduction to his edition of Archimedes he articulated his views concerning the desirability of retaining and cultivating ancient methods and modern techniques side-by-side:

> To preserve ancient authors, the inventors of the sciences,
> from destruction seems an important task for their modern
> followers.... True, their contents can in large part be derived
> more rapidly or constructed more concisely by modern tech-
> niques; yet reading them retains its value. First, it seems
> pleasant to examine the foundations from which the sciences
> have been raised to their present height. Second, it will be of
> some interest to sample the sources from which virtually all the
> discoveries of the moderns are derived.[116]

Such an outlook also stood behind his edition of the Euclid. As he told his readers, beyond the desire to provide an accurate and concise volume, his reason for publishing was "to content the Desires of those who are delighted more with symbolical than verbal Demonstrations." Thus, he adopted the mathematical symbolism that had been introduced by William Oughtred, though not exclusively. To dispense totally with "Conjunctions and Adjectives," he argued, would tend rather to confuse the uninitiated reader and muddle certain of the demonstrations. Barrow also acknowledged the indebtedness of the overall pattern of his book to Pierre Hérigone's method of explicating Euclid, as evidenced in the latter's first volume of *Cursus mathematicus* (Paris, 1634–42; Barrow, almost certainly, used the 1644 edition). This debt was the result more of necessity than of choice, owing to Barrow's imminent departure to the Continent, which would not allow him time to draw a new scheme, even though he wished to do so.

Thanks to the survival of Samuel Hartlib's diary and a portion of the latter's correspondence with James Worthington, some precious details concerning the composition and publication of the *Elements,* as well as the *Euclidis data* (1657), survive. It appears that by June 1654 the Cambridge bookseller and stationer William Nealand was in possession of the manuscript of the *Elements,* for which Barrow had been paid a certain, undisclosed fee. Worthington, however, was unable to inform Hartlib when the book would be published. By October 1655 we discover the reason for the delay. It appears that Nealand, fearing a loss, had decided to adopt the new publishing practice that was slowly emerging in England: publication by subscription. Foremost in the publisher's mind, no doubt, was the great success enjoyed by the ongoing project of publishing by subscription the Polyglot Bible under the editorship of Brian Walton. Worthington wrote Hartlib that "there [were] 300 [who had] subscribed to take of one copie of Barlow's [*sic*] Geometrical treatise." And as if to put the study of mathematics in the proper perspective, Worthington added that "above a thousand" had subscribed for a copy of Epictetus's *Enchiridion,* then being printed at Cambridge. Barrow's book finally appeared in late February 1655/6 (the old manner of calendar reckoning explains why the date 1655 appears on the title page), as Worthington was able to enclose a copy to Hartlib with a letter sent on 26 February. Hartlib, in turn, promised to insert a notice of the book in the next Frankfurt catalogue and distribute copies among his Continental friends such as Jungius.[117]

Barrow did not sit idle after the delivery of the manuscript of the *Elements* in the summer of 1654. In the months preceding his Continental tour, he was furiously at work completing his edition of Euclid's *Data.* Worthington, for his part, seems to have been zealously protective of Barrow's time, for when Hartlib, who was evidently impressed with the mathematical talents of the young Cambridge don, tried to procure Barrow's opinion concerning certain mathematical papers by Nicholas Mercator he was then circulating among English mathematicians (including Seth

Ward), Worthington informed him that Barrow's cluttered schedule and work habits prevented his immediate compliance. In a letter of 14 February 1654/5, Worthington told Hartlib that he "could not strictly tie [Barrow] to any day of returning [Mercator's] mathematical papers, because he [Worthington] would not have him straitened in his thoughts; but desire him to return them some time next week." Two weeks later Worthington was still protective of Barrow's time: "Mr Barrow is preparing for his travels. He thinks of beginning his long journey this month. He is therefore strained in time as to the account and wary perusal of Mercator's papers. He would do nothing slightly. He has scarce time to finish something, which he intends to add upon Euclyd, which he is now upon, and would dispatch – I wish he could be encouraged to stay."[118] The *Data* was eventually published in 1657 – together with a new edition of the *Elements* – and carried a dedication to James Stock, which was penned and sent by Barrow from Italy. Thereafter, the two works were frequently reprinted together.

The Euclid became an instant bestseller, and would remain so for more than a century. Many commented on its educational value. Sir Jonas Moore, for example, insisted that Barrow's text should "be every where follow'd" by the boys at the Mathematical School of Christ's Hospital, London. John Aubrey made a similar stipulation for the students of his utopian school: "Let them always have Barrow's *Euclid* in their pockets." Roger North thus articulated the feelings of most contemporaries when he stated that Barrow's text was "much the best."[119] Half a century later John Keill greatly expanded upon this view, providing a high encomium to the labors of that "great Geometrician" Barrow, which he valued far above those of Tacquet and Dechales. Keill stated that Barrow had generally

> retained the Constructions and Demonstrations of Euclid himself, not having omitted so much as one proposition. Hence, his Demonstrations become more strong and nervous, his Constructions more neat and elegant, and the Genius of the ancient Geometricians more conspicuous, than it is usually found in other Books of this kind. To this he has added several Corollaries and Scholias, which serve not only to shorten the Demonstrations of what follows, but are likewise of use in other Matters.

The only cause for objection that Keill found – and obviously he had to justify his own edition – was that "Barrow's Demonstrations are so very short, and are involved in so many Notes and Symbols, that they are rendered obscure and difficult to one not versed in Geometry. There, many Propositions which appear conspicuous in reading Euclid himself, are made knotty and scarcely intelligible to Learners by this Algebraical Way of Demonstration."[120]

Barrow attached little significance to his Euclid. As early as January 1655/6 Worthington informed Hartlib that Barrow viewed his labors as "but a meane worke," an opinion that Barrow himself confirmed years later when he told Collins that the book had been composed in great haste and that if it were to be reprinted, he would like to incorporate certain corrections.[121] The Euclid, in other words, was conceived of as help for his students and was later transformed into book form for monetary reasons, especially in view of Barrow's impending travel. Indeed, the chance publication of the Euclid conceals the fact that during the early 1650s Barrow had embarked upon a most ambitious and intense study of ancient and modern mathematics. Details concerning these studies are scanty. However, from the little we do know it is clear that c. 1653–4 Barrow was experiencing his own *anni mirabiles,* anticipating that of his most famous disciple, Isaac Newton, a decade later in terms of both age (both were 23–4 at the start) and intensity of labor; for it was during these years that Barrow not only composed his rendition of the Greek mathematicians, but also laid the foundations for all his future work.

When we assemble the preciously few threads attesting to Barrow's studies during this period, it becomes clear how analogous (and independent) were the foundations on which Barrow and other contemporaries such as Wallis and Huygens built their mathematical edifice, as well as the goals they pursued. In this respect it is important to mention Barrow's work on Archimedes, Apollonius, and Theodosius. Until now it has been taken for granted that Barrow's edition of these three geometers, published in 1675, represents his study of them during the 1660s, perhaps even the content of the lectures he had delivered as Lucasian Professor of Mathematics. Yet Barrow completed his work on the manuscripts more than two decades earlier and, as we shall see, it was only Collins's solicitations that induced Barrow to impart his manuscripts, although he refused to amend them in any way for publication. The manuscripts, now in the archives of the Royal Society,[122] were written during 1653. The dating of the manuscripts enables us to perceive Barrow's amazing capacity for work, intensity of labor, and facility of pen; in Pope's words, "his indefatigability in study, immense Comprehension, and accurate Attention to what he sought after." This friend and biographer of Barrow bore witness in later years to how little Barrow slept and how close he kept to his books – habits Barrow had obviously developed much earlier: "I have frequently known him, after his first sleep, rise, light, and after burning out his Candle, return to Bed before Day."[123] The amazing productivity resulting from such a Spartan regime is corroborated by the speed of composition of the Apollonius (which, in the 1675 edition, totals some 104 quarto pages): less than a month, 14 April to 10 May 1653 to be exact. The work on the Greek geometers appears to have been completed by the end of November 1653, Archimedes' *De sphaera & cylindro* being the last in the cycle.

These manuscripts represent Barrow's rapidly increasing command over some of the more difficult, and exciting, texts of Greek geometry. It is highly suggestive that twenty years before Barrow began his own studies, Evangelista Torricelli chose to mention his mastery of precisely these three authors when he found an opportunity to initiate a correspondence with Galileo.[124] Barrow probably composed these manuscripts – which consisted of the contraction of most of the known works of Archimedes, the first four books of Apollonius's *Conics,* and the *Sphaerics* of Theodosius – in a manner similar to that of his subsequent rendition of Euclid; his objective was a lean, symbolic text, rendered into Latin from the original Greek. Quite possibly, Barrow undertook the task in order to penetrate the ideas and techniques utilized by these authors. Whatever the motive, there is no indication that Barrow contemplated publishing the work during the 1650s.

For Barrow, as for other highly gifted mathematicians of the time, the study of these ancient texts was not an end in itself. Barrow, in fact, was pursuing a course very similar to, and yet independent of, the one taken by John Wallis at the same time at Oxford. For example, the latter devoted the three years immediately preceding the publication of his important *Arithmetica infinitorum* (1656) to an intense study of the quadrature of the circle, employing for this purpose the new powerful tool of the method of indivisibles.[125] This problem, which had absorbed mathematicians and amateurs for centuries, received fresh impetus during the late 1630s and 1640s, not the least as a result of the controversies initiated by John Pell, a former member of Trinity College who was teaching on the Continent during the 1640s but was in England during 1652–3, drawing from Cromwell's government the handsome annual salary of £200 as "Professor of Mathematics."[126] Thus, it was natural for Barrow to direct his attention to such studies. Not only did he master the relevant Greek texts he edited – and Wallis himself had argued, in a later tirade against Vincent Leotaud, that Archimedes' *De spaera & cylindro* was "the foundation of all those demonstrattions whether of the ancients or the moderns"[127] – but quite clearly he versed himself in the work of Willebrord Snell, Tacquet, and Christiaan Huygens on this and related subjects. In fact, Barrow may have even anticipated both Gregory and Newton in producing a "proof" for the impossibility of geometrical quadrature; for it was the "futility" of this attempt that, according to Roger North, caused Barrow to discourage North's brother, John, from applying himself to a rigorous study of mathematics. The Lucasian Professor "represented to him what pains he had taken, and particularly that he had spent more time upon one proposition, which was to prove an arc of a circle equal to a straight line (in order to square the circle), than most men spend in qualifying themselves for gainful professions, and all that he got was a demonstration that it was impossible to be done."[128]

Other indications of significant achievements may be found elsewhere. A few months after Barrow's departure to the Continent, Worthington wrote Hartlib of Barrow's slight regard for his Euclid, adding that Barrow "hath undertake to doe something upon Archimedes which shall awaken all the world."[129] Unfortunately, Worthington did not elaborate. He certainly was not referring to Barrow's edition of Archimedes' *Opera,* for it had been completed two and a half years before and made no claim to profound originality.[130] The master of Jesus College may have had in mind the above-mentioned quadrature of the circle; however, the expression "awaken all the world" may have indicated a greater discovery still. Is it possible that as early as 1655 Barrow had come across his method of tangents and much else that would later be incorporated into his geometrical lectures? Quite certainly, he had his method for some time before he wrote to Collins in 1663:

> If you remember, Mersennus and Torricellius do mention a general method of finding the tangents of curve lines by composition of motions, but do not tell it us. Such a one I have sometime found out, and did think to send it to you, it being only one theorem very easily and simply demonstrated; but wanting leisure to dress it, I will attend till you call for it, if you think such a curiosity worth your regarding.[131]

Now in addition to Torricelli's *Opera geometrica* (containing a method of tangents similar to the one developed by Roberval) and Mersenne's *Cogitata physico-mathematica,*[132] both published in 1644, Barrow was also well versed in Hérigone's *Cursus mathematicus,* which had served him well when working on his edition of Euclid. The second (1644) edition of Hérigone's work also included, in the appendix, a stripped version of Fermat's method of tangents. Barrow, therefore, may well have worked out, and extended, such methods. The published form of Barrow's method indeed was both similar to, and more elegant than, the method of Fermat. Unfortunately, those notebooks into which Barrow was in the habit of transcribing his mathematical ideas appear not to have survived, and consequently it is impossible to determine the exact chronology and extent of his mathematical knowledge.

Nevertheless, sufficient material exists to alert us to the fact that Barrow arrived at mathematical maturity in his early twenties, and not in his late thirties as is commonly assumed. His correspondence during the 1660s reveals quite clearly that he was constantly drawing upon material derived from his earlier notebooks, just as Newton would do decades later. For example, in a letter Barrow sent Collins on 1 February 1666/7, in which he related the area under a hyperbola to logarithms, he told Collins: "The matter itself...is of manifold use, and I have sometimes thought of it, but what I have forgot, and I think to little purpose: only concerning the dimension of the hyperbola I find this written in a note book, where I used to cast some things that came to my head."[133] Obviously, when the time came for Barrow to publish his geometrical lectures, he felt it necessary – and was prompted by Collins – to refer occasionally to work published after his own discoveries. But this should not lead us either to deny his genius or to obscure the strong possibility that Barrow had already accomplished much of the work he would be later renowned for before his departure for the Continent in 1655.[134]

Continental travel

The series of disappointments that befell Barrow in 1654 and early 1655 occurred at the very time when the English universities lay open to violent attacks by religious extremists, who demanded that their structure, monopoly over higher education, and prevailing curriculum be abolished and revenues confiscated. Oxford and Cambridge managed to weather this storm. Nevertheless, Barrow's perception of insecurity – in both the personal and public realms – led him to opt for a period of extended travel on the Continent. And with such an end in mind, he applied for one of the three traveling fellowships available to fellows of Trinity College. He was successful, perhaps benefiting from the support of his tutor James Duport, who had been elected vice-master in 1655. In any case, in fulfillment of the terms of the fellowship (which consisted of a three-year stipend), Barrow was required to report at regular intervals on his progress as he toured the European centers of learning. And it is to this requirement that we owe the existence of the few letters (both in prose and in verse) devoted to his itinerary. On 4 May 1655 Barrow obtained the necessary passport; the following month he left England accompanied by his friend – and co-member of the Trinity scientific "club" – Thomas Allen of Gonville and Caius College.[135]

Their first destination was France. In Paris, Barrow sought out his father, whom he had not seen for some years and to whom he gave a much needed monetary gift. Barrow appears to have spent the following weeks assiduously touring the city, a task aided by Barrow's fluency in French. His letters to Trinity describing the French capital display elegance and wit, as well as keen observation. Of the political events of the day, Barrow relates the aftermath of the Fronde and the brewing Jansenist controversy, providing a brisk account of the persecution of Antoine Arnauld before the theology faculty of the Sorbonne. Unfortunately, compared with the excitement in the politicoreligious arena, the scholarly sphere is barren, and Barrow laments the intellectual dearth that seems to have overtaken Paris: "I have already praised Arnauld and anxious inquiry has discovered no one else, if I except Roberval as eminent in mathematics: no Petaus, no Sirmonds, no Mersennes, no Gassendis are to be found; more's the pity! To any sight-seer wandering through the numerous Colleges nothing reveals itself as eminent except roofs, nothing as conspicuous except walls."[136] Indeed, he informs his friends back home that Cambridge and Trinity College have nothing to envy in terms of either the architecture or the learning to be had at Parisian colleges.

The mention of Gilles Personne de Roberval is intriguing, and the nature of their possible relations invites speculation. Roberval, who was appointed professor of mathematics at the Collège Royal in 1651, was an ingenius, if odious, mathematician, who claimed discovery of the method of indivisibles, carried out important work on the cycloid, and had worked out a method of drawing tangents. Such accomplishments, as well as the fact that he was a friend of Fermat, lead us to wonder whether Barrow – who was acquainted (at least by reputation) with Roberval's *Aristarchi Samii de mundi systemate*, which he had referred to in one of his Cambridge orations[137] – had sought him out. More intriguing still is the question whether Barrow met the young Christiaan Huygens, who was also in Paris between July and September 1655. Again, Barrow knew of Huygens's works before he left England, and later in life he would relate to Collins on more than one occasion that he esteemed the writings of the Dutchman.[138] The likelihood of such a meeting is enhanced by the fact that Christiaan's father had strong ties with England, and therefore the son would have been welcomed among the English exiles at Paris.[139] Huygens, for his part, did meet Roberval during his stay at Paris.

Sometime in February 1656, Barrow departed for Florence, where he would remain for more than eight months, in large part because of the plague, which made travel to Rome impossible. While sitting out the plague, Barrow spent much of his time in the magnificent Medici library, which, in addition to its rich collection of ancient manuscripts and printed books, included an exquisite collection of more than ten thousand coins and medals. The keeper of this collection was an Englishman by the name of Fitton, with whom Barrow quickly formed an intimate friendship. Fitton became his teacher and guide in the study of numismatics, which indirectly benefited Barrow's classical studies but which also had a more immediate, practical end; it enabled Barrow to act as an agent for English coin collectors. One such collector was a young London merchant, James Stock – described by Barrow as his "accomplished friend and matchless patron" – to whom Barrow dedicated his 1657 edition of *Euclidi data*. Altogether it is unclear whether the two met on the Continent or whether Barrow père had been a former business associate of the merchant. Stock appears to have contributed generously to Barrow's meager finances. Barrow, in turn, reciprocated by acquiring for Stock a valuable collection of coins and medals. (Barrow regretted that his own poverty prevented him from acquiring some for himself.) In the sensitive issue of finding appropriate coins, Barrow was assisted by a paper, "Of the Value and

Rarity of Medals," written by Fitton, and later (in Constantinople) by the expert advice of another friend, a certain Frenchman, who gave him practical guidance about how to seek out and purchase only genuine and valuable coins. Unfortunately for Barrow, Stock's sudden death in 1658 would not only deprive him of a generous patron, but also leave on his hands a large collection of coins he had to dispose of otherwise than intended. In addition to Stock, Barrow acted on behalf of another merchant, Abraham Hill, who would become Barrow's lifelong friend and first biographer.[140]

His interest in ancient coins naturally kept Barrow abreast of his beloved study of classics and ancient history, for few, if any, libraries in Europe were better equipped for such studies than the Laurenziana. However, in no way did Barrow neglect his scientific studies. From a comment he made years later it appears that he formed a friendship with Carolo Renaldini, the Florentine mathematician, who was engaged at that time in composing his *Algebra*. Although no other evidence survives to document this period, it is tempting to speculate on possible contacts with other Italian men of science in Florence and its environs, either through Barrow's own initiative or through Renaldini. Barrow's own previous studies of, and partisanship toward, the new astronomy and mathematics would have made him naturally curious about the Italian Galileans, many of whom by the mid-1650s were to be found around Florence. Chief among them was Galileo's "last pupil," Vincenzo Viviani, who, like Barrow, was a keen student of ancient geometry; during the second half of the 1650s he was engaged in an attempt to reconstruct the fifth book of Apollonius's *Conics,* an effort that culminated in the publication in 1659 of his *De maximis et minimis.* This attempt to reconstruct the lost books of Apollonius was given new impetus in 1654 by the posthumous publication of Francesco Maurolico's *Emendatio et restitutio conicorum Apollonii Pergaei,* which attempted the reconstruction of Books 5 and 6 of the *Conics.* Indeed, during the time Barrow was in Florence much excitement was generated by an Arabic manuscript housed at the Laurenziana, purported to contain all eight books of the *Conics* but which, in fact, included only seven.

At about the time Barrow arrived in Florence, Giovanni Alfonso Borelli was appointed professor of mathematics at Pisa. Immediately after his inauguration, Borelli, who had already prepared a précis of the first four books of the *Conics* before he left Messina, embarked on a design to edit and translate the Arabic manuscript, and the edition finally appeared in 1661. Barrow would certainly have found much in common with Borelli. Not only did the two share a strong interest in Apollonius – Barrow himself had prepared a digest of the first four books of Apollonius three years earlier – but after his arrival at Pisa, Borelli began editing another Greek mathematician with whom Barrow was well acquainted:

Euclid. Such efforts resulted in the publication of Borelli's *Euclides resti-tutus* in 1658, followed five years later by the *Euclide rinnovato*.

Other members of the Florentine group included the mathematician Carlo Roberto Dati, author of the *Exhortation on the Study of Geometry* who, like Viviani, had done work on the cycloid; the biologist Francesco Redi, then serving as pharmacist to the duke of Medici; and Marcello Malpighi, who taught medicine at Pisa between 1656 and 1659. Although these men were later incorporated into the Academia del Cimento, Barrow was no longer in Florence when its first meeting took place in June 1657. Nevertheless, his eight-month sojourn in Florence raises the question of the benefits Barrow might have derived from the personal contacts he most plausibly had formed with the Italian Galileans – or even from the papers of Torricelli at the Laurenziana – for later, during the 1660s, he certainly exhibited intimate knowledge of the achievements of Italian mathematics. Conversely, one may speculate whether Barrow was able to reciprocate with his own store of mathematical knowledge. Unfortunately, as we shall see later, whatever notes Barrow may have taken in Florence, and whatever books he may have purchased there, were all consumed by fire before his return to England. Consequently, the precise nature of his acquaintance with the results of Torricelli and Cavalieri while he was in Florence, or the possibility that he served as a conduit of such ideas back to England, must remain a matter of conjecture.[141]

This period of leisure study came to an end in November 1656, at which time Barrow embarked on a ship headed for Turkey. By Barrow's own account, opting east was the result more of chance than of choice. Despairing of proceeding to Rome, finding the roads to Venice blocked and not wishing to return to France, Barrow set sail from Leghorn on a ship bound for Constantinople. It was during this voyage that a corsair of Algerian pirates attempted to mount the ship. Barrow, in a display of strength and courage for which he became renowned, joined the crew in defeating the pirates. Later, when asked why he did not leave the fighting to experts, he replied, "Holy Liberty is dearer than my vital breath," adding that the prospect of enslavement by infidels was worse than death.[142]

Following a brief stay on the tiny island of Melos – where the wounded were nursed and the ship repaired – they proceeded to Smyrna, a major trading post of the Levant Company. Here Barrow remained for some seven months, enjoying the unreserved hospitality of the English consul, Spencer Bretton, and the friendship of the learned chaplain of the colony identified by Barrow only as a "second Polycarp." He appears to have been Robert Winchester, a former Trinity man who graduated B.A. in 1635 and later spent many years in the Middle East as chaplain to the Levant Company. Leaving Smyrna, Barrow's ship made its way across the Aegean Sea until it arrived at Constantinople in the summer of 1657. In the Turkish capital, Barrow was to enjoy even more lavish entertainment;

in fact, according to the story related years later by Thomas Barrow to John Aubrey, the English ambassador at Constantinople, Sir Thomas Bendish, "made him stay with him and kept him there a yeare and a halfe, whether he would or no," while the merchant Jonathan Dawes most generously assisted his finances.[143]

From the time he left France in February 1655/6, Barrow appears not to have written any "official" letters to his college – with the exception of a verse poem – as stipulated in the terms of his traveling fellowship. More unfortunate still, none of his private letters from this time have survived. At least one letter to Trinity from Turkey must have been written, however, for his permit was about to expire in May 1658 and Barrow obviously requested a renewal of his leave. The college records show that an extension was indeed granted by the college on 30 March 1658, and a letter sent off, gently admonishing him at the same time for his long neglect in writing. It was in response to this admonition that Barrow composed his last epistolary account to his college, which contained a summary of his travels from the time of his arrival to Italy. At the same time, Barrow also sought a dispensation from Cambridge, for the university statutes required that an M.A. of seven years' standing proceed toward a higher degree in divinity and dispute in the schools. Such a dispensation was granted by Congregation – chaired by Vice-Chancellor Worthington – on 24 February 1657/8. However, Barrow's plans for future travel were frustrated by the end of 1658. The immediate reason for the change of plans appears to have been news of the death of his benefactor James Stock. Writing to Abraham Hill on 17 December 1658, Barrow lamented, "News...must indeed of neccessity be very ungrateful to me, as it hath plunged me into some streights and splitt all my designs of future travel, whereof his assistance would have been the main support for the medals I have bought." That same day Barrow prepared to leave Constantinople for Venice, on his way back to England.

Although precise knowledge of the nature of Barrow's studies and dealings during his long sojourn in Turkey is lacking, certain clues nevertheless exist. The above letter to Stock, for example, indicates that, assisted by the advice of the previously mentioned Frenchman, Barrow continued to improve his knowledge of numismatics as he carried on his role as agent for both Stock and Hill. Barrow must have also paid particular attention to the life and manners of the local Greek church, for he shared the deep interest of seventeenth-century Anglican divines in the doctrine and ancestry of the Eastern church – "a model of an independent, continuous episcopal church" – as part of their ongoing polemic with the Church of Rome.[144] It was also in Turkey that Barrow both improved his knowledge of Greek and began his intensive study of the early church fathers, in particular the "divine" St. Chrysostom, who was to become his favorite church father. His stay in Turkey also prompted him to write a treatise "concerning Turkish history, manners, territory and religion," which, he hoped, would appease his colleagues at Trinity College and make amends for his long silence. It is doubtful if he ever learned Turkish; his discouraging attempts at Cambridge to study Arabic most likely dissuaded him from trying the Semitic languages yet again. At any rate, neither his letters nor his writings indicate that he made any such attempt. Nor did Barrow's close study of Islam make him any more tolerant of its faith. As he expressed in his later sermons, Islamic theology was "a mass of absurd opinions, odd stories, and uncouth ceremonies, compounded chiefly of the dregs of Christian heresies, together with some ingredients of Judaism and Paganism, confusedly jumbled or unskillfully tempered together." Clearly, he carried back with him nothing but disdain for what he viewed as the idolatry, foolishness, and cruelty of Islam.

Even the return trip had its share of escapades. As soon as the ship anchored at Venice, and its passengers were safely on shore, it caught fire; all its cargo (including Barrow's belongings) were destroyed. After a short stay, Barrow proceeded north. Following a quick tour of Germany and Holland, he was back in Cambridge in September 1659. Much had changed both in England and at the university during the previous four years. Most important, the death of Oliver Cromwell a year earlier had catapulted England into a new phase of political and religious unrest that eventually led to the restoration of monarchy. Meanwhile, at Trinity, John Arrowsmith, the master, had died at the end of February 1658/9, a few months after the death of the protector. As early as March, the fellows of the college had petitioned for the appointment of John Wilkins as new master. However, nearly six months elapsed before the appointment was made, a delay that may be attributed to the general political turmoil of the period, to the resumption of attacks on the universities, and, apparently, to Wilkins's initial reluctance to take the Engagement, which he had thus far been able to avoid before his installment as master. At any rate, on 3 September 1659, probably days before Barrow's return, Wilkins took up residence in college. Barrow must have been pleased to have such a man in the Master's Lodge, for the two had a good deal in common, in terms of both their scientific interests and their religious opinions. Small wonder, then, that a strong friendship developed between the two and that Wilkins would prove to be Barrow's chief patron in the following decade.

Barrow's first act following his return to Cambridge was to seek ordination. This act was intended not only to satisfy the Trinity statutes, which demanded that all fellows be ordained within seven years after becoming master of arts, but to reaffirm his commitment toward his intended vocation: theology. The Episcopalian ordination was conferred upon him by the venerated Bishop Ralph Brownrigg sometime before the bishop's death in December 1659,[145] and Barrow intended to settle down to the routine life of a college fellow. Shortly after his return to Cambridge he accepted charge of his first student. Within a few months, however, the prospects of a retired, studious life were to be thwarted by the advent of the Restoration.

The professor

The restoration of Charles II filled Barrow with much joy and enthusiasm, which he gave voice to in both prose and verse. A long Latin poem extolling General George Monck, "restorer" of the monarchy, and an even longer poem, part of the official Cambridge volume congratulating the king on his happy return to power, are both fruits of this celebratory mood. Similar expressions of joy may be found in Barrow's inaugural lecture as Greek professor later that year. However, Barrow's gratification with the reinstatement of his preferred form of religion and political order was soon dampened by the departure from Trinity of his two closest friends. First to leave was the master, John Wilkins, who was forced to relinquish the post he had held for less than a year because of a promise made by Charles I nearly two decades earlier, to Henry Ferne, who now expected the mastership in return for his loyalty to the Stuarts. Shortly thereafter, John Ray began his intense deliberations concerning whether to conform to the new religious order. He had already taken Episcopalian ordination from Bishop Sanderson and would probably have been persuaded by his friends to conform fully, for he believed, as would Barrow later, that "his life's work was in the University." However, the Act of Uniformity of 1662 obliged Ray to declare "that none were bound by the Solemn League and Covenant" (an oath that Ray himself had not taken!) and that there "lies no obligation upon [him], or on any other person, from the oath." Yet Ray preferred to vacate his fellowship than take the oath – although personally he was willing to conform – for he insisted that he could not account for the mind and actions of anyone but himself.[146]

Barrow's jubilation at the Restoration involved another element; it renewed his hopes that now his choice of vocation could finally materialize. As noted earlier, for some time after he had graduated B.A., Barrow pursued the study of medicine as a substitute for the projected study of theology. But after a while, "upon deliberation with himself and conference with his uncle...thinking that profession not well consistent with the oath he has taken when admitted Fellow, to make Divinity the end of his studies, he quitted Medicine and applied himself chiefly to what his Oath seemed to oblige him."[147] However, at all times, even at the height of his mathematical studies, his soul aspired toward less earthly goals. In fact, for Barrow his mathematical studies brought him closer to God, as evidenced by the motto he selected for the title page of his Euclid.

Drawn from Hierocles, it read, "A knowledge of mathematics is the purification of the rational mind." Barrow also revealed to Hill that even while he was in the midst of his work "on Archimedes, he could not forbear to prefer and admire much more Suarez for his book *De Legibus.*" Clearly, Barrow's form of rational theology was already in an advanced stage, for it anticipates his subsequent preaching that the foundation of religion lies on ground as solid and rational as mathematics.

But most revealing is the "prayer" Barrow penned on the front page of the manuscript of his Apollonius:

> How great a Geometrician art thou, O Lord! For while this
> Science has no Bounds, while there is for ever room for the
> Discovery of New Theorems, even by Human Faculties; Thou
> art acquainted with them all at one View, without any Chain of
> Consequences, without any Fatigue of Demonstrations. In
> other Arts and Sciences our Understanding is able to do almost
> nothing; and, like the Imagination of Brutes, seems only to
> dream of some uncertain Propositions: Whence it is that in so
> many Men are almost so many Minds. But in these Geometrical
> Theorems all Men are agreed: In these the Human Faculties
> appear to have some real Abilities, and those Great, Wonderful
> and Amazing. For those Faculties which seem of almost no
> force in other Matters, in this Science appear to be Efficacious,
> Powerful, and Successful, &c. Thee therefore do I take hence
> occasion to Love, Rejoice in, and Admire; and to long for that
> Day, with the Earnest Breathings of my Soul, when thou shalt
> be pleased, out of thy Bounty, out of thy Immense and Sacred
> Benignity, to allow me to behold, and that with a pure Mind,
> and clear Sight, not only these Truths, but those also which are
> more numerous, and more important; and all this without the
> continual and painful Application of the Imagination, which
> we discover these withal.[148]

Undoubtedly, it was such sentiments that prompted John Dunton to provide the following aside: "There is without doubt but a very few Men, who amongst those reasons which induce'em to wish for Heaven, give this, *of expecting the Happiness of a Perfect Knowledge of the Mathematicks there.* Thus Mr. Barrow having wearied himself with these Speculations, resolved to addict himself only to the study of Divinity."[149]

Barrow indeed made a concentrated effort to fulfill his aspiration. In 1661 he proceeded Bachelor of Divinity and was chosen to preach the prestigious commencement sermon that took place on 30 June. The sermon, wrote Worthington a fortnight later, "was practical, and much commended."[150] The following January, Barrow delivered another sermon at St. Mary's Church in Cambridge. Barrow's uncle and namesake, who had been helpful a decade before in resolving Barrow's mind concerning the nature of his studies and probably in introducing Barrow to Henry Hammond as well, attempted once again to help his nephew direct his studies toward theology. Barrow was thus given the highly visible position of preaching, at Westminster Abbey, his uncle's consecration sermon as bishop of Sodor and Man on 5 July 1663. Finally, two weeks later we find Barrow preaching again at St. Mary's in Cambridge. This was to be his last sermon for some six years.

It is obvious that Barrow had made his mind clear concerning his desire to become a divine. Yet for the moment that desire was not fulfilled. Instead, Barrow left us with a sharp verse indicative of his deep frustration, and his only bitter expression to survive:

> Thy Restoration, Royal Charles, I see,
> By none more wish'd, by none less felt, than me.[151]

Whewell was perplexed as to the meaning of such frustration, not understanding "what his sufferings were." Barrow, however, was not vying for high places or riches: only to be able to pursue *his* calling. And Hill's statement that Barrow's friends were expecting some great things for him following the Restoration should be read in the context of such aspirations.

Although his aspirations to pursue theology were temporarily frustrated, Barrow nevertheless reaped an immediate benefit from the Restoration; Ralph Widdrington relinquished the Regius Professorship that he had "usurped" some six years earlier. Duport was first to be offered the position, since it was he who had been forced to relinquish it in 1654, but he declined, and subsequently Barrow was elected with no competition. As noted previously, his inaugural oration was filled with joyous regard for the restored monarchy. Also evident in the oration was relief over the state of the universities and learning in general. Barrow, after all, possessed distinct memories of the attacks on the universities and on learning both in 1653–4 and in 1659, and at least on one occasion during the 1650s he had countered such attacks on Oxford and Cambridge with a passionate defense. Not surprisingly, then, his "Gratulatio Regi," his verse praising the restoration of Charles II, verbalized his relief that the patronage of the king would eliminate "the envy and madness of a maligne rabble." This notion was reiterated with even greater force in his inaugural, in which he rejoiced both in the repression of "the madness of the meaner sort hostile to learning" and in the collapse of "the power of worthless men coveting our wretched remnants of sacred revenues."

The greater part of the oration, however, took the customary form of an inaugural. In accordance with set expectations, Barrow produced an elegant composition providing a vivid encomium of his learned predecessors as Regius Professors. He even went so far as to place Erasmus within the Cambridge tradition, citing the latter's rescue of Greek studies from barbarism: "You were the first for a thousand years, as a veritable pioneer, to teach the human race to speak Latin readily, to read Greek correctly and to write it elegantly." He proceeded to commend Sir Thomas Smith, Sir John Cheke, Andrew Downes, and, not surprisingly, James Duport, whom he extolled in the highest terms. Barrow even made some good-natured, if piqued, allusions to the manner in which his immediate predecessor had obtained the professorship and had subsequently neglected his lecturing duties: It was Widdrington, Barrow quipped, whose "one oration has commended him more than a thousand of mine could do; if his health and other occupations had permitted him to lecture as often as he lectures well, no one would have earned higher distinction in this office." Barrow concluded with an elaboration of the benefits that the study of Greek bestows upon virtually every discipline and profession, taking his cue from the battle cry of the great scholars of the seventeenth century: "The study of Greek is a foundation which must be laid before any one can raise a structure of solid erudition."[152]

Shortly after his election, Barrow seems to have initiated a move to effect an official change in the statutes governing the Regius Professorships. Always a strict observer of statutes, Barrow sought to legitimize officially – not only for his own benefit but for the benefit of future Regius Professors as well – the private arrangement settled upon by James Duport following his own election in 1639. Ever since their foundation under Henry VIII in 1540, the five Regius Professorships of theology, Greek, Hebrew, law, and medicine had carried a salary of £40 p.a. Under the statutes of Queen Mary, it was ordered that the salaries of the first three professors be paid out of the revenues of Trinity College, while the salaries of the other two professors continue to be paid by the Crown. According to the revised statutes, the election of a Trinity man to one of these three professorships would result in an immediate suspension of all benefits associated with the college fellowship – save the title and the right for commons. In the following century, however, the stipend of both Greek and Hebrew professors remained unaltered; yet from 1605, the professor of divinity enjoyed the fruits of the rectory of Somersham, which was annexed to the chair by James I. Thus, when in 1639 Duport refused to accept his election as a professor in view of the meager remuneration attached to the post, the master and seniors of Trinity College responded by an interpretation of the statutes that would allow Duport to forfeit only the statutory stipend but still collect his share of the dividends from surplus income, maintain his seniority as a fellow, and retain pupils. It was this "private" arrangement that Barrow – always scrupulous to preserve the letter and spirit of the law – sought to legalize. The result was a letter patent signed by Charles II in 1661 that allowed all future professors of Greek and Hebrew – who were also fellows of Trinity College – to retain both fellowship and privileges at the college. In addition, the teaching load was much reduced, the professors now being required to teach only twice, instead of four, times per week, and not at all during the long vacation.

These changes effected by Barrow intended to augment both the prestige of the professorship and the quality of instruction. Obviously, Barrow was to benefit personally from such changes, yet his efforts to reform the statutes were far from self-indulgent. Barrow seems to have had in mind three objectives. First, he wished to prevent abuses he knew were bound to occur whenever private arrangements, which technically violated the statutes, were tolerated in order to compensate for inadequate provisions of the position.

Second, and perhaps more important, Barrow was convinced that only generous terms of employment, such as ample remuneration and engaging terms of employment, would attract high-quality professors and ensure their continued tenure in office and dedication to teaching; most important, only such financial security would free them from the pursuit of external sources of income. Similarly, overburdening professors with instruction would serve only to deter them from performing their own research. Barrow as well as many of his contemporaries realized that it was precisely such enterprising stipulations that had enabled the University of Leyden to gain prominence, and that a similar vision lay behind both the foundation of the Savilian and Laudian Professorships at Oxford and the abortive attempt by Fulke Greville, Lord Brooke, to endow a professorship of history at Cambridge in the late 1620s.

Third, Barrow and some like-minded contemporaries intended to redirect the nature and method of university teaching. Fully aware of the momentous transformation in university instruction that had occurred since the foundation of the Regius Professorships in 1540, they realized that university professors were no longer solely responsible for the instruction in Greek and Hebrew, since most colleges had their own lecturers and tutors in such languages. Hence, the objective of the professors should also be redirected toward expanding the foundations that students had previously acquired in their respective colleges. It was precisely this progressive vision of expanding and intensifying the compass of learning that inspired Barrow both as Regius Professor of Greek and, later, as Lucasian Professor of Mathematics.

A grasp of such disinterestedness is crucial to our understanding of Barrow, a person who sought neither riches nor dignities. As will be later shown, his manner of living was, and would always remain, modest to the extreme. During his entire career as both Regius and Lucasian Professor, he exercised the license to retain pupils only thrice, and only at the instigation of close friends who sent their sons to Trinity, specifically asking Barrow to take charge. Barrow's real concern was to accommodate the potential needs of his successors and provide them with both the ample income and leisure that would ensure that they would not be forced to resort to private arrangements – as Duport had – in order to compensate for the inadequate official benefits. As Barrow firmly believed, and repeatedly expressed in various ways, encroachment on the official statutes of professorships or colleges would prove the first step toward abuses and corruption.

If Barrow believed that his enthusiasm for the Greek language, and his progressive vision of the content of teaching, would be shared by Cambridge students, he was quickly disappointed. The audience during his first year as professor was small. The subject matter, Sophocles' *Electra,* was certainly unusual – and trying – for an audience accustomed to more straightforward expositions upon rhetorical or moral texts, such as the orations of Demosthenes, the rhetoric of Aristotle, and the characters of Theophrastus or, alternatively, Homeric verse. Barrow was surprised – and disappointed – to learn that this author "who was so thoroughly pleasing to the ancients seems insipid to [his] own age." As a consequence, his auditory was reduced to only a handful of enthusiasts.

Unquestionably, Barrow's failure to attract large audiences may be attributed to his overestimation of the interests, abilities, and needs of both students and tutors. On the one hand, most students had no wish to acquire more than a tolerable grounding in Greek, which would suffice for an occasional reading of the New Testament – or Homer – in the original. And this need was satisfied by the colleges. Consequently, by implementing his conviction that the lectures of the Regius Professor significantly advance the scope and content of the college instruction, Barrow had overestimated his prospective pool of students. After 1660 relatively few advanced students actively sought to excel in the study of Greek or oriental languages. As the future Archbishop William Sancroft found to his sorrow upon returning to Emmanuel College, the enthusiasm for languages he remembered from his own student days of the 1630s no longer existed. Writing to his former tutor on January 1662/3, he lamented, "Hebrew and Greek learning being out of fashion everywhere...I find not that old genius and spirit of learning."[153] Of course, individual towering scholars, the likes of Thomas Gale, Joshua Barnes, and Richard Bentley, would bestow honor upon Cambridge, but the more general passion for erudition that had been characteristic of the first half of the seventeenth century was rapidly declining.

Barrow must have been crestfallen. Certainly he used the occasion of his first anniversary oration (obligatory for every Regius Professor) to unleash one of his wittiest and most scathing addresses. He was grateful for the opportunity imposed on him to begin each new course of lectures with a prefatory address, so began Barrow's *Oratio sarcasmica in Schola Graeca,* for "it brings hither friends who have not been seen for a whole year past and are not likely...to be seen again until another year has passed." He then proceeded to chide Cambridge students:

> You did me the former kindness that there might be no necessity
> for me to sweat anxiously in polishing up my lectures or stand
> in awe of your keen judgement, but that I might take it easy
> and just prattle on with whatever came into my head, without
> fear of gossip or criticism.... Since you bade me that long fare-
> well a year ago I have sat on my chair incessantly alone...like a
> Prometheus chained on his rock.... I speak, not to mountains
> and woods, but to these walls and benches, murmuring my
> Greek sentences, figures, phrases and etymologies searched out
> from every source, just like an Attic owl segregated from other
> birds.... Continue to stay at home, if you are wise...[m]uffle
> yourselves in your snug blankets or sit by your cosy fireside....
> Why leave domestic delights...to seek out a fool of a lecturer,
> raking in a dunghill of obsolete antiquity, gaping after childish
> critical subtleties, repeating nasty mythological stories, sweep-
> ing up the worthless rubbish of the grammarians?[154]

Not surprisingly, during his second year as Greek professor, Barrow abandoned both Sophocles and tragedy, espousing instead the rhetoric of Aristotle. This decision was reached, no doubt, after much delibera-tion, and even Worthington, who always welcomed experiments for the advancement of learning, applauded both the change and the choice. In a letter to Hartlib he wrote: "Mr. Barrow hath begun this term his Greek lectures. He is off from Sophocles, and reads upon one of Aristotle's best pieces, viz. his Rhetoric, which was thought more considerable and use-ful to discourse upon than Sophocles his Electra."[155]

Barrow's decision to teach Aristotle should be interpreted on two levels. First, it was sensible in view of the simultaneous exposure students would have to both an esteemed model of the Greek language and a key rhe-torical text. But more important, the decision of a distinguished man of science to lecture on Aristotle was intended to mitigate the general assault on the Stagirite and at the same time caution against excessive adherence to Cartesianism. Undisputedly, Barrow was aware of the obsolete nature of Aristotelian natural philosophy. Yet he – probably as much as Duport – feared the implications of such an attitude toward Aristotle, the ancients, and polite literature in general (certainly the lack of interest in his pre-vious set of lectures was indicative of precisely such an attitude); quite clearly, his intent was to demonstrate that not *all* of the Aristotelian cor-pus was useless or obsolete.

The underside of this choice of Aristotle was Barrow's attitude toward Descartes. Despite his respect for Descartes as both a mathematician and a philosopher, Barrow found the latter's scorn for the "humanities" repugnant. Certainly, Barrow was neither ready nor willing to replace one autocrat of knowledge with another. Moreover, as noted earlier, by the early 1650s Barrow had begun to realize the implications of Cartesian rationalism – especially as exemplified by Hobbes – on revealed religion; and like More and Cudworth, he was willing to utilize more than ever before what was useful in Aristotle in order to preserve religion, morals, and learning. Barrow deemed it essential, therefore, to exhort younger members of the university, who were particularly prone toward novelties, to temper their enthusiasm for the merits of Descartes's philosophy and not to dismiss en bloc the entire Aristotelian corpus. As he told his audience, "If any of you have been driven by the breeze of novelty, or enchanted by the charm of a perverted fashion, to revolt from your loyalty to him [Aristotle], I must strive by advice and exhortation to lead you back" to the comprehensive corpus of writings of that "perpetual dictator of the Republic of Letters" where "not one particular science kept asunder from the rest, but all science that is worthy of a free intellect bound together in close association, joined up on a uniform thread." Thus, with a mixture of pointed criticism at the frail basis on which the Cartesian edifice was constructed, and with carefully directed puns, Barrow railed against the wholesale adoption, by youngsters, of a "new fashioned" philosophy that

> blunts the apprehension with too much 'Meditation' breaks the force and retards the alacrity of the growing intellect. . . . Aristotle's philsophy is not based on arbitrary figments of the mind,[156] does not resort to insensible causes, does not take refuge in absurd hypotheses, does not feed the mind with chimeras nor vex it with tortures nor whirl its fancies into giddiness. It does not endeavour to shape, as by some lathe, smooth globules, to scrape together the dust of a subtile matter, to twist together minute channels of aetherial spaces, to search the very bowles of the earth, in short, to penetrate new worlds, but to contemplate this one.

The stricture was circumscribed, however, and Barrow was careful to conclude with the wish that undergraduates "'let their tongues be polished by Aristotle', even if they recoil from having their minds instructed by him."[157]

It is possible that Barrow's disillusionment as Regius Professor of Greek, combined with the lack of any immediate theological employment, prompted him to entertain notions of leaving Cambridge. For despite his love of the Greek language, the fiasco of the Sophocles lectures taught Barrow a sobering lesson: the Cambridge students did not share his advanced vision and were not inclined to pursue advanced lectures on uncommon books. Barrow, for his part, was disinclined to dissipate his time by becoming a mere grammarian. In his inaugural lecture as Lucasian Professor he referred back to his tenure as Regius Professor of Greek and reflected upon feelings he had most certainly entertained two years earlier: "I who upon Constraint had engaged my self in the Province of Greek Professor, then desireable to none by reason of the small or even no Reward for the vast Labour attending it," but now that other men can – and wish to – fill the position, "I conceived there was nothing to hinder me from entirely following my own Inclination... by withdrawing my-self from the Drudgery of *Grammar* to the more desireable Exercise of *Mathematics.* .... I did not enslave myself to Grammar Learning, nor suffered my Ears to be bored through with the Languages."[158] He there-fore consented to stand as a candidate for the Gresham Professorship of Geometry, which became vacant on 27 June 1662, following the unex-pected death of Laurence Rooke. The person who persuaded Barrow to apply for the position and who helped procure his election three weeks later was John Wilkins, who would become Barrow's chief patron during the 1660s.

The responsibilities of the new Gresham Professor were moderate to the extreme. Barrow was required to lecture twice every Thursday: in Latin, between eight and nine in the morning, and in English, between two and three in the afternoon. As to the topic of these lectures, he was "to read... every Trinity term arithmetic, in Michaelmas and Hilary terms theoreti-cal geometry, in Easter term practical geometry." The salary attached to the position was £50 p.a.; in addition, Barrow was provided with rooms at the college. Barrow, however, did not relinquish his Greek professor-ship following this election. Such an action would have resulted in the severance of all ties with Cambridge, for the statutes of Trinity College did not allow a regular fellow – as opposed to a fellow who was Regius Professor and was therefore exempt from all tutorial and administrative responsibilities – the simultaneous possession of a nonuniversity secular appointment carrying such generous remuneration. Barrow, it appears, was weighing his options. Or, still another possibility, rumors concerning the possible endowment of a professorship of mathematics at Cambridge may have already been circulating. As it turned out, the modest demands on his time at Gresham College even allowed Barrow to substitute for his friend Walter Pope, Gresham Professor of Astronomy, who was trav-eling on the Continent and whose Friday lectures Barrow delivered from June 1663 to May 1664. Hill tells us that "among other of his Lectures were divers of the Projections of the Sphere, which he lent out also, and many other Papers we hear no more of," and we also know that he de-livered a series of lectures on perspectives that Collins intended to pub-lish during the 1670s. In addition, Barrow appears to have utilized some of his earlier work on Archimedes in his geometrical lectures at Gresham

College. Thus, according to Kippis, Barrow's lecture on Archimedes' method, published in Latin in 1678, was translated from English, suggesting that it was a Gresham lecture. Barrow himself announced his intention, in a letter to Francis Willughby of 26 March 1662, to lecture (presumably at Gresham College) upon Archimedes' *De aequiponderantibus.*[159]

The set of rooms that Barrow occupied at Gresham College were the same in which a historic meeting on 28 November 1660 resulted in the foundation of the Royal Society; for it was in Laurence Rooke's rooms that the early meetings of the society took place. Yet Barrow's own relations with the Royal Society were most curious. On 17 September 1662 he was elected fellow of the society. He was almost certainly proposed by John Wilkins – who, that very day, had proposed for election two other Cambridge dons, Henry More and Ralph Cudworth – and was admitted fellow at the meeting of 29 October 1662. Yet, despite his prominence as a man of science and the successor of Rooke, Barrow's association with the Royal Society was nominal. There is little evidence that he attended or took part in any of its meetings, even when he was teaching at London, let alone after he had returned to Cambridge. The society, for its part, tried to induce him to participate more actively. At the meeting of 3 December 1662, for example, it was ordered "that Mr. Barrow be desired to give an account of an echo at Cambridge"; but this is the last we hear of such a request. On 30 March 1664, when the Royal Society inaugurated its ambitious plan to reform the branches of the arts and sciences, Barrow was nominated a member of two committees (although there is no evidence that his prior consent had been obtained). The first was in charge of considering and improving "astronomical and optical matters" – and here he was in the company of other illustrious "absentees," including Christiaan Huygens and Johann Hevelius, as well as such active members as Paul Neile, Seth Ward, and Robert Boyle. The second committee was responsible for "collecting all the phaenomena of nature hitherto observed, and all experiments made and recorded." There is no evidence that Barrow ever did anything in this matter either. His inactivity was also expressed by the fact that he never paid his dues to the society, not even when on 29 October 1666 a meeting of the council passed a resolution dispensing Barrow from half of his weekly payments. The society, however, was reluctant to expel one of the greatest living English scientists, and Barrow remained on the list of fellows until his death.[160]

It is difficult to account for Barrow's aloofness. It seems likely that his behavior – shared by other Cambridge fellows such as More and Cudworth – was motivated by the tension that existed during the 1660s and 1670s between the Royal Society and the universities. Contrary to many assertions, however, this tension was not necessarily the result of an entrenched hostility toward the new science, although many dons shared the worries of active members of the Royal Society concerning the implications

of certain aspects of the new science on revealed religion. More important was the threat – real or imagined – that the Royal Society posed to
the monopoly of Oxford and Cambridge over English higher education.
Many devoted sons of the universities – and Barrow appears to have been
one – expressed their concern not only for the excessive rhetorical zeal
of such enthusiastic propagandists of the society as Thomas Sprat and
Joseph Glanvill in denigrating university learning, but for the devastating consequences they feared that exclusive devotion to the "new philosophy" – as preached by the champions of the Royal Society – would have
on all good learning. And Barrow, who held equal respect for *litterae
humaniores* and scientific studies, may have opted to dissociate himself
from a body that publicly scorned much of the education provided by the
universities, and much of the learning cherished by its members.[161]

While serving as Gresham Professor, then, Barrow was probably pon-
dering various options. Again, Hill informs us that Barrow was offered the
position of keeper of the Cottonian Library and that he actually tried it
out before turning down the position. The offer came from Sir John Cotton, who on 13 May 1662 succeeded his father, Sir Thomas, as third bar-
onet. Sir John not only was a true scholar but was, as were his father and
his children, a Trinity man and himself a pupil of Duport. Such a Trinity "connection," added to the universal recognition of Barrow's scholarship, obviously commended Barrow to Sir John. "About this Time," the
biographer adds, Barrow was also "offer'd a living of good Value; but
the Condition annexed, of teaching the Patron's Son, made him refuse it,
as too like a Simoniacal Contract." It is possible that this offer also came
from Cotton. In a letter from Barrow to Cotton, dated 27 March 1663,
Barrow warmly thanked Cotton for his kindness and indicated his willingness to accept an offer made by Cotton.[162] In the end, however, Barrow,
preferred Cambridge. As he declared in his inaugural lecture, "Might I
have my free Choice, I would a hundred Times sooner always chuse [Cam-
bridge], than to live in the Pomps of a Court, the Tumults of a City, or
Solitude of a Country."[163] By the end of 1663, he could forgo all thoughts
of leaving Cambridge.

On 11 June 1663, a month before his death, Henry Lucas drew up his
will directing his executors to provide for an endowment of a mathematical professorship at Cambridge. Lucas, who had studied for a while at
St. John's College, became secretary to Henry Rich, first earl of Holland –
who succeeded Buckingham as chancellor of Cambridge in 1628 – and
retained the post until he was beheaded in 1649. Lucas represented Cambridge in both the Short and Long Parliaments and was instrumental in
safeguarding the income and privileges of the university.[164] We know very
little of Lucas's life, so it is difficult to determine what prompted him to
found the chair. Certainly, judging by the impressive library of some 3,200
volumes he bequeathed to Cambridge – for the use of his professors – his

interests were as far removed from mathematics or science in general as could be: With the exception of a handful of books by Galileo, Gassendi, Harvey, Gilbert, and Bacon, the library was strictly one of a cultivated humanist in the tradition of James Duport. More likely, since Lucas died a bachelor, he had been induced to erect a "memorial" to his name by endowing a professorship that would rival the mathematical chairs founded by Sir Henry Savile at Oxford in 1619.[165]

The dream of such a professorship had indeed been long cherished by Cambridge men, who looked with envy at the spate of professorial endowments, the botanical garden, and the magnificent Bodleian Library at Oxford. In contrast, no willing benefactors could be found for Cambridge. The search for such benefactors became ever more crucial since the institutional structure of both Oxford and Cambridge made it impossible for the university, as a body, to finance any such ventures. (It was exactly this situation that Barrow personally encountered the following decade, when he attempted to prod the university into building an edifice that would rival the Sheldonian at Oxford.) Far more than the colleges, the meagerly endowed university had to rely on the good services of former alumni who could be persuaded to support the university *rather* than – or in addition to – their respective colleges. Cambridge's failure to emulate the Savilian example, however, was in no way due to idleness. As early as 1621 the newly elected Savilian Professor of Geometry, Henry Briggs (formerly of St. John's College, Cambridge), wrote a letter to his friend Samuel Ward, master of Sidney Sussex College, enclosing a copy of the Savilian statutes and declaring his strong wish to help Cambridge obtain a similar endowment: "I should be very glad to see Cambridge as well provided for the stipends of these and other professions. . . . If you heare of any that are able and willinge to bestowe somuche monye so well, if you please to sende me worde, I will very willingely bestowe some parte of this (6 weekes) vacation in survayinge and plottinge the grounde, for the love I owe to my mother Cambridge."[166]

Neither Briggs nor Ward, however, appears to have found a benefactor. Five years later another disappointment occurred when Sir Francis Bacon's will, stipulating the endowment of a lectureship in natural philosophy, foundered since the former lord chancellor died deeply in debt. Another great man of English science, William Harvey, had also intended to endow a chair of experimental philosophy with a laboratory and gardens at Cambridge, but, as Sir Charles Scarburgh recalled years later, the Puritan rise to power made him change his mind: "He used to say with tears in his eyes: 'if I dedicated my property as I had determined to the promotion of the discovery of truth and to the public good I might just as well have made Anabaptists, fanatics and every kind of robber and parricide my heirs.'" Finally, as noted previously, Thomas Hill, master of Trinity College, persuaded Sir John Wollaston to support yearly a lecturer in mathematics as a first step toward endowing a professorship, but again the plan came to naught.[167]

It was left to Lucas, then, to implement what a generation of Cambridge men had aspired to. In fact, it is quite possible that Lucas had intended to establish the chair even before the Civil War. In 1639 Hartlib recorded in his diary that "one of Sir W. Boswell's private friends is erecting a Mathematicua Professor Honorarius as it were in Cambridge setling a stipend for a 100 lib. per annum with a house of purpose. Only he advances somewhat slowly in it. The founder will have the Professor of that place to take his corporal oath every 3 years [and] to denominate the fittest successor unto him." The identification of Lucas as Boswell's friend is plausible in view of the likeness of purpose and terms as expressed between the 1639 scheme and the actual endowment, as well as the possible relations between the persons involved. Boswell was an exact contemporary of Lucas at Cambridge – a fellow of Jesus College from 1606 to 1629 – and the two rose together in the circle around George Villiers, first duke of Buckingham. Furthermore, Lucas's friend and executor, Thomas Buck – who according to Barrow played an important role in the eventual establishment of the Lucasian chair – had been an undergraduate at Jesus College, whence he matriculated in 1609, perhaps even as Boswell's pupil.[168] Be this as it may, Buck, who for many years was a powerful figure at Cambridge, serving as a university bedel from 1626 until 1670 as well as university printer, appears to have played a crucial role in bringing the foundation to fruit. According to Barrow in his inaugural oration, Cambridge owed "the first Fruits of this huge Benefit, in Part to him: our great Benefactor being excited by his Admonition, persuaded by his Advice, and drawn by his Exhortation, both to institute his Mathematical Profession, and to endow and adorn your Library with a most choice Treasure of Books."[169] And it was Buck, together with the other executor, the London lawyer Robert Raworth, who were entrusted with the selection of the first professor. Their choice was Isaac Barrow.

According to Abraham Hill, it was Wilkins's intervention that helped Barrow procure the position. It is indeed plausible that Wilkins, the former master of Trinity College, had recommended Barrow. But it is also likely that Buck and Barrow – two staunch Royalists – already knew of each other in the early 1650s. As an esquire bedel and university printer, Buck certainly would have known the Regius Professor of Greek.

One of the most striking features of Barrow's inaugural oration was his emphasis, yet again, on his dedication to teaching. Mention has already been made of Barrow's ever willingness to assist the studies of his students, as well as his diligence in providing good textbooks for newcomers. He exhibited the same spirit as a professor:

While I was a private Person nor otherwise obliged, being
enamoured only with the Loveliness of the Thing, I showed
such hearty Desires and Endeavours to have these Sciences in
the highest Degree recommended to you; it cannot now be
doubted but by Reason of my publick Office, and more solemn
Engagement, I will more diligently apply myself to their Promotion according to my slender Ability.... But what the Laws
do strictly require of me, that I have always shewed the greatest
Readiness to perform, so that I did not only willingly admit,
but earnestly invited you of my own Accord, to familiar Meetings, so that I not only once or twice every Week opened the
Doors of my private Chamber, but daily publish'd the inward
Secrets of my Heart, and unfilded my Breast to all Comers. If
it be then your Pleasure, ye Lovers of Study, come always...
what you may do by your Right, you shall make me do willingly, nay gladly and joyfully. Ask your Questions, make your
Enquiries, bid and command.... If you meet with any Obsta-
cles or Difficulties, or are retarded with any Doubts while you
are walking in the cumbersome Road of this study of Mathematics, I beg you to impart them, and I shall endeavour to
remove every Hindrance.[170]

Barrow was indeed the epitome of a devoted and dedicated teacher.
Perhaps even more now than as professor of Greek, he intended to convey to his students his own passion for the subject as well as prod them to
the frontiers of the respective fields he lectured on. His attitude did not
go unrewarded. From Barrow's letters to Collins it appears that he was
scarcely master of his time, for his students took his invitation literally,
attending in great numbers his lectures and seeking him out in private.[171]
His dedication to students can also be gathered from his books. His Euclid, we mentioned earlier, was dedicated to three of his (and Duport's)
pupils at Trinity. In a similar manner he intended to dedicate his geometrical lectures to John Moore, at the time a young student from Clare
College, and was dissuaded only when a "friend" (Newton?) suggested
that the verse dedication Barrow had composed was not "appropriate"
for the subject at hand.[172]

Undoubtedly, Barrow's performance contributed a great deal to the
future cultivation of the mathematical sciences at Cambridge. Having
said this, it is worth emphasizing a point that will be treated more fully
later – namely, that as historians of science we tend to forget the special
circumstances surrounding the composition and publication of Barrow's
lectures. We read and appraise his mathematical, geometrical, and optical
lectures according to the same criteria applied to the labors of his great
contemporaries Gregory, Newton, and Leibniz. We forget that Barrow's

entire published scientific corpus consists of lectures he delivered at Cambridge and that in no way was he a professional mathematician writing for the benefit of like-minded specialists. His lifelong passion was theology, and when he finally dedicated himself to his chosen calling, he merely consented to the publication of his lectures; he refused either to polish or to expand them. And as Dugald Stewart correctly observed, Barrow always gives the impression that he has given us only half of what he knows; unfortunately, we shall never know the extent and depth of the other half. And Barrow himself made these points crystal clear in his preface to the optical lectures. The passage deserves to be quoted at length, for it imparts Barrow's own estimation of the lectures and his attitude toward their publication:

> This small Work, such as it is, you will upon perusal immediately perceive by many Tokens, not to have been designed to be made Publick, tho' I had often been importuned by several to do so. With the repeated Instances of whom, but not without Fear and Reluctance, I was at length prevailed upon; chiefly because it seem'd to me a laudable Ambition, and even incumbent upon me as Part of my Duty, to endeavour, if not effectually to promote Learning, by setting an Example to those who should succeed me in the Professorship with which I am the first that was honoured. And likewise because I have a small Opinion of its containing something that will both profit and please. Notwithstanding, desire you, who are well versed in Matters of this Kind to remember what those Things are which you are perusing; not such as were designed for you alone; not published of my own accord; nor containing the exquisite Thoughts of a Mind wholly employ'd upon the Subject; but that they are School Lectures, read by Duty, and sometimes spoken too hastily, that I might end my designed Talk within the Hour; and lastly, that they were accommodated for the Instruction of a promiscuous Multitude of Scholars, to whom I was obliged to explain many Things which to you need not have been done. . . . For I know that for your Satisfaction I ought to curtail many Things, and substitute better in their room; to transpose the greater Part, and to correct and reform the Whole: Which to attempt, I neither had the Inclination or Leisure, no nor Power to execute. . . . [But] when I resolv'd to publish, I couldnot bear the Pains of reading over again a great Part of these Things; either from my being tired with them; or not caring to undergo the Pains and Study in new modelling them.[173]

Barrow handed the lectures over to Collins to be either published or abandoned.

Publications

The exact chronology and sequence of Barrow's Lucasian lectures has never been worked out. This confusion is especially problematic because the dating influences our understanding of the genesis of his scientific ideas. Having delivered his inaugural oration on 14 March 1663/4, Barrow devoted the remainder of the spring to reading the first six of his mathematical lectures. The new academic year began in the fall of 1664, and Barrow appears to have set a routine of delivering nine lectures per session. Certainly, it was during this fall term that he delivered the second set of mathematical lectures, Lectures 7–15 (dated 1665) in the printed *Lectiones mathematicae*.

Barrow, it seems, then suspended these philosophical lectures, promising his auditors – who thus far sat "patiently" through such orations "of less Consequence" – better things to follow the winter vacation: "I wish what is to come may be more worthy your so benign and candid Attention, and more acceptable to your Desires; that it may more delight your ingenious Curiosity, and set a greater Edge upon your own poor Industry."[174] During the following term, in the winter and spring of 1665, Barrow delivered the first five of the geometrical lectures. He began by reminding his auditors of the promise he had made at the end of the previous term: "I am now entering into a new Field, whether more pleasant or fruitful, I cannot truly say, but yielding a most copious Variety which consequently is agreeable; and as it comprehends, for the most Part, the *Original of Mathematical Hypotheses,* from whence Definitions are formed, and Properties flow, it must Necessarily be very useful too."[175] This homogeneous group of lectures logically follow topics expounded upon the previous term. Probably, in order to complete the sequel of nine lectures, he devoted the remainder of the term to his four lectures on Archimedes. By the summer, the university had closed down because of the plague, and classes did not resume until April 1666. Barrow then returned to his initial topic and read the last series of mathematical lectures, numbers 16–23 of the printed version. He alluded to the plague in his opening remark – "I am glad to find you in Health, and wish you a long Continuance of it"[176] – and proceeded with the topic of proportion. Barrow's optimism, however, was premature, for by the end of June the plague had returned and the students had been dismissed, not to be recalled until Easter 1667.

"It is now long since we have laid down and explained some general Properties of Curves, from a certain common Principle of Generation..."[177] Thus Barrow greeted his auditors as he embarked upon the second cycle of his geometrical lectures. This remark suggests that the lectures took place following the plague and before the delivery of the optical lectures in 1668–9. Unfortunately, it is not possible to pinpoint their exact dating, assuming, of course, that my conjecture concerning Barrow's sequence of lectures is valid. Nor can it be stated with confidence exactly when Barrow resumed his professorial duties after the resumption of classes. The lectures could have been delivered in late spring 1667, but it is also plausible that no professorial lectures were delivered at all for the remainder of the academic year and that Barrow devoted the academic year 1667–8 to the delivery of these geometrical lectures. At any rate, as I shall argue presently, by March 1668 John Collins had obtained from Barrow a copy of these lectures.

Admittedly, there is no solid documentary evidence confirming this chronology. In fact, some scholars even doubt that the geometrical lectures were delivered in Cambridge at all, in light of the fact that the title page of the *Lectiones geometricae* does not include the wording *"habitae Cantabrigiae"* carried by Barrow's mathematical and optical lectures. Others have speculated that if Barrow read the lectures, he did so in 1669, before his resignation from the Lucasian chair. The circumstantial evidence, however, is quite suggestive. It is certain that the mathematical lectures were delivered first. It is equally certain, for reasons to be elaborated upon shortly, that the optical lectures were the last to be read by Barrow. The geometrical lectures, therefore, assuming that they were delivered, must have been given before the optical lectures.

To my mind, it is clear that Barrow did deliver such lectures at Cambridge. To begin with, by the 1660s Barrow was not in the habit of composing mathematical treatises except for lectures, and the few references to the geometrical lectures in Barrow's surviving correspondence indicate that the geometrical lectures were no exception. Second, John Collins, who made it his business to inquire after, and attempt to publish, every mathematical work composed by Barrow, never mentions another set of lectures that Barrow may have delivered at Cambridge. And finally, given the number of years Barrow served as Lucasian Professor, the three sets of lectures fit the duration of his tenure.

Additional confirmation that the optical lectures followed the second set of the geometrical lectures may be derived from Barrow's opening remarks. He articulated his awareness of the fact that his previous lectures were becoming too difficult for his auditors, an awareness that effected a change of course:

> Since, in pursuing my original purpose, I realised that a number
> of matters came up which demanded too much attention, and
> were therefore unsuitable for an unprepared audience, and that
> this disadvantage could not be avoided in the treatment of
> Pure Geometry, I determined to abandon this subject for the
> moment, and to stray forthwith into pleasanter fields, bright
> with the flowers of Physics and sown with the harvest of Mechanics, the fields of what they call Mixed Mathematics.[178]

Finally, when Isaac Newton chose optics to inaugurate his Lucasian professorship, he stated in his opening lecture that he was proceeding from where Barrow had left off, thereby intimating that the latter had concluded his tenure with his set of optical lectures.[179]

Equally revealing is the letter sent by Collins to Edmund Boldero, the Cambridge vice-chancellor, that precipitated the publication of Barrow's Lucasian lectures. The letter purports to represent the wish of members of the Royal Society to see various of Barrow's works in print. Although the letter is not dated, Boldero was selected vice-chancellor in November 1668; hence, the letter was probably sent late that month or in early December 1668. Inter alia, Boldero was informed that Barrow had intended to deposit with the previous vice-chancellor, James Fleetwood, his "Anniversarie" optical lectures in accordance with the statutes of the Lucasian Professorship. The letter appears to refer to a complete manuscript. Yet even as late as Easter Eve 1669, Barrow had not yet completed delivering all the optical lectures.[180] Most likely, then, Barrow had composed the entire series in advance; he delivered the first half during the spring or fall of 1668 and intended to deliver the second half the following spring.

The issue of dating aside, the letter is important because it indicates the manner in which Collins overcame Barrow's reluctance to publish. Collins resorted to a semiofficial letter, laboring on the latter's sense of duty as Lucasian Professor and exemplar to future incumbents of the chair, in order to bring pressure to bear upon the publication-shy Barrow. The ploy proved successful. By 30 December Collins could write triumphantly to Gregory that Barrow "doth hereafter intend a Treatise of Opticks for the Presse but is not yet ready."[181] Upon forwarding the lectures on 23 February 1668/9, Barrow had informed Collins that the "lectures, having...got a warrant to go abroad, fly to you, and to your protection I commit them." Barrow was referring, most likely, to the imprimatur given by Edmund Boldero, the Cambridge vice-chancellor, Peter Gunning and John Pearson, masters of St. John's and Trinity Colleges, respectively, although the published imprimatur is dated a month later: 22 March 1668/9.[182] The cheerful Collins thus informed Gregory on 15 March 1668/9 that "Mr Barrows Opticks suddainly to goe to the Presse," and the printing of the volume was completed by the end of November 1669.[183]

It would appear, then, that Barrow's optical lectures, though printed first, were actually delivered after the geometrical lectures, that months before Collins sent his letter to Boldero he was already in possession of the manuscript of the geometrical lectures; for the letters exchanged between Barrow and Collins earlier that year, in March and May 1668,[184] contain detailed discussions of the geometrical lectures, and in particular the eleventh lecture. Here, too, we are struck by the persistence with which Collins cajoled Barrow into publishing the lectures. To counter Barrow's doubts about the significance of his lectures, the haste in which they had been composed, and consequent errors that they probably contained, Collins came up with the idea of showing the manuscript to a member of the Royal Society, most likely Lord Brounker. Barrow, for his part, consented to abide by the decision. Collins obtained the favorable reading and Barrow capitulated. He was resolved, he wrote Collins in May, "if you so much differ in opinion from me as to think any of those things fit to be published, I would." Thus came about the publication of Barrow's lectures.[185]

The man who masterminded Barrow's publishing ventures was justly titled the "English Mersenne" for his role both in disseminating Continental mathematical knowledge in England and in providing Europeans with news of the latest English accomplishments. A lover of all things mathematical and a firm advocate of the necessity of publishing as much and as quickly as possible, Collins tirelessly urged members of his ever growing circle of friends and correspondents to entrust their scientific labors into his hands for publication. Certainly, his efforts on Barrow's behalf determined the course of the publishing history of nearly all of the latter's scientific publications. They also shed much light on the nature of scientific publication in England during the latter half of the seventeenth century, and for both reasons it is instructive to pay particular attention to Collins's role.

Collins embarked on the publication of Barrow's works cognizant of the difficulties – even dangers – inherent in the publication of scientific works in England. The London publishing world was still reeling from the effects of the recent plague and the even more catastrophic fire of 1665–6. The latter brought many English publishers and booksellers, who were said to have incurred losses of between £150 and £200,000,[186] to the brink of ruin. This situation proved particularly disastrous for the local scientific market, which, even before the fire, was relatively small – and expensive – compared with the Dutch, French, and Italian markets. As early as February 1666/7, for example, Collins revealed to Wallis that recent scientific treatises, such as Johann Heinrich Rahn's *An Introduction to Algebra,* rendered into English by Thomas Brancker and John Pell, and Henry Briggs's *Arithmetica logarithmica,* had been overprinted, the result being that the latter had been remaindered and turned out on the streets at 1s 6d per copy. Even Barrow's Euclid (a bestseller under most conditions) had been bulk purchased by Collins and Henry Sutton the instrumentmaker "at 1s. a book in quires."[187] Booksellers, lamented Collins on another occasion, "having sustained great Losses, and finding but small or slow sale at home, and wanting a Correspondency and Vent abroad, lye under great Discouragements."[188] Collins, however, was not to be discouraged in his efforts to persuade printers and booksellers to print scientific books: On one occasion he was able to find a bookseller who would "print any book, if he [be] but sure to sell eighty or an hundred books for ready money," and Collins was hoping to turn this readiness to his advantage.[189]

In his attempt, then, to seek out mathematical projects from his friends and arrange for their publication, Collins was endeavoring almost single-handedly to reverse the prevailing publishing bias against scientific books. Oblivious to commercial failures and the difficulty of finding even enterprising – not to mention qualified – printers, again and again Collins would seek to convince reluctant publishers of the economic viability of "good" books written or edited by his friends, producing as ammunition testimonials from Oxford and Cambridge dons to this effect. His long-term hope was that the promise of financial success would encourage the publishers to increase their involvement in scientific publishing, perhaps even to expand their "list" to include far riskier works that were neither textbooks nor potential bestsellers. From Collins's correspondence it becomes clear that his publishing agenda included, in addition to the works of such Englishmen as Wallis, Baker, Strode, and Horrox, many unpublished manuscripts of French and Italian mathematicians that could find no willing publisher on the Continent for similar commercial reasons. Such ventures, he believed, would contribute to the advancement of mathematical learning as well as to the glorification of English science. Undoubtedly, it is to Collins's untiring efforts that we owe the publication of many works that otherwise would never have been submitted to a publisher – let alone committed to print – and chief among these are the works of Isaac Barrow.

Nor did Collins's efforts as self-appointed agent of English and Continental mathematicians cease with a commitment of publication. He sought to ensure that his friends obtained the best possible terms from the publishers; he oversaw the books into press; he corrected proofs, drew diagrams, and even helped publicize the books. Again and again in his letters he apologizes for his tardy response because of one or another self-imposed publishing task. Often we find him, therefore, excusing a delay in replying to a letter or to requests because of the large constraints on his time. In just this way he apologized to Sluse in 1670: "At present all the spare time I have is spent in correcting of Dr. Wallis or Mr. Barrow's books at the press, and in drawing the schemes."[190] Small wonder, given such a major commitment of his time and resources, that when Collins's economic prospects became uncertain during the early 1670s, he was tempted to become first a stationer, then a publisher, so that he could combine his insatiable love of mathematics with his need for a livelihood.[191]

Fortunate for all was Collins's seemingly invincible optimism and enthusiasm, for few, if any, of his projects ever proceeded smoothly. With regard to Barrow, the first projects to suffer the vicissitudes of the publishing industry were the optical and geometrical lectures. Although the

Lectiones opticae and the *Lectiones geometricae* were printed in 1669 and 1670, respectively, owing to the grave financial difficulties of the publishers copies of both books were unavailable for sale for some time. According to a letter of 1 November 1670 from Collins to Gregory:

> It so happens that neither his booke of Opticks, nor that of Geometrick Lectures are yet exposed to Sale; the Stationer being poore, lurketh, is runne in Debt for Paper and Printing, and unable to pay, and I myselfe have disbursed £15 for the Plates, the which I know not when I shall be reimbursed: by this meanes the Printer would deliver out no more Bookes than those Coppies the Author was to have for his paines, and a few for my selfe.

However, not before the middle of December was Collins able to release Barrow's author copies from the hands of William Godbid, the printer, and still the books did not go on sale.[192] In fact, before that date, it was only because of Collins's efforts that some complimentary copies were forwarded to such correspondents as Gregory, Wallis, Sluse, and Huygens.

As for the books themselves, it appears that they became available to the public only two years later. Collins wrote Gregory on 28 May 1672, "Dr Barrowes bookes doe just now begin to peepe abroad being in the hand of other Stationers."[193] The new stationer was Walter Kettilby, who apparently purchased the entire stock of both optical and geometrical lectures from Godbid and sold them as he had received them, save for inserting his name on the title page. On the whole, Kettilby opted to sell the sets of lectures paired together – resulting in the 1672 combined edition – although he may have sold a few copies of the geometrical lectures separately.[194] But Kettilby's finances were not much better than those of the other booksellers, the result being that in order to reduce his heavy debt he sold the entire remaining stock of Barrow's books in bulk. Collins himself tells us what happened. The optical and geometrical lectures, he calculated, "did cost the first undertaker [Dunmore and Pulleyn] 4s. 8d. or near 5s., and yet at last a great number of both books, together with the plates, were sold by Sir Thomas Davis, Lord Mayor, formerly a stationer. . . into whose hands they came for a debt, at the rate of 1s. 6d. a pair. . . to Mr. Scot. . . . or otherwise they would have turned to waste paper."[195] The person who rescued Barrow's Lucasian lectures from the shredder was Robert Scot, a bibliophile and lover of learning who not only would issue a new edition of the optical and geometrical lectures, but would also publish the next Barrovian venture.[196]

It was Collins who directed Scot's attention to Barrow's Lucasian lectures at about the same time that he sought to convince the bookseller to publish Barrow's edition of the Greek geometricians. Perhaps in order to encourage Scot, as well as boost sales of a new edition, Collins proposed the addition of an appendix to the geometrical lectures, which would include "some elegant additions de Maximis et Minimis."[197] The result was the 1674 edition of the optical and geometrical lectures, which again paired the existing printed text of the two sets under a new title page, with the addition of two and a half pages of supplementary theorems.

Despite his abandonment of mathematics, Barrow was distressed over the fate of his lectures. Collins informed Gregory in 1670 that the fact "that his Labours are yet not to be sold, doth much trouble" Barrow, who grieved primarily about his inability to bestow complimentary copies on his friends. Certainly, distribution delays, to use modern jargon, may explain the scarcity of references to the lectures in the early 1670s, and Barrow's further disillusionment with mathematics. His biographer, Hill, informs us that after Barrow heard that only Gregory and Sluse had read through the geometrical lectures, "the little relish that such things met with did help to loosen him from these Speculations, and the more engage his inclination to the study of Morality and Divinity." Collins, too, informed his correspondents from 1670 onward that Barrow was considering mathematical studies to be barren.[198]

Such feelings, however, should not necessarily be interpreted to mean that the books did not sell. The financial difficulties of the booksellers were not caused by low sales of Barrow's lectures but preceded their publication (although the venture may have aggravated existing difficulties). Their subsequent need to sell the stock in a hurry, and at bulk rate, was the outcome of a desperate need to generate cash flow, probably preventing a profit in the long run. Unfortunately, it is difficult to speculate on the sales of the various editions, since we have no information concerning their size. Regarding the market itself, it appears that Barrow himself may have "flooded" the market when he finally received his author copies before Christmas 1670. Collins told Gregory on 15 December that he had received fifty copies of both optical and geometrical lectures from the printer to give Barrow. However, Barrow received them unbound, and not only did he have to pay for the binding, but he needed to purchase an additional thirty copies of each set of lectures since he intended "to give away at least 80 of each kind bound"![199] Needless to add, many of those Barrow had in mind as recipients of complimentary copies would have been the prospective buyers of the first edition. Even so, the survival rate of books carrying the imprint of the second, and even more the third, editions assuredly shows signs of decent sales. Six copies of the combined edition and four of the geometrical lectures – bearing Walter Kettilby's imprint – are recorded by Wing, and ten copies of the combined 1674 edition are similarly recorded. Compared with the records found by Wing of other books, including theological bestsellers, the survival rate hints at both a sizable printing run and at least a moderate distribution of copies.

True, Collins would refer some years later to losses incurred to book-sellers who published Barrow's lectures and Wallis's works.[200] However, such statements were made in order to explain to Thomas Baker why Collins was unable to find a publisher who would print the mathematical works of the latter. Certainly, from what we know, the problems of Bar-row's publishers were aggravated – not caused – by such ventures. After all, Dunmore and Pulleyn, the original publishers, did not even have a chance to try to sell the volumes, and Kettilby did not invest in printing, but bought an existing edition en mass. The trouble with both was their preexisting debt and inability to allow time for sales. Scot, who had both money and time, certainly did not think that in producing a new edition he had made a bad investment.

Odd as it may seem, the ordeal with Barrow's Lucasian lectures did not dampen Collins's zeal. Just the opposite. Gratified by the favorable reac-tion of the few people to whom he himself had sent copies of the book, Collins began to prod Barrow for more work. In particular, Collins was interested in the manuscripts of those editions of the Greek geometers that had been prepared by Barrow years before and that Collins had hoped to include in a major publishing venture to print as many ancient mathematical texts as possible. Not only were Barrow's texts intended to play an important role in this enterprise; the abbreviated method he had employed in the edited texts – as well as in his Euclid – was to be the standard format used in the rest of the projected volumes, which would include the last three books of Apollonius's *Conics,* those Archimedian fragments not dealt with by Barrow, as well as the works of Pappus and Serenus.

As a matter of fact, Collins had tried to convince Barrow to publish his rendition of Apollonius, Archimedes, and Theodosius as early as 1665. At that time Barrow wrote an interesting reply:

> For your proposition concerning Archimedes and Apollonius, I cannot well tell what to answer. I have been offered, by a friend, to be at charges of printing them for me, which would yield me, I suppose, a considerable benefit; for I think I could put off many here. But till I be necessitated by some engage-ment, I shall hardly ever induce myself to take the pains and spend the time requisite for the reviewal of them; although within two or three months I think I could perform that. If the stationer, you mention, should make me a round offer, and propose fair conditions, I might perhaps be moved; till such occasion I am likely to supersede.[201]

Most likely, the plague and the fire of London put an end to such schemes. Nevertheless, when Collins and Scot became fast friends around 1670, old hopes revived and the previous scheme was put on a new footing.

Collins moved swiftly. In early October 1670, while Barrow was waiting for copies of his geometrical lectures, Collins requested both a copy of Barrow's edition of Apollonius's *Conics* and a copy of the perspective lectures Barrow had delivered at Gresham College some eight years before. His purpose was yet another project: a comprehensive treatise on optics in English. As Collins wrote to Sluse in October 1670, "One of our stationers here [John Martin] is very desirous to print in English a Treatise of Perspective, Catoptrics and Dioptrics.... Mr. Barrow is willing to communicate his Lectures of Perspective, which he read in English, but is not willing to own them as an author." Sluse, for his part, was asked to recommend suitable treatises on catoptrics and dioptrics that could be joined to Barrow's Gresham lectures, since the latter's *Lectiones opticae* was considered "somewhat too difficult for the vulgar." Barrow, though willing to impart the lectures, was modest about their value: "I see part of them written so ill, and so confusedly, that I fear you will hardly be able to make anything of them."[202] This project, however, came to nought, and Collins proceeded with his original Barrovian agenda of publishing the ancient geometers; as far as this project can be reconstructed, initially Collins intended to publish each of the ancient geometers separately, with Apollonius heading the list.

In the same letter in which Barrow trivialized the value of his perspective lectures, he likewise slighted his rendition of the *Conics*: "I have also... sent you my Apollonius, which yet is little worth your looking in, having in it nothing considerable, but its brevity." But Collins was not to be dissuaded. He was adamant not only about publishing Barrow's edition of the first four books of the *Conics,* but about adding the last three books – published by Borrelli in 1661 – trimmed and tailored to Barrow's method. With this objective, Collins proceeded by sending Barrow's manuscript to press in March 1671/2, at the same time seeking to enlist Edward Bernard, Savilian Professor of Astronomy at Oxford and a skillful Semitic scholar, to complete the task begun by Barrow. Bernard, Collins revealed to Gregory on 25 March 1671, had discovered in the Bodleian Library "two intire Coppies of the first 7 Bookes of Apollonius.... the 3 latter bookes when translated [from the Arabic] and put into Dr Barrowes method, may probably be printed with Dr Barrowes Comment on the first foure and be sold together."[203] Bernard seemed eager to see Barrow's edition published. On 14 March 1671/2 he thus urged Collins to proceed, insisting that "Dr Barrow's work must by no means tarry from the press, nor yet be put into the hands of untoward booksellers." Despite this support, however, in the end Bernard refused to participate in the project, fearing it would have an adverse effect upon true scholarship. According to Wallis, who articulated his co-Savilian Professor's reasons in a letter of November 1672, Bernard firmly believed that the complete texts should be published in the original language before any "epitome" be made.[204] In the face of such convictions, Collins proceeded with Barrow alone. By 20 February 1672/3 the text and most plates of the Apollonius had been printed, and the entire edition was completed a week or two later. In March, some of Collins's correspondents, including Wallis and Strode, received copies in the form of unbound sheets, presumably so they could inform Collins of errors.[205] And now that the Apollonius was nearly complete, Collins was ready to follow through with the rest of Barrow's unpublished manuscripts: Archimedes, Theodosius, and "about 30 Lectures of Mathematicall Sciences in generall."[206]

Conceivably, Bernard's refusal to edit the last three books of Apollonius thwarted the original plan to publish a complete, and separate, edition of the *Conics*.[207] For the time being, Collins left the Apollonius in the proof stage and concentrated on the Archimedes and Theodosius, which he now intended to adjoin to the Apollonius. Ultimately, his efforts proved to be no less frustrating than those involved in publishing the *Conics*. The text of the Archimedes, with the exception of one or two plates, was printed by 5 March 1673/4. However, it was exactly now that Collins's labor really began. For a start, he wanted to accomplish with Archimedes what had been denied him with the Apollonius – namely, to append texts and fragments that Barrow had not edited, such as the Arenarius, the copy of which Barrow claimed was "so corrupt, that without more time than he [could] allow to it, he [could] make nothing of it." Again, Barrow suggested Bernard as the person best equipped to check whether there existed in Oxford a more accurate manuscript of the Archimedian text and, if so, perhaps be persuaded to edit it. Collins also sought Wallis's advice and contribution, hoping the latter would add "the mechanics at the end of the edition of Rivaltus . . . [as well as] the Lemmata at the end of Borellius' Explication of the three latter books of Apollonius." Wallis declined. Initially begging lack of time, he added, "I think it better if Dr. Barrow would do it himself, that so the whole may be more uniform, as done by the same hand."[208]

Frustrated by the disinclination of both Savilian Professors to contribute substance to the project, Collins found some consolation in the figure of Henry Aldrich. In 1675 the future dean of Christ Church was a young don acclaimed both as a mathematician and as a Greek scholar who also served as Busby's lecturer in mathematics at Christ Church. Aldrich, we learn, "hath fitted Serenus de Sectione Cylindri for the press in Dr Barrow's method," and Scot was apparently willing to publish "another volume of the Ancients," which once again raised Collins's hopes that such a volume would include (in addition to Serenus and Pappus) the last three books of Apollonius.[209]

During this period (1673–5) the printed proofs of the various pieces in the projected volume were distributed by Collins among friends – Gregory, Bernard, Kersey, and Wallis – who were requested to check text and diagrams both against Barrow's own manuscript and for content. This proofreading was especially important in the case of the Archimedes, because Collins had been informed, apparently by Wallis, that the circulating printed version "prove[d] very faulty, chiefly through the errata of the MS Exemplar." By 10 September 1674 Collins had announced to Gregory that Scot was now intent on publishing all three Barrovian texts "next term." Gregory, for his part, was asked "to communicate as soone as you can an account of what errors you have met withal in the Theodosius and Apollonius and I hope next week to find an opportunity of sending you...Dr Barrow's Archimedes, craving your observations."[210]

Two weeks later Collins forwarded the proofs of the newly printed Archimedes. This time he asked Gregory "as soon as [he could] be pleased to transmit an account of what errors of the Press [he had] already observed in the Apollonius, or shall hereafter observe in the Archimedes." Progress, however, was slow. By 1 May 1675 Scot, it appears, had given up hope of publishing in the present volume either Serenus or the last three books of Apollonius and was "resolved to publish what is already done," namely, the three authors prepared by Barrow. Collins responded by once again urging Gregory and his students to finish their editorial work: "If your scholars have found out the errata in Dr Barrow's Archimedes and Theodosius vouchsafe to impart the same." The entire edition was finally printed by 1 June; but Collins was now visibly worried, for as late as September Wallis had informed him that even the present version was defective, including "many faults...which were not committed at the Presse." In a somewhat frantic gesture, Collins requested additional corrections from Gregory "with an explication of the symbols used throughout the whole" and turned to Newton for help. Writing Newton in July 1675, Collins enclosed an account of the errors pointed out by Wallis, expressing his hope "that out of your respects to the Doctor you would be pleased to find out the rest."[211]

In the midst of this turmoil, Collins embarked on yet another Barrovian project. As he informed Gregory on 31 December 1674, Scot was now "desirous to reprint Dr Barrow's Euclid, to which the Doctor consents, but hath no time to send up a correct copy, or to make any corrections or alterations of that in print. And if you or your scholars will vouchsafe to be helpful in this case, you will oblige the Doctor and bookseller who is willing to make a recompense." Recalling the distress caused by the faulty manuscript of Archimedes, Collins was intent on ensuring that Euclid "[should] have better success." To this purpose, he enlisted not only Gregory and the latter's students to review the former edition of Barrow and suggest corrections and improvements, but Newton as well. Newton indeed delivered. On 3 August 1675 Collins informed Gregory that Euclid "was carefully corrected and sent by Mr Newton."[212] Yet no edition of Euclid was ever published by Scot; the 1678 edition was published by John Williams and John Dunmore. Since that edition included new material previously in Collins's hands – namely, Barrow's lecture on Archimedes' method[213] – he was almost certainly responsible for the edition.

The experience of these years left its toll on Collins. Certainly, a more jaded and pessimistic tone can be detected in his correspondence of the late 1670s. In the letter to Thomas Baker previously alluded to, for example, he wrote, "The truth of it is, Mathematical learning will not here go off without a dowry; the booksellers have lost so much by the works of Drs. Wallis and Horrox, the Optic and Geometric Lectures of Dr. Barrow, &c., though by Mr. Gregory and others esteemed the best things extant, that it is no easy task to persuade booksellers to undertake any thing but toys that are mathematical."[214] Nevertheless, Collins never ceased to attempt to publish the literary remains of his friend. Immediately following Barrow's death, Collins was already planning to contact the elder Barrow to ensure that both the mathematical lectures and the Gresham perspective lectures be preserved and eventually published. He expressed similar hopes for other fragments, such as a "little tract of his about constructions for equations, but imperfect."[215] As it turned out, the perspective lectures were never published. Nor were the various fragments and spinoffs from the geometrical lectures. Only the mathematical lectures were eventually published.

It is difficult to determine the nature of Collins's involvement with the publication of Barrow's *Lectiones mathematicae,* which first appeared in 1683, shortly before Collins's death. The book was printed by John Playford, the younger, for the bookseller George Wells. If Collins was responsible for persuading Wells to undertake the edition, he was certainly not around to see the manuscript through the press. In fact, the publication of the lectures was chaotic. The "first edition" of 1683 was a duodecimo volume composed merely of Barrow's prefatory oration and the first eight lectures, although the title page promised the lectures for 1664, 1665, and 1666. The following year an octavo "edition" was printed by Wells, this time containing Lectures 9–23, as well as the four lectures on Archimedes, with three separate title pages for each set of lectures (9–15 and 16–23 for 1665 and 1666, respectively), and the Archimedes lecture of uncertain date. Finally, in 1685, another octavo edition was published, again carrying the all-inclusive title page of the 1683 edition, but again the text included only the prefatory oration and first eight lectures. It was as if the publisher, realizing the disarray, provided those who purchased the 1684 edition with the missing first part of the book.

Master of Trinity College

Barrow's integrity was a recurring motif in his life, and the issue becomes crucial to our understanding of his resignation of the Lucasian

Professorship. Modern historians, in contrast to Barrow's contemporaries, have failed to appreciate the force and sincerity of the motives that animated this decision. Indeed, some of the most influential Newtonian scholars, seeking to explain the act immediately responsible for Newton's own appointment to the position, have attributed this resignation to Barrow's alleged outright careerism, thereby contrasting him with the "virtuous" Newton, whose predicament it was to live amongst place seekers. "Man of the world and fighter that [Barrow] was," insists Whiteside,

> it must frequently have been a sore point to him that the King's confirmation in January 1664 of Lucas' statutes, aimed expressly to prevent the Lucasian professor taking an active part in academic administration or politics, forbade him. . . from holding any university or college position of any kind. . . . Barrow's resignation of the chair in 1669 may reasonably have been but a necessary preliminary to his being considered, when next it was vacant, for the Mastership of Trinity.

Westfall, referring with approval to this passage, carries the argument a step further. Barrow, dissatisfied with the Greek and Lucasian Professorships, was "clearly hungry for more": "It is known. . . beyond doubt that Barrow was a man ambitious for preferment."[216] Not only are such speculations not substantiated; they demonstrate a lack of understanding of Barrow and, more important still, considerably downplay the evidence for Barrow's motives, namely, that he aimed at a higher – but *not* a more profitable – calling. To understand these priorities, which, though suspect today, were in full accordance with contemporary beliefs, it is necessary to reconstruct the web of events that precipitated Barrow's action.

Barrow, we have seen, experienced at least two "vocational" crises concerning the nature of his true calling. It appears that by 1669 he had experienced yet another crisis, even more acute than the previous two. One year shy of his fortieth birthday, Barrow seems to have been struck by a deep sense of doubt about his accomplishments, not in the field of science, but with regard to God and his salvation. Something of this anxiety can be sensed from Barrow's letter to his friend Tillotson, when he presented the latter with a copy of his *Lectiones geometricae* in 1670:

> While you, dear man, expound to the people the mysteries of sacred truth, closing the mouths of petulant sophists and, at the same time, waging successful war on behalf of God's law; lo, I am tied miserably to these hooks which you see, wasting my time and intellect. The explanation of my hard lot is manifest, but I will be modest about this unwanted offspring.[217]

Barrow's early biographer furnishes yet another example of his anguish: "He was afraid, as a clergyman, of spending too much time upon Mathematics; for. . . he had vowed in his ordination to serve God in the Gospel of his Son, and he could not make a bible out of his Euclid, or a pulpit out of his mathematical chair."

The problem was intensified by Barrow's own advocacy, in principle, of undivided devotion to one's "calling" – and for just this reason he had insisted that university professors be debarred from any other offices or employments. How much more so, then, when one was involved in the pursuit of the highest of callings? Barrow, therefore, took the obvious next step and resigned, not only his professorship, but also active interest in secular studies. As Walter Pope, who was living with Barrow during much of the period between 1669 and 1672, described it, "At that time he applied himself wholly to Divinity, having given a divorce to Mathematics, and Poetry, and the rest of the *Belles lettres,* wherein he was so profoundly versed, making it his chief, if not only business, to write in defence of the Church of *England,* and compose Sermons."[218] In so doing Barrow simply followed, with greater sincerity than many, a long-trodden tradition of abandoning one's "secular" studies for the pursuit of divinity. One example should suffice to illustrate this point. On 14 October 1522, five days before he was consecrated bishop of London, Cuthbert Tunstal penned the dedication of his famous arithmetical treatise *De arte supputandi* to Thomas More, in which he expressed convictions corresponding to those that would guide Barrow a century and a half later:

> But now that I...have nevertheless been nominated...I not
> only resolved that I would devote what is left of my life to
> sacred literature putting all wordly writings entirely aside, but
> at first I thought it fitting that those papers...should be thrown
> away. For I did not consider them worthy to come into the
> hands of learned men, nor did I think it right that any part of
> my life should for the future be filched from sacred studies to
> polish them up.[219]

Such sentiments may explain why Barrow, too, was unwilling to polish and revise his lectures for publication.

Barrow, then, made a clean and complete break with his career in order to achieve that leisure and peace of mind he so longed for. In so doing, he incurred a considerable loss of income, for he was now left with only his fellowship at Trinity. Yet Barrow felt no remorse. As he wrote John Mapletoft after he had made up his mind, "In sooth I never find any regrett for my being a poore meane fellow."[220] It was equally obvious, however, that it would be difficult for Barrow to remain "a poore meane fellow" for long, for he was coming under increasing pressure from friends to take the final step and become an ecclesiast. This Barrow was unwilling to do. His reluctance as well as his overall predicament were the same that faced other distinguished scholar-divines such as Joseph Mede and Henry More, both of whom, despite their piety and dedication to serving God, would not fit into the church hierarchy. Indeed, this was the dilemma facing many intellectuals. Impatient with church administration and the demands it would make on their leisure to study, all such scholars believed themselves unfit for the task of curing souls or constantly preaching to country folk, since their sermons were learned and academic. Barrow and his like believed that they could best serve God and church with their pens and by educating future generations of soldiers in the wars of the Lord. Obviously, such a career could be pursued only at the universities. This was why Mede and More preferred to remain "mere" fellows at Christ Church, declining many outside offers, while others such as John Pearson, who eventually received church dignity, were ultimately judged to be better divines than they were bishops.

Even before his resignation,[221] Barrow dedicated himself with renewed vigor and enthusiasm to his new vocation. His first major effort, in accordance with college statutes, was a treatise on the Creed, a topic that had been treated most exhaustively by the master of Trinity, John Pearson, in his own masterpiece, *Exposition of the Creed* (1659 and later editions). The two works, however, were quite dissimilar. As Osmond rightly observed, while Pearson "had delivered his Lectures during the Interregnum to an adult congregation with a highly developed curiosity in dogmatical theology; Barrow was dealing with a more juvenile congregation, and at a time when it seemed all-important to emphasize the practical implications of the faith."[222] The series was ostensibly delivered in anticipation of Barrow's becoming college preacher. Barrow, however, was not appointed to the position until 21 December 1671. In the meantime, he appears to have divided his time between Cambridge and Salisbury, where he lived with his close friend the bishop, Seth Ward; and in 1670 he was also appointed royal chaplain.

Barrow's friends, however, never ceased in their efforts to secure him a church position. In fact, Ward was so keen to keep Barrow with him at Salisbury that he proposed to appoint his younger friend archdeacon of Salisbury, when the position became vacant on 26 August 1670 following the death of Joshua Childrey. Barrow declined the offer, and John Sherman was appointed in his stead. Sherman's elevation, however, freed the following year the prebend of Yetminster; this gift Barrow consented to accept, and was duly installed on 16 May 1671.[223] The reason for his change of heart was familial. According to Pope, Barrow wished to raise the sum of £500 in order to give it to his half-sister "for a portion, that would procure her a good husband." Quite likely, Thomas Barrow's age and shaky finances prevented him from providing his daughter with an adequate dowry, and Elizabeth, who was now in her midthirties, was facing the prospect of remaining a spinster.

Another person to labor on Barrow's behalf was his uncle, by now bishop of Bath and Wells, who bestowed a small sinecure on his nephew. However, Barrow used the income from this gift solely for charitable purposes and continued to do so until his death, when the living was bestowed on the famous Hebraic scholar Humphrey Prideaux. Barrow resigned the prebend once the terms of the lease he had set out expired.[224]

A man with such qualifications and determined friends, however, could not be long expected to remain an obscure college fellow. And since Barrow showed no desire to become a church functionary, his friends directed their attention toward acquiring for him an appropriate university appointment, which, given the positions he had already filled, could be nothing else than the Regius Professorship in Divinity or a mastership of a college. In the process Barrow gained two very powerful allies. The first was George Villiers, second duke of Buckingham, chancellor of Cambridge, a Trinity man, and, most important, the patron of John Wilkins, who, no doubt, mobilized the duke on Barrow's behalf. The second was Gilbert Sheldon, archbishop of Canterbury, and here Bishops Barrow and Ward had been instrumental. This combination of patrons was unusual, since Buckingham and Sheldon were usually to be found on opposite sides of the religious and political spectra. It is yet another tribute to Barrow's character that two such different figures joined efforts on his behalf. The coalition was also essential since a "package deal" had to be worked out. If Barrow was to be made master of Trinity College – as appears to have been the plan by 1671 – a promotion had to be found for the incumbent master, John Pearson. This was a prickly matter, for during the past few years, Buckingham and the "cabal" had systematically thwarted Sheldon's efforts to confer a bishopric upon Pearson, and it took the Barrow case to remove such resistance. Charles II was glad to award the mastership – which was a gift of the Crown – to his own Chaplain and, in his own words, "the best scholar in England." By a strange coincidence, it was the death of Barrow's friend and patron John Wilkins, bishop of Chester, that cemented the exchange. Wilkins died on 19 November 1672 and was buried on 12 December. Eight days later Isaac Barrow, senior, was already in the position to write a thankful note to Archbishop Sheldon: "I was very willing to take this occasion to present my humble duty, & most thankfull acknowledgment of y^r Graces favour to my Nephew Dr Barrow in y^t plentifull provision you have made for him."[225] Pearson was consecrated bishop of Chester on 9 February 1672/3, and Barrow received his letter mandate as master of Trinity College six days later.

Even before returning to Cambridge, Barrow displayed his high regard for his college statutes. Although the mandate from the king provided him with a dispensation to marry, Barrow regarded marriage inappropriate for good mastership. He therefore went through the additional expense of having a new letter patent drawn, one that removed the dispensation. On Thursday afternoon, 27 February 1672/3, Barrow arrived back at Cambridge from London[226] to take charge of the college that had been his home for more than a quarter of a century. Trinity was the largest of the Cambridge colleges, with some four hundred members.[227] It was also the wealthiest. By the end of the century, each fellowship was worth, all told, some £150–200 p.a., and the mastership carried a gross remuneration of roughly £500 p.a.[228] Intent upon providing an example worthy of emulation, Barrow, whose needs were always modest, gave up certain luxuries enjoyed by his predecessors, such as the use of a coach. To judge from all reports, Barrow's five-year tenure proved a happy one for both master and college. As Barrow himself commented in the summer of 1673, "Trinity College is, God be thanked, in peace (I wish all Christendome were so well)."[229] Admissions remained high and discipline stern; college lectures and other exercises were faithfully kept; and it seems that little, if any, dissension plagued the society, all of which was to change radically during the terms of the next three masters. In retrospect, however, two major issues appear to have occupied the mind and energy of the master. The first concerned the sticky problem he inherited from Pearson: how to cope with the growing burden of external interventions – royal and otherwise – in college affairs, particularly in the elections of fellows. The second involved Barrow's efforts to initiate what would prove to be his greatest contribution to Trinity: the building of the magnificent Wren Library, which, unfortunately, he did not live to see completed.

External intervention in elections for headships and fellowships in Cambridge colleges did not come into being with Charles II, who merely emulated practices widely used during the 1640s and 1650s. After the restoration, however, the recourse of students and scholars to outside support reached epidemic proportions, owing to the new precedents set during the 1660s. It all began innocently enough when those ejected fellows and college heads who had not been able to find adequate provisions elsewhere began, as a matter of course, to demand the return of their university positions. In subsequent years, however, Charles II increasingly realized that fellowships and scholarships constituted a cheap medium for rewarding loyal subjects and, potentially more alarming still, the sons and kin of those who had been loyal to the Stuarts. Thus, by 1674 intervention had reached such massive proportions that Edmund Boldero, master of Jesus College and the incumbent vice-chancellor, had managed to convince the chancellor, the duke of Monmouth, to act on behalf of Cambridge and bring a halt to such a state of affairs. Monmouth was able to obtain from the king a letter in which Charles declared that having been made aware

> that the number of letters lately requiring or recommending students to be elected to scholarships or fellowships is prejudicial to the freedom of election allowed by the statutes of the colleges, [he was] declaring therefore...that he [would] not expect compliance from the heads of colleges with his said

letters, unless the persons recommended [were] found qualified for the preferment; leaving them otherwise to make their selections according to their statutes.[230]

By this late date, however, the damage was already done, and the dangerous precedent set. Trinity College was particularly vulnerable to such abuses because of its size and wealth and because it was a royal foundation.

This disturbing state of affairs can be detected in the pattern of election to Trinity fellowships after the Restoration. In 1661 thirteen new fellows were elected, four of whom were Westminster scholars. In the next triennial election, in 1664, seventeen fellows were added, including seven Westminster scholars. In contrast, 1667 witnessed the election of only eight fellows. But an extraordinary, and totally out of sequence, election occurred the following year, 1668, when no less than fifteen fellows (four of whom were Westminster scholars) were added. Clearly, a total of fifty-three fellows within seven years was phenomenal, not to say alarming. Small wonder, then, that in 1670 and 1671 only one Westminster scholar was elected per year, that no elections were held in 1672, and that only three Westminster scholars were elected in 1673.

Such was the background to the first major open fellowship election to occur in six years – that of 1674 – and, as it happened, the only one to be held during Barrow's mastership. Obvious to all was the rigorous competition as well as the widespread attempts of various contenders to garner all the outside support they could muster, especially in the light of past events. It was a well-publicized fact that all three Westminster scholars elected fellows the previous year owed their fellowships to letters mandate received from Charles II![231] Thus, between February and May 1674 we find at least four scholars producing letters from the king demanding their appointment as fellows (Nathaniel Rashleigh, Purbeck Richardson, Christopher Wyvil, and William Davis); a fifth, William Ellis, who was already an M.A., and therefore ineligible for election, produced a letter from the king instructing the college to bestow on him the vacant office of library keeper.[232]

Barrow was aware of what was about to follow. As early as 1673, when there was a possibility of an election, he had sought to enhance the chances of a deserving student by writing the student's father to obtain letters of recommendation as they "would have done better [in the eyes of the College Fellows] than mandate." But since no election took place, Barrow advised the father confidentially to procure a letter mandate after all and send it to Barrow "to be produced when needful, that is, in case the places voidable be stopped by other mandates, or any go that way to skip over your son's head."[233] This episode illustrates the dilemma Barrow faced. Continuous outside intervention in college affairs, he believed, could not but have a devastating effect on the college, resulting in disregard for both merit and statute, and hence the failure of many worthy students to obtain fellowships; the options for such students would be either to cut short their courses of study or to migrate to other colleges, as was indeed happening with increasing frequency. Barrow decided, therefore, to resort to a gamble and force the election of deserving students, no matter how many vacancies there actually were or how many students had been elected by virtue of obtaining letters mandate. His strategy was to turn the technicality used in the king's letters on its head. Since the king's orders stipulated that supernumerary fellows be elected – that is, a student be elected before a vacancy became available and be instituted later when such a vacancy did occur – Barrow sought to effect the election of other scholars the same way. As he wrote the duke of Monmouth:[234]

> It having pleased his Majesty to grant his letters mandatory... for electing divers persons into Fellowships, beyond the number of present vacancies; so that, in obedience to those commands they must be obliged to pre-elect some of those persons. And whereas there are in the said College many Bachelors of Arts capable of Fellowships.... It is therefore... for the encouragement of merit and study, humbly requested that they may be allowed, with his Majesty's leave, to pre-elect, together with those whom his Majesty hath recommended, some others of the said Bachelors, who shall, by trial in the statutable way, appear best deserving that preferment.

Barrow was allowed to have his way. The result was staggering: the election of an unprecedented number of fellows, twenty-three, twelve of whom were Westminster scholars. The master himself was placed under much pressure to lend his support to various candidates. But again it appears that Barrow's integrity and concern for the college enabled him to execute his office with a clear conscience. As early as July 1673, for example, one of his closest friends, John Mapletoft, asked him to enhance the election of William Davies. Barrow refused: "Your project concerning Mr. Davies I cannot admitt. Trinity College is, God be thanked, in peace... and it is my duty, if I can, to keep uproars thence. I do wish Mr. Davyes heartily well, and would doe him any good I could; but this I conceive neither faisible nor fitting."[235] Whatever the nature of the proposal, Davies went ahead and obtained a letter from the king, which ensured his election.

Whether Barrow intended to make a remonstrance beyond a desire to effect the election of deserving scholars who would not have been able to procure letters from the king is unclear. It is also impossible to speculate how he intended to cope with future scheduled elections. Certainly, his unconventional act signified only a short-term solution to a major problem. Be this as it may, Barrow died in May 1677, and no new fellows –

not even Westminster scholars – were elected before 1679. For example, only one of the fifteen scholars who were elected in 1672 became a fellow (in 1679), while of the eleven scholars elected the following year, only three would be elected fellows in 1680. A revealing testimony to the problem, and an "ingenious" attempt by one of these unfortunate scholars to better his chances, is provided in a memorandum of Adam Ottley, the future bishop of St. David's, who entered Trinity on 29 January 1671/2, was elected scholar in 1674, and graduated B.A. early in 1676. Because the statutes of the college stipulated that a bachelor be elected into a fellowship within three years of obtaining his degree or leave the college, the predicament of the immense backlog created by the previous elections no doubt weighed heavily upon Ottley. Thus, the mandate he requested from the king contained not the usual formula for an election, but one "permitting him to jump the queue"! The fellows were much divided over such an enterprising solution, and eventually resolved to send Ottley to Trinity Hall instead.[236]

But perhaps the most serious, as well as consequential, event connected with the 1674 elections was the appointment of John Montague as fellow. He was the fourth son of Admiral Edward Montague, first earl of Sandwich, who died in battle three months after his son entered Trinity College on 12 April 1672. The following year John was created M.A., *jure natalium,* and now in September 1674 he wished to be elected fellow. Thomas Ross, Montague's former tutor and presently the king's librarian and groom of the private chamber, mobilized both Samuel Pepys and the duke of Monmouth to act on Montague's behalf. On 26 September the king sent a letter to Barrow instructing the master and seniors of Trinity to elect the young nobleman, notwithstanding Montague's "*not being a scholar,*" and grant him all necessary dispensations so that he could sit for examination.[237] Montague was elected. He was also created doctor of divinity in 1682 (at the age of 27!) by royal mandate and the following year appointed master of Trinity, a post he held for seventeen years. Small wonder that it was under the mastership of Montague – as congenial a person as he may have been – that the college entered a stage of rapid deterioration, for the master kindled neither discipline nor learning.

These events substantiate my previously stated argument that Barrow was a realist, not a ritualist, in his advocacy of the preservation of both letter and spirit of the statutes. He was correct in his judgment that breach and encroachment of regulations would only encourage further and more blatant violations, and that even estimable motives could not justify the creation of a precedent that would encourage subsequent abuse and corruption. Certainly, the events that occurred immediately after the ordeal of the above-mentioned elections justify such views, as well as shed light on certain events that have been often quoted, yet little understood, by historians.

The need to constantly play a delicate balancing act – to preserve the independence and prestige of fellowships at Trinity College without incurring royal displeasure – made Barrow sensitive to any attempts at further encroachments upon college statutes. It is against this background that two other events of 1674–5 must be viewed. Toward the end of 1674, Francis Aston completed seven years from the time he had been elected a fellow of Trinity. According to the statutes, he was expected either to take holy orders or to vacate his fellowship. Aston, an accomplished virtuoso, a friend of Newton, and future secretary of the Royal Society, had for some years been traveling on the Continent and was keen to retain his fellowship without taking orders. He therefore sought to obtain a dispensation, and to such an end he enlisted the powerful support of Sir Joseph Williamson, secretary of state. Barrow, for his part, was adamant about not allowing such a dangerous precedent – regardless of Aston's learning and character – especially at a time when so many students were openly resorting to external pressures to bear upon college affairs. He therefore wrote a firm letter to Williamson, remonstrating that if Aston received the dispensation, what would prevent any young fellow with friends in high places from doing the same? Barrow further elucidated the objectives and duties of a fellowship, stressing that serving as a sinecure was not one of them.[238] Barrow had his way.

Similarly, when the law fellowship at Trinity became vacant in 1673, and Newton, who was equally desirous to be dispensed from ordination, stood as a candidate, Barrow did not hesitate to thwart his friend's wish and decide in favor of Robert Uvedale – not so much, I believe, because of the latter's "seniority," as because it was obvious that Newton had no intention of dedicating himself to the study of law. Barrow's integrity and strong sense of propriety, coupled with a deep concern over dangerous precedents, caused him to deny even a friend an unlawful request.[239] Barrow did, however, recognize the rationale behind Newton's wish. Although he may not have known the extent of Newton's heterodox religious beliefs, he was aware of the incongruity between scientific and theological vocations and was more than willing to lend his strong support in order to procure for Newton a royal letter of dispensation that would emancipate not only Newton, but all subsequent Lucasian Professors from the onus of taking holy orders.

Barrow's good sense was lost on his successors, but his love for his college and determination to advance learning found a more durable expression in the new library at Trinity College. According to Roger North, Trinity came to possess such a magnificent library because the Cambridge heads of house refused to lend a hand to building the Cambridge version of the Sheldonian:

> They say that Dr Barrow pressed the heads of the university to build a theatre, it being a profanation and scandal that the

> disputations and speeches should be had in the university
> church, and that also be deformed with scaffolds, and defiled
> with rude crowds and outcries. The matter was formally con-
> sidered at a council of the heads, and arguments of difficulty
> and want of supplies went strong against it. Dr Barrow assured
> them that if they made a sorry building, they might fail of
> contributions, but if they made it exceeding magnificent and
> stately, and at least exceeding that of Oxford, all gentlemen of
> their interest would generously contribute.... But sage
> occasion prevailed, and the matter at that time was wholly laid
> aside. Dr Barrow was piqued at this pusillanimity and declared
> that he would go straight to his college and lay out the founda-
> tions of building to enlarge his back court, and close it with a
> stately library, which would be more magnificent and costly
> than what he had proposed to them.[240]

Even though the story may be somewhat inaccurate, since Barrow made
the *public* appeal to build a theater only in June 1676 and construction
work on the library had begun some four months earlier, still Barrow was
correct in his estimation that it would be possible to generate widespread
excitement and raise substantial funds. In his last year, he himself did
much of the paper work involved in raising the huge sum required; the
final cost ultimately reached in excess of £16,000. Hill describes the care
Barrow took in his public relations: "He writ out quires of paper, chiefly
to those who had been of the College, first to engage them, and then to
give them thanks, which he never omitted." Barrow, then, wrote to past
members of the college, public figures, and his personal friends, both
soliciting their contributions and urging them to use their influence to
recruit relatives and friends for a such purpose.[241]

Barrow was also able to obtain the good services of his friend Christopher Wren, who rendered his architectural skills gratis, even after he
was asked to alter his original design.[242] Undoubtedly, had Barrow lived
longer, his actual contribution to the building would have been larger
than the £100 he gave in 1676; equally plausible, Barrow would have bequeathed his books to the new library had he not died intestate. Still, his
vision and enthusiasm made the venture possible, and the building stands
as a monument to Barrow's achievement.

Early in April 1677 Barrow traveled to London to take part in the annual
election of Westminster scholars. On 13 April he preached a sermon at
Guildhall Chapel, and shortly thereafter he contracted "malignant fever."
According to Pope, he attempted to cure himself by fasting and taking
opium, a combination Barrow had once successfully administered to himself in Constantinople, but to no avail. He died on Friday, 4 May 1677,
in his forty-seventh year. Many grieved his death. To Collins, Thomas

Baker revealed that the news "extracted tears" from his eyes; and John Locke, then in Paris, mourned the death of a "very considerable friend." A letter of condolence sent by Sir Heneage Finch, the lord keeper, to Thomas Barrow stated that the father "had too much cause to grieve, for no Father lost a better Son." And Newton, who was rarely given to expressing emotions, told Conduitt years later that "no one had greater cause to regret Barrow's death than he."[243]

Barrow was buried at Westminster Abbey three days after his death. Later, a poignant epitaph was composed by John Mapletoft and corrected by Thomas Gale. The character of Barrow portrayed in that elegy was faithful to the man:

> He was a Godlike, and truly great Man, if Probity, Piety,
> Learning in the highest degree, and equal Modesty, most holy
> and sweet Manners, can confer that Title. . . . He imitated God,
> whom he had served from his Youth, in wanting few things,
> and doing good to all, even to Posterity, to whom, tho dead, he
> yet Preaches.

Other contemporaries were unanimous in their appreciation of his integrity, modesty, and willingness to help others. For John Evelyn he was "that excellent pious & most learned man, Divine, Mathematitian, Poet, *Traveller,* & most humble person."[244] Indeed, Barrow seems to have been one of the very few leading scholars of the early modern period who had no enemies. This trait is particularly conspicuous, given the fractious nature of the world of seventeenth-century mathematics in which he obtained eminence. In fact, it was this quality of good-naturedness that caused Roger North to observe that Barrow was the only great virtuoso who was free of arrogance.[245]

Since only two of Barrow's sermons were published during his lifetime, naturally his reputation as a theologian was posthumous. In rapid succession John Tillotson arranged the publication of Barrow's sermons (in three folio volumes published between 1678 and 1680 and for the manuscript of which the publisher Barbazon Aylmer paid Thomas Barrow the staggering sum of £450).[246] In 1680 Tillotson also issued Barrow's masterpiece, *A Treatise of the Pope's Supremacy.* This influential work – the only one that Barrow, on his deathbed, requested that Tillotson publish – was, according to Anthony Wood, put to good use during the reign of James II, when Matthew Tindal, a fellow of All Souls College, Oxford, declared himself a Catholic in 1687. He "was esteemed a zealous brother, and was on the point of being a Carthusian, but reading Dr. Isaac Barrow his book and by conversation with some of his house, he denied the popish religion." Tindal, it seems, was rescued from Catholicism only to become a deist.[247]

The enormous reputation of "that very great Man Dr. *Barrow,*" as William Whiston referred to him,[248] was to last for nearly two centuries. His name was to be found in almost any constructed "pantheon" of the great English divines. Locke named Barrow, along with Whichcote and Tillotson, one of the greatest practical divines, while Gilbert Burnet preferred Sanderson, Farrington, and Hammond as Barrow's companions in such studies. James Thomson, for his part, adjoined the triumvirate of Barrow, Stillingfleet, and South to his catalogue of English sages, which included Bacon, Boyle, Locke, and Newton.[249] Philip Dodridge singled out for recommendation Barrow and Tillotson among the divines of the established Church of England, thereby continuing the tradition of praise of Barrow by Nonconformist divines, which had begun with Richard Baxter.[250] Finally, we may quote Dugald Stewart:

> [Barrow's] theological works (adorned throughout by classical erudition, and by a vigorous, though unpolished eloquence) exhibit, in every page, marks of the same inventive genius which, in mathematics, had secured to him a rank second alone to that of Newton. As a writer, he is equally distinguished by the redundancy of his matter, and by the pregnant brevity of his expression; but what more peculiarly characterizes his manner, is a certain air of powerful and of conscious facility in the execution of whatever he undertakes. Whether the subject be mathematical, metaphysical, or theological, he seems always to bring to it a mind which feels itself superior to the occasion; and which, in contending with the greatest difficulties, puts forth but half its strength.[251]

Today, of course, Barrow is remembered principally for his mathematics, and in this domain, too, his reputation was high. His textbook on Euclid was continuously reprinted, in both English and Latin, and some fifty editions had been published by the middle of the eighteenth century. His optical lectures, as Shapiro documents in his chapter, were relatively well known - if not always properly appreciated - during the eighteenth century. Perhaps the testimonial of a highly educated amateur in matters optical, James Logan of Philadelphia, best conveys some of the prevailing sentiments regarding Barrow's reputation as a mathematician:

> After Kepler divers wrote on the Subject, with some further Improvments, but they generally trod very much in the same path, till Dr. Barrow's excellent Optic Lectures were published in 1669. These struck out new Lights to the whole Science of Optics, and in Dioptrics particularly, he gave or pointed out certain Analogies, wch. had not been done before, for finding the ffoci, but it was generally by compounded Ratios.[252]

Notes

1. John Aubrey, *Brief Lives,* ed. A. Clark, 2 vols. (Oxford, 1898), 1:87–93; Charles H. Cooper, *Athenae Cantabrigienses,* 3 vols. (Cambridge, 1858–1913), 2:98, 545; *Admissions to Trinity College, Cambridge* (London, 1911), 2:38, 102, 146. A translation by Isaac Barrow, M.D., of Actuarius's *Compendium of Urines* is in British Library (BL) Ms. Sloan 1635, fols. 30–61. His will, dated 15 February 1616/17), is mentioned by Baker, Cambridge University Library (CUL) Ms. Mm 1. 37, p. 145.
2. Percy H. Osmond, *Isaac Barrow, His Life and Times* (London, 1944), p. 7; Aubrey, *Brief Lives,* 1:88.
3. *The Oxinden Letters,* ed. Dorothy Gardiner (London, 1933), pp. 116, 184, 248–51, and passim; Henry R. Plomer, "The Oxinden Letters," *The Library,* 2d ser., 6 (1905), 29–44.
4. Paul A. Parrish, *Richard Crashaw* (Boston, 1980), pp. 20–1.
5. Aubrey, *Brief Lives,* 1:87; *The Theological Works of Isaac Barrow,* ed. Alexander Napier, 9 vols. (Cambridge, 1859), 1:xxxix.
6. The main sources for the biographical details of Barrow's life are Abraham Hill, "Some Account of the Life of Dr. Isaac Barrow," in Napier, ed., *Theological Works of Isaac Barrow,* 1:xxxvii–liv; Aubrey, *Brief Lives,* 1:87–94; *Biographia Britannica,* ed. A. Kippis, 6 vols. (London, 1747–66), 2:629–38; Walter Pope, *Life of. . . Seth, Lord Bishop of Salisbury,* ed. J. B. Bamborough (Oxford, 1961), pp. 136–81.
7. *The Autobiography of Sir John Branston* (Camden Society, vol. 32, 1845), p. 124; Victoria County History, *A History of the County of Essex,* ed. W. R. Powell, 8 vols. (London, 1956–87), 2:534; William Hunt, *The Puritan Moment* (Cambridge, Mass., 1983), p. 257; A. G. Matthews, *Calamy Revised* (Oxford, 1934), pp. 70–1; J. T. Cliffe, *The Puritan Gentry* (London, 1984), p. 80.
8. Christoph J. Scriba, "The Autobiography of John Wallis, F.R.S.," *Notes and Records of the Royal Society,* 25 (1970), 24–5.
9. Ibid., pp. 25–7, 44 n18. For a general discussion of the grammar school curriculum, see Foster Watson, *The English Grammar Schools to 1660* (rpt. London, 1968); T. A. Baldwin, *William Shakspere's Small Latine & Lesse Greeke,* 2 vols. (Urbana, Ill., 1944).
10. Aubrey, *Brief Lives,* 1:88.
11. See p. 256.
12. Another former member of Peterhouse was his mother's uncle, Bowes Buggin, who received his M.A. in 1633 and who may have also influenced the choice of a college for his nephew. *Admissions to Peterhouse,* p. 26.
13. Aubrey, *Brief Lives,* 1:89.
14. J. D. Twigg, "The Parliamentary Visitation of the University of Cambridge, 1644–5," *English Historical Review,* 98 (1983), 513–28.
15. Joseph Hunter, *The Rise of the Old Dissent Exemplified in the Life of Oliver Heywood* (London, 1842), p. 44; John Tulloch, *Rational Theology and Christian Philosophy in England in the Seventeenth Century,* 2d ed., 2 vols. (Edinburgh and London, 1874), 2:55.
16. Thomas Hill, *The Strength of the Saints to Make Jesus Christ Their Strength* (London, 1648), sig. A2ᵛ–A3; idem, "A Letter to the Seniors of Trinity-College," in *The Best and Worst of Paul* (Cambridge, 1648), unpaginated.
17. *The Diary and Correspondence of Dr. John Worthington,* ed. James Crossley, 3 vols. (Chetham Soc., Manchester, vols. 13, 36, 114, 1847–86), 1:22.
18. See p. 257.
19. Anthony Tuckney, *Death Disarmed* (London, 1654), p. 43. Sheffield University Library, Hartlib Papers, 15/6/22, Thomas Smith to Samuel Hartlib, 20 November 1648. Wollaston had been a recurrent trustee of Gresham College, which may explain his

willingness to encourage mathematical learning at Cambridge; see Ian Adamson, "The Foundation and Early History of Gresham College, 1596-1704" (Ph.D. thesis, Cambridge University, 1975), pp. 288, 293. Hill's sermon "The Beauty and Sweetness of an Olive Branch of Peace and Brotherly Accommodation Budding," in *Six Sermons* (London, 1649), is dedicated to Wollaston. The first lecturer seems to have been John Smith, the Platonist; see p. 19. If not before, the lectureship must have been terminated by 1652, the year in which Wollaston retired from public life and ceased to be a member of the Gresham committee and the year in which Smith died.

20. J. C. T. Oates, *Cambridge University Library, A History: From the Beginnings to the Copyright Act of Queen Anne* (Cambridge, 1986), pp. 249-51.

21. *The Laws or Statutes of the University of Cambridge* (Cambridge, 1828), p. 7; William Harrison, *Description of Britain* (1577), quoted in *Education in Tudor and Stuart England*, ed. David Cressy (London, 1975), p. 117; George Dyer, *The Privileges of the University of Cambridge*, 2 vols. (London, 1824), 1:289, quoted in Mark H. Curtis, *Oxford and Cambridge in Transition, 1558-1642* (Oxford, 1959), p. 97.

22. Roger North, *General Preface & Life of Dr John North*, ed. Peter Millard (Toronto, 1984), p. 114.

23. J. Looney, "Undergraduate Education at Early Stuart Cambridge," *History of Education*, 10 (1981), 9-19.

24. Hill, "A Letter."

25. *The Works of the Pious and Profoundly Learned Joseph Mede*, ed. John Worthington (London, 1677), p. iv.

26. *Conway Letters*, ed. Marjorie H. Nicolson (New Haven, Conn., 1930), pp. 393, 395, 397-8.

27. J. H. Monk, "Memoir of Dr. James Duport," *Museum Criticum*, 2 (1826), 672-98.

28. See p. 56.

29. Monk, "Memoir," p. 680; Matthews, *Calamy Revised*, passim.

30. Aubrey, *Brief Lives*, 1:89.

31. Monk, "Memoir," p. 695; James Bass Mullinger, *Cambridge Characteristics in the Seventeenth Century* (London and Cambridge, 1867), p. 183; Osmond, *Isaac Barrow*, p. 28. Charles Webster, *The Great Instauration* (London, 1975), p. 135 n126.

32. James Duport, *Musae subsecivae* (Cambridge, 1676), pp. 47-8.

33. Ibid., pp. 65-6, 315-18; see also Don Cameron Allen's introduction to his edition of John Hall's *Paradoxes* (Gainesville, Fla., 1956).

34. "Rules to Be Observed by Young Pupils & Schollers in the University," Trinity College, Cambridge, Ms. O.10A.33, pp. 1-15, printed by G. M. Trevelyan in *Cambridge Review*, 64 (1943), 328-30. The quotations are from p. 330 (emphasis added).

35. H. W. Garrod, "Phalaris and Phalarism," in *Seventeenth Century Studies Presented to Sir Herbert Grierson* (Oxford, 1938), p. 361. Duport and Temple were obviously not alone in espousing such convictions. John Eachard commented in 1670 that "we are now in an Age of great Philosophers and Men of Reason.... And that Greek and Latin, which heretofore (though never so impertinently fetched in) was counted admirable, because it had a learned twang; yet, now, such stuff, being out of fashion, is esteemed but very bad company." *The Grounds & Occasions of the Contempt of the Clergy and Religion* (Oxford, 1670), reprinted in *An English Garner* (Westminster, 1903), p. 263.

36. J. Glucker, "Casaubon's Aristotle," *Classica et Mediaevalia*, 25 (1964), 296.

37. Thus, he may not have had firsthand knowledge of Descartes, but he got it from works such as Jean Baptiste Du Hamel's *De consensu vetus & novae philosophiae*. By contrast, his hostility toward Hobbes was grounded firsthand on his reading of *De cive* (1647), *Leviathan* (1651), and *Of libertie and necessitie* (1654). He owned copies of Barrow's works, which were obviously presentation copies, but more interesting, Duport possessed two copies of James Gregory's *Optica promota* (1663) and a few mathematical

textbooks. Other books included John Wilkins's *The Discovery of a New World* (1638), Gassendi's *De motu impresso a motore translato* (1642), Ismael Boulliau's *De natura lucis* (1638); three works by Francis Bacon including *The Advancement of Learning* and *Sylva sylvarum,* and a like number of Robert Boyle's works: *Certain Physiological Essays* (1661), *New Experiments Physico-Mechanical* (1662), and *Seraphick Love* (1659). Other English works included Walter Charleton's *The Immortality of the Human Soul* (1657), Thomas Sprat's *History of the Royal Society* (1667), and other pieces of apologists for the Royal Society such as Joseph Glanvill's *The Vanity of Dogmatizing* (1661), *Plus ultra* (1668), and one of his replies to Henry Stubbe. Henry Power's *Experimentall Philosophy* (1664) was included, as was William Gilbert's *De mundo* (1651). Finally, many of the recent English biological works are to be found, such as William Harvey's *De generatione animalium* (1651) and two works by Thomas Willis, the *Cerebri anatome* (1664) and *De anima brutorum* (1667).

38. The following analysis is based, in part, on Duport's "Rules" and Richard Holdsworth's "Directions for a Student in the Universitie," printed in Harris F. Fletcher, *The Intellectual Development of John Milton,* 2 vols. (Urbana, Ill., 1956–61), 2:624–55; the sale catalogue of John Ray's library (1707–8), reprinted in *Sale Catalogues of Libraries of Eminent Persons,* Vol. 11, ed. H. A. Feisenberger (London, 1975), pp. 117–48; John Nidd's inventory, E. S. Leedham-Green, *Books in Cambridge Inventories,* 2 vols. (Cambridge, 1986), 1:576–84; the account books of John Masters – a fellow-commoner at Trinity from 1646 to 1647 – described in *The Flemings in Oxford,* ed. J. R. Magrath, 3 vols. (Oxford, 1904–24), 1:374–91. In addition, the conclusions presented below concerning the nature of university studies at Cambridge in the mid-seventeenth century are based on research carried out for Vol. 4 of the *History of the University of Oxford,* ed. Nicholas Tyacke (Oxford University Press, forthcoming).

39. Richard Holdsworth conveyed the same message to his students at Emmanuel.

40. Trinity College, Cambridge, Mss. R.9.40, R.9.38.

41. BL Ms. Add. 4292, fol. 85ᵛ.

42. Edmund Calamy, *An Abridgment of Mr. Baxter's Life and Times,* 2d ed., 2 vols. (London, 1713), 2:340; quoted in Caroline Francis Richardson, *English Preachers and Preaching, 1640–1670* (London, 1928), p. 161.

43. For Barrow's commonplace book of observations and collections from such Greek historians, see Trinity College, Cambridge, Ms. R.9.38. The manuscript is dated 9 July 1648.

44. Thomas Hearne, *Ductor historicus,* 2d ed. (London, 1705), p. 128.

45. Lord Brooke expected the occupant of the professorship of history he founded to be versed "in cosmography, chronology, and other science requisite for his profession." BL Ms. Harl. 7038 fol. 78ᵛ, quoted in Curtis, *Oxford and Cambridge in Transition,* p. 117. Newton, too, took the notion for granted. In his own proposal "of educating youth in the universities," he recommended the tutor to read to his students of "the Globes & principles of Geography & Chronology in order to understand History." *Unpublished Scientific Papers of Isaac Newton,* ed. A. Rupert and Marie B. Hall (Cambridge, 1962), p. 370.

46. "Rules to Be Observed," p. 330.

47. Thomas Fuller, *The Holy State, and the Profane State,* ed. James Nichols (London, 1841), p. 68; Henry Hammond, *XIX Sermons* (London, 1664), p. 102; *The Autobiography of Richard Baxter,* ed. N. H. Keeble (London, 1974), p. 9.

48. *Works of Mede,* p. 853, *Conway Letters,* p. 231.

49. Hartlib Papers 15/6/22; *Correspondence of Scientific Men of the Seventeenth Century,* ed. Stephen J. Rigaud, 2 vols. (Oxford, 1841), 2:558–61; John Wallis, *Treatise on Algebra* (Oxford, 1685), pp. 121, 177, 209; J. E. Saveson, "Some Aspects of the Thought and Style of John Smith" (Ph.D. thesis, Cambridge University, 1955), app., pp. 9–

63, passim; idem, "The Library of John Smith, the Cambridge Platonist," *Notes and Queries,* 203 (1958), 215–16.

50. Mordechai Feingold, *The Mathematicians' Apprenticeship* (Cambridge, 1984), pp. 52, 156–7; Bodleian Library (Bodl.) Ms. Selden, supra 109, fols. 266, 258–58ᵛ; E. Millington, *Bibliotheca Cudworthiana* (London, 1691).

51. John Hall, *The Advancement of Learning* (1649), ed. A. K. Croston (Liverpool, 1953), pp. 39–40.

52. Napier, ed., *Theological Works of Isaac Barrow,* 9:29; Osmond, *Isaac Barrow,* p. 25; John Ray, *The Wisdom of God,* 4th ed. (London, 1704), pp. 202–3.

53. *The Correspondence of Isaac Newton,* ed. H. W. Turnbull, J. F. Scott, A. R. Hall, and Laura Tilling, 7 vols. (Cambridge, 1959–77), 1:493.

54. Napier, ed., *Theological Works of Isaac Barrow,* 9:43; Osmond, *Isaac Barrow,* p. 39; Isaac Barrow, *The Usefulness of Mathematical Learning,* trans. John Kirkby (London, 1734), p. xxv. For the original, see Isaac Barrow, *The Mathematical Works,* ed. William Whewell (Cambridge, 1860, rpt. 1973).

55. CUL Baker Mss. 35, p. 315; *Notes & Queries,* 2d Ser., 8 (1857), 305. Jolley was admitted a few months before Barrow, in April 1645, while Rycaut arrived in May 1646.

56. Pope, *Life of...Seth,* p. 140.

57. Ibid., pp. 140–1; Osmond, *Isaac Barrow,* pp. 20–1.

58. *Theologian and Ecclesiastic,* 12 (1848), 169, 172, 372; for Hammond, see John W. Packer, *The Transformation of Anglicanism* (Manchester, 1969). Barrow would pay a personal tribute to the memory of Hammond in an epitaph he composed following Hammond's death in 1660: Napier, ed., *Theological Works of Isaac Barrow,* 9: 540–1.

59. *Calendar of State Papers Domestic* (*CSPD*) (Interregnum), 7:246–7, 270, 375; 8:58, 61–2; Charles H. Cooper, *Annals of Cambridge,* 6 vols. (Cambridge, 1842–1908), 3:457–8; James A. Winn, *John Dryden and His World* (New Haven, Conn., 1987), pp. 66–72; Monk, "Memoir," p. 684.

60. Napier, ed., *Theological Works of Isaac Barrow,* 9:19–34; Osmond, *Isaac Barrow,* pp. 22–7.

61. Marjorie Nicolson, "The Early Stage of Cartesianism in England," *Studies in Philology,* 26 (1929), 356–74; Charles Webster, "Henry More and Descartes: Some New Sources," *British Journal for the History of Science,* 4 (1969), 359–77; Alan Gabbey, "Philosophia Cartesiana Triumphata: Henry More (1646–71)," in *Problems of Cartesianism,* ed. T. Lennon, John M. Nicholas, and John W. Davis (Kingston, Ontario, 1982), pp. 171–250.

62. Christ College, Cambridge, Ms. BB.6.7., fol. 19, quoted in Paolo Cristofolini, *Cartesiani e Sociniani* (Urbino, 1974), p. 132 n47.

63. *The Works of Sir Thomas Browne,* ed. Geoffrey Keynes, 4 vols. (Chicago, 1964), 4:260.

64. In fact, so popular was the teaching of Descartes at Christ's Church that a shortage of texts was felt, which John Finch tried to alleviate while on the Continent in 1652 by sending two copies of the *Principia philosophiae* to More. Archibald Malloch, *Finch and Baines* (Cambridge, 1917), p. 13.

65. Barbara J. Shapiro, "The Universities and Science in Seventeenth Century England," *Journal of British Studies,* 1 (1971), 73; Christ Church, Oxford, Evelyn Papers, Letters, No. 79.

66. See p. 277.

67. Trinity College, Cambridge, Ms. O.10.A.33, p. 18.

68. A. H. Maclean, "George Lawson and John Locke," *Cambridge Historical Journal,* 9 (1947–9), 72.

69. BL Ms. Add 22,910, fols. 13–15; Roger North, *The Lives of the Norths,* 3 vols., ed. Augustus Jessopp (London, 1890), 3:15.

70. A copy of the decree is in Bodl. Ms. Rawl. C146 fol. 37. Obviously, it was too late to turn the tide. A few years later, an old alumnus of St. Catherine was informed that "the controversies of the schools are adjusted more by experiments, such as Mr. Boyle's, than by the maxims of Aristotle." J. J. Smith, *The Cambridge Portfolio*, 2 vols. (London, 1840), 1:285.

71. C. L. S. Linnell, "Daniel Scargill, a Penitent 'Hobbist,'" *Church Quarterly Review*, 116 (1955), 257–60; James L. Axtell, "The Mechanics of Opposition: Restoration Cambridge. V. Daniel Scargill," *Bulletin of the Institute of Historical Research*, 38 (1965), 102–11.

72. Crossley, ed., *Diary and Correspondence of Worthington*, 2:286.

73. Ibid., 2:254.

74. Gabbey, "Philosophia Cartesiana triumphata," p. 198.

75. Napier, ed., *Theological Works of Isaac Barrow*, 1:xl–xli.

76. Ibid., 9:79–104; Osmond, *Isaac Barrow*, pp. 28–33.

77. Shapiro, "The Universities and Science," p. 63.

78. *Complete Prose Works of John Milton*, 8 vols. (1953–84), 1:314.

79. Letter to Walter Curle, Bishop of Winchester on 2 February 1635/6, in *The Works of... William Laud*, ed. W. Scott and J. Bliss, 7 vols. (Oxford, 1847–60), 5:117.

80. William Perkins, *Workes*, 3 vols. (London, 1626–31), 3:442, 461; John Morgan, *Godly Learning* (Cambridge, 1986), pp. 235–8.

81. *Wadsworth Remains* (London, 1680), p. 5.

82. W. W. Rouse Ball, *Notes on the History of Trinity College, Cambridge* (London, 1899), pp. 97–8.

83. J. H. Gray, *The Queens' College* (London, 1899), pp. 177–8; Hall, *The Advancement of Learning*, p. 5.

84. *The Rev. Oliver Heywood... His Autobiography, Diaries, Anecdotes and Event Books*, ed. J. H. Turner, 4 vols. (Brighouse, 1882), 1:162.

85. Sheffield University Library, Hartlib Papers 15/6/3, letter of 22 October 1647.

86. BL Ms. 4292, fols. 84–85^v, Walter Needham to Richard Busby, 8 February 1654/5.

87. Walter Charleton, *The Immortality of the Human Soul* (London, 1657), p. 50.

88. Robert Sprackling, *Medela ignorantiae* (London, 1665), Dedication to Francis Glisson, sig. A2^v.

89. Barbara J. Shapiro, "Latitudinarianism and Science in Seventeenth-Century England," in *The Intellectual Revolution of the Seventeenth Century*, ed. Charles Webster (London and Boston, 1974), pp. 300–1.

90. Bodl. Ms. Tanner 39 fol. 19, quoted in Matthew, *Calamy Revised*, p. 416.

91. Thomas Sprat, *History of the Royal Society*, ed. Jackson I. Cope and Harold W. Jones (St. Louis, Mo., and London, 1959), pp. 55–6.

92. Paul Hammond, "Dryden and Trinity," *Review of English Studies*, N.S., 25 (1985), 51–7.

93. *Heywood... His Autobiography*, p. 146; Sheffield University Library, Hartlib Papers, "Ephemerides," 1653 (gg–gg), 5; Webster, *The Great Instauration*, p. 135.

94. Hartlib, "Ephemerides," 1653 (JJ–JJ), 7.

95. Derek J. Price, "The Early Observatory Instruments of Trinity College, Cambridge," *Annals of Science*, 8 (1952), 10; the *Epitome* is in the Library of Jesus College, Oxford, shelf mark APT 46.

96. Charles E. Raven, *John Ray, Naturalist*, 2d ed. (Cambridge, 1950), pp. 47–8; Leedham-Green, *Books in Cambridge Inventories*, 1:576–84.

97. Thomas Birch, *The History of the Royal Society of London*, 4 vols. (London, 1756–7), 2:476, 483, 487, 499; 2:17, 90, 97; Hammond, "Dryden and Trinity," pp. 42–7.

98. BL Ms. Add. 4292, fols. 84–5^v; Joseph Needham, *A History of Embryology*, 2d ed. (Cambridge, 1959), pp. 131–3, 158–62; Raven, *John Ray*, pp. 48, 463. In his letter,

Needham also alluded to his study of Ptolemy, an author Barrow, too, had studied in the early 1650s.

99. Raven, *John Ray,* pp. 35, 147–8; Isaac Barrow, *Geometrical Lectures,* trans. Edmund Stone (London, 1735), pp. 249–50; *The Correspondence of Henry Oldenburg,* ed. D. Rupert and Marie Boas Hall, 13 vols. (Madison, Wis., and London, 1965–86), 13:434; Vols. 9, 11, passim; Rigaud, ed., *Correspondence of Scientific Men,* 1:227. In 1687, Jessop also published his *Propositiones hydrostaticae.* Two other students of Templer, the brothers John and Samuel Fisher, entered Trinity in 1649 and 1650, respectively, and joined the group. The two continued their scientific collaboration with Ray and Jessop after the Restoration as well. Raven, *John Ray,* p. 148.

100. Cambridge University Archives, Acct. 2 (1), fol. 780.

101. J. Lough, "Martin Lister's Travels in France," *Durham University Journal,* 76 (1983–4), 41.

102. E. G. R. Taylor, *The Mathematical Practitioners of Tudor and Stuart England* (Cambridge, 1954), p. 199. Feingold, *The Mathematicians' Apprenticeship,* p. 85.

103. BL Ms. Sloane 1342, fol. 3; Charles Webster, "Henry Power's Experimental Philosophy," *Ambix,* 14 (1967), 150–78. Raven, *John Ray,* p. 45. Naturally, the letter is suggestive of contact not only between Power, Needham, and Ray, but also between the Trinity men and colleagues of Power at Christ's College.

104. BL Mss. Sloane 587, 591, 696; Matthews, *Calamy Revised,* p. 398.

105. Napier, ed., *Theological Works of Isaac Barrow,* 1:xli; 9:iv, 45–6.

106. Ibid., 1:xliii. The two would remain friends until Barrow's death.

107. Arthur Gray, *Jesus College [Cambridge]* (London, 1902), p. 118; Crossley, ed., *Diary and Correspondence of Worthington,* 2:ii:378.

108. After the Restoration – when Widdrington faced ejection – he claimed that far from causing Molle's ejection by vying for the post, his predecessor had actually resigned in his favor; he further claimed that he, Widdrington, continued to pay Molle his stipend for the remainder of Molle's life some seven years later. Conveniently, Molle was dead when the persecuted Widdrington made his defense. Marjorie Nicolson, "Christ's College and the Latitude-Men," *Modern Philology,* 9 (1929), 43.

109. Hartlib, "Ephemerides," 1655 (27–27), 4.

110. Ibid. (26–26), 5; (28–28), 6; (33–33), 5; Ralph Cudworth, *The True Intellectual System of the Universe,* ed. Thomas Birch, 2 vols. (London, 1837), 1:12.

111. Hartlib, "Ephemerides," 1655 (26–26), 5.

112. Crossley, ed., *Diary and Correspondence of Worthington,* 1:82–3.

113. Napier, ed., *Theological Works of Isaac Barrow,* 1:lxii.

114. Ibid., Raven, *John Ray,* pp. 46–7.

115. Isaac Barrow, "To the Reader," in *Euclid's Elements,* ed. J. Barrow (London, 1751).

116. Translation taken from Anthony Grafton, Chapter 5, this volume.

117. BL Ms. Add. 6209, fol. 119ᵛ; Hartlib, "Ephemerides," 1655 (35–35), 2; Crossley, ed., *Diary and Correspondence of Worthington,* 1:79, 81–2. Whatever fears Nealand may have had concerning the commercial success of the book, they proved to be unfounded. When Collins negotiated a new edition of the Euclid in 1665, Barrow told him that Nealand "got (as I Have been told from himself) some hundreds of pounds by it, and did not keep conditions with me in printing it so well as he did promise me." Rigaud, ed., *Correspondence of Scientific Men,* 1:45.

118. Napier, ed., *Theological Works of Isaac Barrow,* 1:lxii. Neither from the Worthington letters nor from other letters sent by Hartlib at the time is it possible to identify what treatise Barrow was asked to evaluate. It may have been the one on the revision of the calendar. See Wilbur Applebaum, "A Descriptive Catalogue of the Manuscripts of Nicholaus Mercator, F.R.S. (1620–87), in Sheffield University Library," *Notes and Records of the Royal Society,* 41 (1986), 28–9.

119. Jonas Moore, *A New Systeme of the Mathematicks* (London, 1681), preface (by John Flamsteed); *Aubrey on Education,* ed. J. E. Stephens (London and Boston, 1972), p. 109; North, *Lives of the Norths,* 3:16.
120. Euclid, *Elements,* ed. John Keill, 3rd ed. (London, 1733), preface, sig. A3ᵛ–A4.
121. Napier, ed., *Theological Works of Isaac Barrow,* 1:lxii; Rigaud, ed., *Correspondence of Scientific Men,* 2:45.
122. Royal Society, London, Mss. 18–20; Abraham Hill and John Tillotson presented Mss. 18 and 20 to the Royal Society in 1683, while Ms. 19 was given by William Jones in 1715.
123. Pope, *The Life of. . . Seth,* p. 154.
124. Galileo Galilei, *Opere,* ed. A. Favaro, 20 vols. (Florence, 1890–1909), 14:387, Torricelli to Galileo, 11 September 1632.
125. J. F. Scott, *The Mathematical Work of John Wallis* (London, 1938), p. 15.
126. *CSPD* (Interregnum), 4:461, 494–5; 5:437; 6:93, 311, 345, 422, 457. There exists no evidence of any teaching by Pell during this period. The salary may have been intended only to recompense Pell, shortly to be dispatched on a diplomatic mission to the Continent. Since Pell was a former member of Trinity College, since he was a close associate of Samuel Hartlib, and since both Pell and Barrow shared the friendship of Herbert Thorndike – reputed by Seth Ward to have been one of the best mathematicians in England – it is not inconceivable that Barrow had himself known Pell and perhaps discussed mathematics with him.
127. Rigaud, ed., *Correspondence of Scientific Men,* 2:487.
128. North, *General Preface & Life of Dr John North,* p. 129.
129. Hartlib Papers, "Ephemerides," 1656 (38–38), 8.
130. Barrow himself, we may recall, later told Collins that his edition of the Greek geometers was but "elementary things." Rigaud, ed., *Correspondence of Scientific Men,* 2:45.
131. Ibid., 2:34.
132. Both works are in Barrow's library catalogue. But, again, the fact that we are told that he had sold his books before his departure to the Continent prevents us from concluding that he possessed – or read – these works previously.
133. Rigaud, ed., *Correspondence of Scientific Men,* 2:47.
134. William Jones had in his possession a volume containing several papers of Barrow – namely, *Compendium pro tangentibus; Aequationum constructio per conicas sectiones; Aequationum constructio geometrica;* and *Additamenta de curvis,* which, according to Kippis, "seem to have been written before his *Geometrical Lectures,*" and may well have also been composed c. 1654. Another short paper, "A General Theorem for Determining the Tangents to Curve Lines, and the Areas of Curve Figures by Motion," was undoubtedly Barrow's compliance with Collins's wishes in 1663 to send him Barrow's method of tangents. See *Biographica Britannica,* 2:637n.x.
135. *CSPD* (Interregnum), 8:584.
136. Napier, ed., *Theological Works of Isaac Barrow,* 9:118; Osmond, *Isaac Barrow,* p. 53.
137. Napier, ed., *Theological Works of Isaac Barrow,* 9:81.
138. Rigaud, ed., *Correspondence of Scientific Men,* 2:33, 70.
139. For the elder Huygens see A. G. H. Bachrach, *Sir Constantine Huygens and Britain, 1596–1687* (Leiden, 1962).
140. Osmond, *Isaac Barrow,* pp. 57–9.
141. *Dictionary of Scientific Biography,* ed. Charles C. Gillespie, 16 vols. (New York, 1970–80), s.v., Borrelli, Viviani, Redi; Eric Cochrane, *Florence in the Forgotten Centuries* (Chicago, 1973), p. 232.
142. Osmond, *Isaac Barrow,* p. 61.
143. Aubrey, *Brief Lives,* 1:91.

144. Hugh Trevor Roper, "The Church of England and the Greek Church in the time of Charles I," in *Religious Motivation,* ed. Derek Baker (Oxford, 1978), pp. 213–40.

145. An amusing anecdote describes Barrow's examination by the Bishop's chaplain:

> Chaplain: *Quid est fides?* Barrow: *Quod non vides.* Chaplain: *Quid est spes?* Barrow: *Magna res.* Chaplain: *Quid est caritas?* Barrow: *Magna raritas.* "On which the chaplain retired in dudgeon and reported to the Bishop that there was a candidate for ordination who would give him nothing but 'rhyming answers to moral questions.' But the Bishop, who knew Barrow by repute, was content."

Osmond, *Isaac Barrow,* p. 73.

146. Raven, *John Ray,* pp. 59–60; Osmond, *Isaac Barrow,* p. 85.

147. Napier, ed., *Theological Works of Isaac Barrow,* 1:xli.

148. Ibid., 1:xlvii; the translation is taken from *The Elements of Euclid,* ed. William Whiston, 3rd ed. (London, 1727), sig. a2–a2^v.

149. John Dunton, *The Young Students Library* (London, 1692), p. 14.

150. Crossley, ed., *Diary and Correspondence of Worthington,* 1:346.

151. "Te magis optavit rediturum, Carole, nemo, Et nemo sensit te rediisse minus." Napier, ed., *Theological Works of Isaac Barrow,* 1:xlv; the English version is taken from the *Biographica Britannica,* 2:632.

152. Napier, ed., *Theological Works of Isaac Barrow,* 9:137–54; Osmond, *Isaac Barrow,* pp. 85–90.

153. George D'Oyly, *The Life of William Sancroft,* 2d ed. (London, 1840), p. 78.

154. Napier, ed., *Theological Works of Isaac Barrow,* 9:155–8; Osmond, *Isaac Barrow,* pp. 91–2.

155. Crossley, ed., *Diary and Correspondence of Worthington,* 2:64, letter to Samuel Hartlib, 26 October 1661.

156. He had already made this point by 1652.

157. Napier, ed., *Theological Works of Isaac Barrow,* 9:158–69; Osmond, *Isaac Barrow,* pp. 92–6.

158. Barrow, *The Usefulness of Mathematical Learning,* p. xxii.

159. Adamson, "The Foundation and Early History of Gresham College," p. 270; Napier, ed., *Theological Works of Isaac Barrow,* 1:xlvi; Sherburne, who probably had seen a copy, claimed that these lectures were prepared for the press. Edward Sherburne, *The Sphere of Marcus Manilius* (London, 1675), app., p. 112; *Biographica Britannica,* 2:635n.x; John Ray, *Philosophical Letters,* ed. William Derham (London, 1718), p. 360.

160. Barrow's only known contribution to the society was a paper on the contemporary value of the Roman *sestertius*. Birch, *History of the Royal Society,* 1:111, 119, 138, 406–7; 2:118; Michael Hunter, *The Royal Society and Its Fellows, 1660–1700,* British Society for the History of Science, Monograph 4 (Chalfont St. Giles, 1982), pp. 91, 106, 180–1.

161. See, in general, Michael Hunter, *Science and Society in Restoration England* (Cambridge, 1981).

162. Bodl. Ms. Smith 25, fol. 13.

163. Barrow, *The Usefulness of Mathematical Learning,* p. xxiv.

164. Ibid., pp. viii–xii; Millicent Barton Rex, *University Representation in England, 1604–1690* (London, 1954), esp. pp. 121, 146–8, 168–9, 176 n31, 181 nn112,115.

165. Lucas's library catalogue is in CUL Ms. Mm.4.27. More than 700 of the books in his library were in French and Italian, and an additional 180 were in English.

166. Bodl. Ms. Tanner 73 fol. 68, Henry Briggs to Samuel Ward, 6 August 1621.

167. T. Bass Mullinger, "The Relations of Francis Bacon, Lord Verulam, with the University of Cambridge," *Publ. Cambridge Antiq. Soc.,* 9 (1896-8), 227-36, 233. L. M. Payne, "Sir Charles Scarburgh's Harveian Oration, 1662," *Journal of the History of Medicine & Allied Sciences,* 12 (1957) 158-64, 163.

168. Hartlib Papers, "Ephemerides 1639," (H-J), 1. In my *Mathematicians' Apprenticeship,* I speculated that Boswell's friend may have been William Harvey, but for the reasons given above Lucas appears to be the more likely candidate.

169. For Buck, see Barrow, *The Usefulness of Mathematical Learning,* pp. xviii-xx (the quotation is from p. xix); H. P. Stokes, *The Esquire Bedells of the University of Cambridge,* Cambridge Antiquarian Society, Vol. 47 (Cambridge, 1911); George J. Gray and William M. Palmer, *Abstracts from the Wills and Testamentary Documents of Printers, Binders, and Stationers at Cambridge, from 1504 to 1609* (London, 1915), pp. 110-14; William M. Baillie, "The Printing of Privileged Books at Cambridge, 1631-1634," *Transactions of the Cambridge Bibliographical Society,* 5 (1969-71), 155-66.

170. Barrow, *The Usefulness of Mathematical Learning,* pp. xxv-vi.

171. Rigaud, ed., *Correspondence of Scientific Men,* 2:43, 46.

172. Osmond, *Isaac Barrow,* p. 132.

173. Barrow, *Geometrical Lectures,* pp. i-iv.

174. Barrow, *The Usefulness of Mathematical Learning,* pp. 288-9.

175. Barrow, *Geometrical Lectures,* p. 1; idem, *Mathematical Works,* 2:159.

176. Barrow, *The Usefulness of Mathematical Learning,* p. 293; idem, *Mathematical Works,* 2:253.

177. Barrow, *Geometrical Lectures,* p. 99; idem, *Mathematical Works,* 2:208.

178. *Isaac Barrow's Optical Lectures,* trans. H. C. Fay, ed. A. G. Bennett and D. F. Edgar (London, 1987), p. 10; Barrow, *Mathematical Works,* 2:11.

179. *James Gregory Tercentenary Memorial Volume,* ed. Herbert W. Turnbull (London, 1939), p. 154.

180. Rigaud, ed., *Correspondence of Scientific Men,* 1:71.

181. Turnbull, ed., *Gregory Memorial Volume,* p. 55.

182. Ibid., 2:67.

183. Ibid., pp. 71, 73.

184. Rigaud, ed., *Correspondence of Scientific Men,* 2:56-67.

185. Ibid., 2:64-5.

186. *The Diary of Samuel Pepys,* ed. Robert Latham and William Matthews, 11 vols. (Berkeley and Los Angeles, 1970-83), 7:297, 309-10; *The Diary of John Evelyn,* ed. E. S. de Beer, 6 vols. (Oxford, 1955), 3:459. Walter G. Bell, *The Great Fire of London in 1666* (London, 1920), pp. 139-40, 224-7.

187. Rigaud, ed., *Correspondence of Scientific Men,* 2:470.

188. Turnbull, ed., *Gregory Memorial Volume,* p. 181.

189. Rigaud, ed., *Correspondence of Scientific Men,* 1:178, letter to Francis Vernon, 14 December 1671. Six years later Collins similarly told Baker that if a stationer be paid "for one hundred books, ready money, at a rate he sells them in his shop, there are stationers, [who] will undertake any book proposed." Ibid., 2:21.

190. Rigaud, ed., *Correspondence of Scientific Men,* 1:142; see also Turnbull, ed., *Gregory Memorial Volume,* p. 76.

191. Rigaud, ed., *Correspondence of Scientific Men,* 1:201; Turnbull, ed., *Gregory Memorial Volume,* pp. 84-5, 244-6.

192. Turnbull, ed., *Gregory Memorial Volume,* pp. 109, 137, 153. Barrow himself wrote Collins that he had hoped to receive copies of the geometrical lectures in time to distribute them among his friends who came to the Cambridge commencement, but in vain.

193. Turnbull, ed., *Gregory Memorial Volume,* p. 233.

194. D. G. Wing, *Short Title Catalogue of Books Printed in England...1641-1700,* 3 vols. (New York, 1945-51), 1:127. B937 gives a separate imprint of the 1672 edition of the geometrical lectures, but since the combined 1672 edition merely represents the putting together of both sets, it could have been possible for anyone to purchase an unbound volume and retain only one set of lectures.

195. Rigaud, ed., *Correspondence of Scientific Men,* 2:22, letter to Baker, 24 April 1677. Similar information had been conveyed to Gregory two years earlier, with the additional piece of information that the plates were being thrown into the bargain. Turnbull, ed., *Gregory Memorial Volume,* p. 330.

196. For Scot, see Plomer, *A Dictionary of the Printers and Booksellers,* pp. 264-5; Leona Rostenberg, "Robert Scott, Importer & University Agent," in *Literary, Political, Scientific, Religious & Legal Publishing, Printing & Bookselling in England, 1551-1700,* 2 vols. (New York, 1965), 2:281-313. It was perhaps Scot of whom Collins said, "A bookseller here will print any book, if he [be] but sure to sell eighty or an hundred books for ready money." But it was also Scot who observed, as Collins quoted, that "England doth not vent above twenty or thirty of any new mathematical book he brings over; and therefore to think of printing the ancient math[ematician]s is, in my [Collins's] opinion, a design very hazardous to stationers." Rigaud, ed., *Correspondence of Scientific Men,* 1:178, 200-1.

197. Rupert and Hall, eds., *Correspondence of Henry Oldenburg,* 11:316; Turnbull, ed., *Gregory Memorial Volume,* pp. 329, 333. The material had been in Collins's hands since 1670, and a copy had been sent to Gregory (ibid., pp. 161-5), but for some reason not all the propositions were published in 1674; in particular, the analogous round-solids problem was left out.

198. Napier, ed., *Theological Works of Isaac Barrow,* 1:xlvi-xlvii; Rigaud, ed., *Correspondence of Scientific Men,* 1:142.

199. Turnbull, ed., *Gregory Memorial Volume,* pp. 137-8.

200. Rigaud, ed., *Correspondence of Scientific Men,* 2:14-15, 20-2.

201. Ibid., 2:45. Collins expressed interest at the same time in Barrow's Euclid, and Barrow told Collins that if he decided to publish the Archimedes and the Apollonius, he would have to annex them to a new edition of Euclid.

202. Rigaud, ed., *Correspondence of Scientific Men,* 1:148, 2:75; Turnbull, ed., *Gregory Memorial Volume,* p. 107.

203. Turnbull, ed., *Gregory Memorial Volume,* p. 179: A year later Collins was still expressing his confidence that Bernard would join the venture. Ibid., p. 218.

204. Rigaud, ed., *Correspondence of Scientific Men,* 1:188, 2:552: "For printing the other [epitome] first would rather endanger the loss of the author [Apollonius] himself; it having been found already in experience, that Commandine having printed the translation before the original, hath so far hazarded the loss of the original, as that to this day it is not published."

205. Turnbull, ed., *Gregory Memorial Volume,* p. 258; Rigaud, ed., *Correspondence of Scientific Men,* 2:449, 555. Strode would later substitute the proofs with the bound volume. Ibid., 2:453.

206. Turnbull, ed., *Gregory Memorial Volume,* p. 247.

207. Bernard did, however, continue to take part in the team effort to check and correct the printed versions of Barrow's works, including the first four books of the Apollonius. See, e.g., Bodl. Ms. Rawl. D. 697, fols. 67ᵛ-68.

208. Rigaud, ed., *Correspondence of Scientific Men,* 2:586-7, 589.

209. Turnbull, ed., *Gregory Memorial Volume,* pp. 317, 305.

210. Ibid., pp. 317, 284.

211. Ibid., pp. 285-6, 299, 305, 333; Rigaud, ed., *The Correspondence of Isaac Newton,* 1:346.

212. Turnbull, ed., *Gregory Memorial Volume,* pp. 294–5, 314, 317; Rigaud, ed., *Correspondence of Isaac Newton,* 1:333, 346.

213. *Lectio...in qua theoremata Archimedis de Sphaera & cylindro per methodum indivisibilium investigata exhibentur.* Kippis stated that the lecture was originally "written in English, but soon after the author's death...[was] turned into Latin." *Biographica Britannia,* 2:635n.x.

214. Rigaud, ed., *Correspondence of Scientific Men,* 2:14.

215. Ibid., 2:24.

216. *The Mathematical Papers of Isaac Newton,* ed. D. T. Whiteside, 8 vols. (Cambridge, 1967–81), 3:xiv, n14; Richard S. Westfall, *Never at Rest* (Cambridge, 1980), pp. 187, 207. Osmond appears to be closer to the mark when he observes that "if we get the impression that he seems in more directions than one to fall short of supreme achievement, it may be attributed less to any lack of intellectual grasp than to a most exceptional absence of self-seeking and self-appraisal. Personal ambition and professional jealousy seem to have been omitted in his composition." *Isaac Barrow,* p. 225.

217. Napier, ed., *Theological Works of Isaac Barrow,* 9:572, translation supplied by Osmond, *Isaac Barrow,* p. 143.

218. Pope, *Life of...Seth,* p. 153.

219. Charles Sturge, *Cuthbert Tunstal* (London, 1938), p. 73.

220. Napier, ed., *Theological Works of Isaac Barrow,* 1:lxx.

221. Barrow told Collins of his theological studies on Easter Eve, 1669. Rigaud, ed., *Correspondence of Scientific Men,* 2:71.

222. Osmond, *Isaac Barrow,* p. 187.

223. John Le Neve, *Fasti Eccleiae Anglicanae,* 3 vols. (Oxford, 1854), 2:626, 659. It is a pity that Pope thought it "not necessary to declare" the reason for Barrow declining the archdeaconship, but it may be speculated that it was owing to his unwillingness to become a church functionary. *Life of...Seth,* p. 152.

224. Le Neve, *Fasti,* p. 660; *Letters of Humphrey Prideaux...to John Ellis,* ed. Edward M. Thompson, Camden Society, N.S., 15 (1875), 62.

225. BL. Ms. Tanner 146, fol. 51.

226. *The Diary of Samuel Newton,* ed. J. E. Foster (Cambridge, 1890), p. 70.

227. John Ivory, *The Foundation of the University of Cambridge* (Cambridge, 1672); Cooper, *Annals,* 3:554.

228. W. W. Rous Ball, *Cambridge Notes* (Cambridge, 1921), p. 97. Westfall estimated the value of Newton's fellowship in 1668 at £60 p.a. *Never at Rest,* pp. 180–1.

229. Napier, ed., *Theological Works of Isaac Barrow,* 1:lxxii, letter to Mapletoft, July 1673.

230. *CSPD,* Charles II, 16:363. The effectiveness of such a dispensation was tested the following year when Boldero refused to comply with a "recommendation" sent by Sir Joseph Williamson unless stronger letters were to follow. Ibid., 17:351.

231. Edward Bathurst obtained both a mandate from the king and a letter from the duke of Buckingham on 26 June 1673 directing Trinity to make him a fellow instantly; this order seems to have raised hopes of an election, and both Charles Fraser and Leonard Welstead, the other two Westminster fellows to be elected, were granted letters from the king in September and October, respectively. *CSPD,* 15:394–5, 526, 587.

232. Ibid., 16:211, 241, 252, 269, 186. Ellis was not elected librarian; the position was given to James Mansfield. Other letters from the king, or other officials, are to be found in the college archives.

233. Ibid., 15:574. The son, Henry Firebrace, was indeed elected in 1674, assisted by the letter mandate his father (who was chief clerk to the king's kitchen) was able to obtain from Charles II.

234. *The Life, Journals, and Correspondence of Samuel Pepys,* ed. John Smith, 2 vols. (London, 1841), 1:154.

235. Napier, ed., *Theological Works of Isaac Barrow*, 1:lxxii.

236. National Library of Wales, Ottley Mss. 1130–4, quoted by D. R. Hirschberg, "The Government and Church Patronage in England, 1660–1760," *Journal of British Studies*, 20 (1980), 119. Equally futile was Bishop Seth Ward's effort to intercede on behalf of his nephew Thomas Ward who graduated B.A. in early 1677. Writing to Archbishop Sancroft c. 1680, Bishop Ward argued that Thomas failed to obtain a fellowship at Trinity College - despite his recommendation by several fellows - "by reason of the interposition of his Majesty's letters." Bodl. Ms. Tanner Letters 38, fol. 125.

237. Bodl. Ms. Rawl. A 191, fol. 66; *The Life, Journals, and Correspondence of Samuel Pepys*, 1:153–5; *CSPD*, 16:369 (emphasis added).

238. Napier, ed., *Theological Works of Isaac Barrow*, 1:liv–lv.

239. It may indeed be true, as Westfall pointed out, that neither did Uvedale intend to pursue the profession of law. But I believe that Barrow's concern, as Edleston pointed out, was more with the incompatibility of a law fellowship "with the efficient discharge of the duties of the Mathematical Professor." Uvedale, at any rate, appears to have at least given the impression that he was seriously intending to follow the study of law. J. Edleston, *Correspondence of Sir Isaac Newton and Professor Cotes* (London, 1850), pp. xlviii–xlix; Westfall, *Never at Rest*, p. 331 and note 151.

240. North, *General Preface & Life of Dr John North*, p. 148. For a fuller account on the building of Trinity College Library see Robert Willis and John W. Clark, *The Architectural History of the University of Cambridge*, 3 vols. (rpt. Cambridge, 1988), 2:531–51; Philip Gaskel, *Trinity College Library* (Cambridge, 1980), pp. 137–41.

241. Napier, ed., *Theological Works of Isaac Barrow*, 1:xlix; for some samples of Barrow's many letters, see BL Ms. Rawl. Lett. 109, fol. 2; Bodl. Ms. Smith 45, fol. 83; Osmond, *Isaac Barrow*, p. 206.

242. Rous Ball, *Cambridge Notes*, p. 136.

243. *The Diary of Robert Hooke*, ed. Henry W. Robinson and Walter Adams (London, 1935), pp. 288–9; Pope, *Life of... Seth*, pp. 178–81; Rigaud, ed., *Correspondence of Scientific Men*, 2:24, 27; *The Correspondence of John Locke*, ed. E. S. de Beer, 8 vols. to date (Oxford, 1976–), 1:504; Louis T. More, *Isaac Newton* (New York, 1934), p. 199.

244. De Beer, *Diary of John Evelyn*, 4:62.

245. BL Ms. Add. 32,545, fol. 13ᵛ, quoted in F. J. M. Korsten, *Roger North* (Amsterdam, 1981), p. 278 n214.

246. A second edition was published by Aylmer between 1682 and 1687. Complete editions or separate treatises were often reprinted during the eighteenth century; Napier's definitive edition of 1859 was the third complete edition to appear in the nineteenth century.

247. *The Life and Times of Anthony Wood*, ed. Andrew Clark, 5 vols. (Oxford Historical Society, 1891–1900), 2:264. Interestingly, Barrow's sermons also effected the famous conversion of Captain Booth from skepticism to Christianity in Fielding's *Amelia*. See Henry Fielding, *Amelia*, ed. Martin C. Battestin (Middletown, Conn., 1983), p. 511. Indeed, Barrow's powerful reasoning could lend itself to eighteenth-century deists as well. Lord Bolingbroke was well versed in Barrow's works and at one point, after quoting Barrow with approval, added, "If we seriously weigh the case, we shall find that to require faith without reason is to demand an impossibility, and that God therefore neither doth nor can enjoin us faith without reason." R. N. Stromberg, *Religious Liberalism in Eighteenth century England* (Oxford, 1954), p. 174.

248. William Whiston, *Historical Memoirs of the Life and Writings of Dr. Samuel Clarke*, 3rd ed. (London, 1748), p. 23.

249. T. E. S. Clarke and H. C. Foxcroft, *A Life of Gilbert Burnet* (Cambridge, 1907), p. 312; Alan D. McKillop, *The Background of Thomson's Seasons* (rpt. Hamden,

1961), pp. 28–9. In the second edition of his masterpiece, however, Thomson dropped the orthodox divines and elevated instead the heterodox Shaftesbury.

250. Isabel Rivers, "Dissenting and Methodist Books of Practical Divinity," in *Books and Their Readers in Eighteenth-Century England,* ed. Isabel River (Leicester and New York, 1982), p. 138.

251. *The Collected Works of Dugald Stewart,* 9 vols. (Edinburgh, 1854), 1:90–1. Stewart also maintained that Barrow's mathematical lectures displayed "*metaphysical* talents of the highest order."

252. In a letter written in 1740 Logan added that the *Lectiones opticae* were "the most Compleat Sett of Lectures on the Subject that had ever appeared before." Edwin Wolf, *The Library of James Logan of Philadelphia, 1674–1751* (Philadelphia, 1974), p. 39.

2

The Optical Lectures *and the foundations of the theory of optical imagery*

ALAN E. SHAPIRO

Introduction

Isaac Barrow completed a major phase of the Keplerian revolution in geometrical optics by creating a mathematical theory of optical imagery. In the opening years of the seventeenth century Kepler had introduced the new concepts of pencil of rays and of image formation by pencils and thereby laid the foundation, even if largely qualitative, of the modern theory of image formation in lenses and optical systems. He then applied these new ideas to the optical system of the eye; and with his discovery that vision occurs by means of a real inverted image formed on the retina, he created an entirely new theory of visual perception. Telescope design and construction, studies of vision, and a search for the true law of refraction were vigorously pursued in the decades following Kepler's pioneering research, but the mathematical investigation of optical imagery languished. Not until midcentury was a solution provided to even such an elementary problem as the determination of the focal point of thin spherical lenses. Barrow's principal achievement was to determine the location of the image after any reflection or refraction in plane and spherical surfaces. By synthesizing and reformulating a number of Keplerian concepts, he created the mathematical foundation of a general theory of optical imagery and, with his concept of oblique pencils of rays, began the exact study of astigmatism and caustics.

Barrow's *Optical Lectures*[1] is devoted almost exclusively to a single subject, the mathematical theory of image location. Barrow strays from it only in the opening two lectures on physical models of light and in some brief remarks on color. The *Lectures* are the work of a mathematical physicist, as Barrow himself indicates in his conclusion:

I have here delivered what my thoughts have suggested to me concerning that part of optics which is more properly mathematical. As for the other parts of that science (which being rather physical, do consequently pretty frequently abound with plausible conjectures instead of certain principles), there has in them scarce anything very probable occur'd to my observation different from what has been already said by Kepler, Scheinerus, Descartes, and others after them.[2]

Thus, he altogether omits applied optics and such topics as the eye and vision, or the telescope and microscope, which were among the principal areas of seventeenth-century optics. Barrow feared that his strictly theoretical approach limited the value of his *Lectures*. While it was being readied for the press he wrote to John Collins:

[H]ad I known M. Huygens had been printing his Optics, I should hardly have sent my book. He is one that hath had considerations a long time upon that subject, and is used to be very exact in what he does, and hath joined much experience with his speculations. What I have done is only what, in a small time, my thoughts did suggest, and I never had opportunity of any experience. So that I have great reason to believe what he hath done, with so much advantage in all respects, will be much more perfect.[3]

Barrow need not have worried about the imminent appearance of Huygens' *Dioptrica,* which was published only in the following century, but his concern was otherwise well founded.[4] With their common Keplerian heritage and their commitment to pursuing reflection and refraction in spherical surfaces rather than following the well-trodden Cartesian path of conoids, Barrow and Huygens traversed similar ground in the mathematical theory of images and frequently made the same discoveries. Barrow's approach, however, was more general (if not for the most part simpler and more elegant), especially as he extended it to the refraction of oblique pencils. Despite the restricted scope of his *Lectures* and its failure to treat applied optics, it contains many notable results: the modern (i.e., Keplerian) theory of images in spherical mirrors and refracting surfaces, the general solution of the image points in thick lenses, and a derivation of the radius of the primary rainbow. All of this made his *Optical Lectures* the most advanced treatise on geometrical optics yet published, a status it would maintain until the end of the century.

The location of an image

Barrow was best known in the late seventeenth and eighteenth centuries for what I shall call Barrow's principle of image location: An image is located at the place from which the rays entering a single eye diverge; that is, the image is perceived at the place of the real or virtual geometrical image. It is ironic that in our own century Barrow's principle, with which his name was for so long properly associated, is erroneously attributed to Kepler, who would have rejected it.[5] The question of the location of an image actually involves two distinct questions: (i) Where is the image perceived to be located? (ii) Where is the place to which rays from given points of an object diverge or converge? (I call these the "perceived image" and the "geometrical image," respectively.) The first is a question of physiological optics and psychology, whereas the second is a question of geometrical optics and physics. Barrow identified the solution of the first question with that of the second, considering them to be a single question of geometrical optics, and never discussed physiological or psychological mechanisms of judging distance. Kepler, in contrast, directly addressed both questions, but he identified some aspects of the second question with the first. Barrow's identification of the perceived and geometrical location of an image seems to rest on the naive or unquestioned assumption that an image is perceived in the same way as an object and that when an object sends rays to our eyes, we judge it to be in that place from which the rays actually originate, or in its true place. As Barrow explained it:

> Images are clearly nothing other than light from objects so
> reflected or refracted that it is again collected in one place and
> in such a situation as it had when it flowed from the original
> object and proceeded in a direct path to the eye; whereby it
> happens that images represent objects similarly but as if they
> were located elsewhere.[6]

It may seem strange that I wish to attach Barrow's name to a naive principle, but I intend to show that by means of his naive conception of the problem he was able to develop a sophisticated mathematical theory of imagery.

Through the course of the seventeenth century, geometrical optics had developed into a rather refined mathematical science. For the sake of those not familiar with the concepts of caustics, image points, and astigmatism, I shall briefly illustrate them by means of the simple problem of describing the refracted rays and image of a point seen across a plane refracting surface. Barrow participated in the mathematical development of all of these concepts. By approaching the earlier development of geometrical optics from the sophisticated level that it had attained by the latter part of the seventeenth century, we shall gain deeper insight into its history. Let *A* (Figure 1) be a radiant point in a denser medium that is viewed by an eye in the rare medium above the refracting surface *EF*. From point *A* to the refracting surface drop the perpendicular *ABC*, which

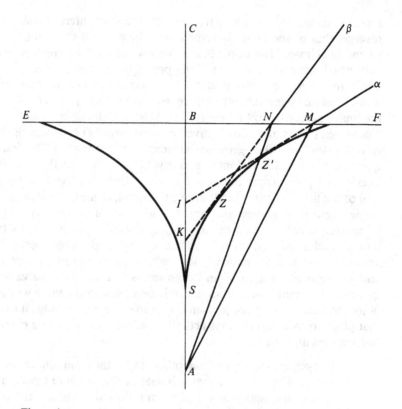

Figure 1

is the axis. This perpendicular line (which in spherical surfaces passes through the center) is important in geometrical optics; it was called by medieval opticians the "cathetus" and by Barrow the "axis" and occasionally the "radiant axis." At the surface the rays, such as *AM, AN,* will be refracted, and if the refracted rays αMI, βNK are produced backward, they will in general no longer possess a single center as they did at *A,* but rather two. The generation of two points of convergence, or foci, is known as "astigmatism," and it requires that the process of image formation be considered in three dimensions. If a narrow oblique pencil of incident rays (i.e., a cone of rays whose axis makes a large angle with the axis of the optical system) is considered, the refracted rays lying in two particular surfaces will intersect in the immediate neighborhood of two image points. Let us first consider refracted rays in the plane of incidence *ABN,* which is the traditional concern of geometrical optics; all of these rays will be tangent to a curve, *ESZZ',* called the "caustic."[7] (Jakob Bernoulli introduced the terms "catacaustic" and "diacaustic" to distinguish curves tangent to reflected and refracted rays, respectively.)[8] Thus the refracted ray βZK of the incident pencil *MAN* will be tangent to the caustic at *Z*, where no other rays intersect it; but very many neighboring rays, which are tangent to the caustic in its vicinity, will intersect it immediately above and below the point *Z*, and particularly densely about that point, which limit point is called the "tangential" focus or image point.[9] Determining the tangential image point – or, as Barrow called it, the "relative image" – is therefore equivalent to determining a point on the caustic. Now let us turn to those incident rays lying on the surface of the cone generated by rotating the ray *AN* about axis *AB*, for their refractions will all intersect on the axis at point *K*, which is called the "sagittal" focus or image point.[10] The rays from this image point will enter an eye at β in a plane perpendicular to that of the paper. No other refracted rays converge to an image point. If now we confine our attention to a very small, central portion of this pencil of rays that lies about the axis, or what is known as a pencil of "paraxial rays" (i.e., rays that are parallel to the axis or make a very small angle with it),[11] all of its refracted rays will intersect in the immediate vicinity of the cusp of the caustic *S*, and the two image points will coincide. Barrow called the paraxial image point, which always lies on the axis, the "absolute image." The simple rules of elementary geometrical optics that we all learned in school are based on paraxial rays, which possess perfect focal properties.

The Lectiones XVIII

As the Lucasian Professor of Mathematics at the University of Cambridge, Barrow was required to lecture once each week during the term and to submit annually at least ten of these lectures to the vice-chancellor for deposit in the university library for public use. Barrow had intended to submit his optical lectures to the vice-chancellor in 1668, but he yielded to Collins's importunities and allowed him to publish them.[12] From Barrow's "Letter to the Reader" it is obvious that he was relieved to have his lectures taken off his hands and would have nothing more to do with them. He advised the knowledgeable reader not to expect a polished scholarly treatise,

> but school lectures, wrested from me by the necessity of my
> office... and prepared for the instruction of an ordinary class
> of students for whom it was important that I not omit many
> things which will seem to you elementary.... To satisfy you
> I know indeed that it would be beneficial to cut many things,
> substitute better ones, rearrange the greater part, and submit
> the whole to be reforged; but I had neither the inclination
> nor leisure to take pains with it, nor the opportunity to carry
> it out.... In fact after I resolved to publish them, I could
> not even endure reading a great part of them again, for I was

either seized by an aversion to them or shunned undertaking the study to revise them; but as weakly mothers frequently do, I entrusted my expelled offspring to the nursing care of not unwilling friends to be either brought out or abandoned, as they saw fit. One of these friends...our colleague, Mr. Isaac Newton, a man of extraordinary genius and remarkable skill, reviewed the copy, noting a number of things to be corrected and also added some things from his own stock, which you will see annexed here and there to our own things with praise. The other...Mr. John Collins took care of the publication at very great trouble to himself.[13]

Though Barrow exaggerates, he did lose interest in his lectures, and they do in some ways read like school lectures that require revision. They are often repetitive and tedious, and the lectures on spherical mirrors, especially, could have been substantially condensed by not treating so many cases separately. Yet their contents, especially the principal concepts of image location, are so sophisticated that they were surely above the head of any student.

Despite Barrow's claim that the lectures were published essentially as delivered, he continued to make revisions and additions through the summer of 1669, just a few months before they were published. In late February 1669 Barrow sent Collins the manuscript of his lectures. "In the fourteenth lecture," he advised him, "I have tore out some leaves, concerning the determination of images in all kinds of lenses, for all cases, which I shall send you somewhat more exacted. I have also a lecture or more behind, which shall be sent in due time; also somewhat of preface."[14] It is not apparent exactly which of the lectures were still to follow, but during the next six months Barrow sent a steady stream of revisions and new material to Collins. Some of these papers were intended for the geometrical lectures, which were originally to be appended to the optical lectures, as announced on its title page. As the number of geometrical lectures accumulated and Barrow continued to press him for expeditious publication, Collins decided – even while the optical lectures were being printed – to publish the two sets of lectures separately. It was only on 20 July that Barrow sent the "remainder of the Opticks"; and on 20 August, while reading proof, he was still promising that the "Synopsis" (the second "epistola") would be sent "in good time."[15] On Easter Eve Barrow asked Collins to return the last three lectures of the "book"; he wanted to copy them, for "I have not, indeed, yet read them in the schools, so that they would ease me of making new ones."[16] If, as I suspect, Barrow was referring to his optical and not geometrical lectures, then he planned to deliver some of his optical lectures in the spring of 1669. Since we know that he intended to deposit a set of optical lectures earlier in 1668, no doubt those from 1667 and 1668, we can conclude that they were most likely delivered at various times in 1667, 1668, and perhaps 1669.[17] The *Lectiones XVIII* were published in early November 1669.[18]

The core of the *Lectures* are Lectures IV–XIII, in which the image of a point after any reflection or refraction at a plane or spherical surface is determined. The first three lectures are introductory, with the first two presenting Barrow's physical models and the third the elementary geometrical optics of reflection and refraction. Lecture XIV, on the image points of thick lenses, is an important application of Lectures XI and XIII to two surfaces and in February 1669 was still in an unsatisfactory state requiring revision. Lecture XV is a slight one, a qualitative description of extended images, that was rescued from obscurity by an interesting appendix on the number of images in binocular vision. Lectures I–III, X (which applies the results of VIII and IX for convex mirrors to concave ones), and XV are the most likely candidates for those that were later added to the original set of lectures sent to Collins in February 1669. The final three lectures, which may be those returned to Barrow and read in 1669, extend the principles of the core lectures on the images of points to those of lines.

In composing his *Lectiones XVIII* Barrow exhibited a broad command of the optical literature from the great medieval Arab mathematician Ibn al-Haytham (Alhazen) to the Jesuit mathematician Andreas Tacquet, whose *Opera mathematica* was also published in 1669. Barrow cited the early contributors to modern optics, Kepler, Scheiner, Descartes, and Hobbes, as well as more traditional seventeenth-century writers, Herigone, Fabri, Vossius, and Maignan. The catalogue of Barrow's library that was compiled after his death shows that he also had many uncited optical works of less importance (except to a historian), such as those by Euclid, Pecham, Maurolycus, Aguilon, Zucchi, and Eschinard. His library, however, contained two uncited works of more importance – Bonaventura Cavalieri's *Exercitationes geometricae sex* (1647) and James Gregory's *Optica promota* (1663) – which it is difficult to believe he did not read. Cavalieri would seem to have deserved a historical citation for deriving the focal points of paraxial rays for thin lenses, the first significant step in the imagery of spherical surfaces since Kepler. Gregory's *Optica promota* contains a number of suggestive ideas that Barrow seems to have developed mathematically, most notably that the image is located at the point of divergence of a pencil of rays entering one eye. This similarity, however, may simply reflect independent development of common Keplerian ideas, for Gregory had only the highest praise for Barrow's *Lectiones XVIII* when it appeared.

Barrow and Newton

In his "Letter to the Reader" Barrow effusively praised the young Newton and acknowledged his efforts in proofreading his lectures. It is

uncertain when Newton's earliest contacts with Barrow occurred, but –
to confine myself to their optical interests – Newton must have attended
Barrow's optical lectures beginning in 1667. By 1667 Newton was sufficiently well versed in geometrical optics to appreciate the mathematical
sophistication of Barrow's approach. He had, among other things, designed his reflecting telescope and carried out research on Cartesian refracting surfaces and the grinding of nonspherical lenses.[19] The predominant influence on Newton in this period was Descartes. Only with the two
constructions by Newton that Barrow inserted in his *Lectiones XVIII*
do we see a Barrovian influence. One construction (appended to Lecture
XIV) – to determine the image point of a thick lens – derived from research that Newton had carried out well before he had heard or read
Barrow's lectures.[20] There can be no doubt, though, that Barrow's work
inspired the other construction (XIII:26) – to determine the tangential
image point for the refraction of an oblique pencil of rays at a spherical
surface. Newton had not earlier utilized pencils of rays. His initial exposure to this Keplerian concept occurred in 1666 through Gregory's *Optica
promota,* but Barrow was the first to develop the concept of oblique pencils – indeed, it is almost his hallmark.

When Newton was appointed as Barrow's successor to the Lucasian
chair, he chose to continue the theme of Barrow's optical lectures. In his
inaugural lecture in January 1670 he paid direct tribute to Barrow by observing that it would be difficult to follow his lectures since they "brought
together such a great variety of optical topics and a vast quantity of discoveries with their very accurate demonstrations."[21] Newton, to be sure,
had no difficulty in following Barrow, for he devoted most of his lectures
to his new theory of light and color, which was completely independent
of Barrow. He also devoted a substantial portion of his lectures to traditional geometrical optics, which does follow directly in Barrow's footsteps. He further pursued the refraction of oblique pencils and recognized
that there were in fact two image points; and he extended the method of
oblique pencils to a general solution for the radius of the rainbow of
any order – both notable achievements. In general, as we shall see, Newton completely adopted Barrow's theory of optical imagery. Although he
would never publish his own *Optical Lectures,* he did later incorporate
Barrow's principle of image location into the *Opticks.*

If Barrow's geometrical optics so strongly influenced Newton, what
can be said of the impression that Newton made on Barrow? Quite simply, that Barrow was always most impressed by Newton and continually
did what he could to further his reputation and career. Besides arranging
for Newton to succeed him as Lucasian Professor, he encouraged the
publication of his *Optical Lectures* and carried his reflecting telescope to
London in December 1671 to present to the Royal Society. Since Barrow
was so kind to Newton, it has been argued that Newton "betrayed" Barrow

by allowing him to publish some "guesses" on color, which Newton had already shown to be erroneous.[22] If Newton had told Barrow about his theory of color in 1669, it probably would not have been very convincing. At that time he had not yet fully articulated the theory and had only written some notes on it in his "Of Colours." It is easy for us today to forget how difficult Newton's theory is and how few people initially accepted it after he first published it in 1672.[23] In fact, when Newton had carefully formulated his theory in his *Optical Lectures,* he did take Barrow into his confidence and show it to him in late 1671 – the only person I know of who did see the *Lectures* in this period. Barrow was converted and declared it "one of the greatest performances of Ingenuity this age hath affoarded."[24] To anyone who has actually read Barrow's *Lectiones XVIII,* it is obvious that his "guesses" on color are not at all related to his main theme, optical imagery, and were inserted at the end of his derivation of the radius of the primary rainbow merely to lighten the unrelenting sequence of difficult mathematics. Finally, Barrow was not subjected to ridicule for publishing these "guesses," since they were well within the mainstream of contemporary ideas of color.[25] It was, after all, Newton's ideas that were so radical and subject to ridicule.

Physical theory of light and color

The method adopted by Barrow to establish the principles of geometrical optics was to deduce its theorems from a set of six hypotheses, which in turn were justified by a physical theory of light. He required that the hypotheses (which, he tells us, can also be considered axioms, theorems, or laws) "be admitted as agreeing with experience and not in any way opposed to reason."[26] They were in fact noncontroversial, universally accepted optical principles, except perhaps for the sine law of refraction, which, despite being known for more than a generation, was just becoming widely accepted in Barrow's own day – "by now, I believe, received by most of the better opticians."[27] The physical theory whereby the hypotheses were deduced and elucidated, on the contrary, need not be accepted, since it was only "probable." Barrow himself did not put great stock in it and recognized that other physical theories could be assumed as well.[28] Since Barrow in fact derives the principles of geometrical optics from the hypotheses, which are all phenomenological, and not from his physical theory, the latter plays no role beyond its heuristic function in the first two lectures.

The mechanical theory of light propounded by Barrow is an unusual hybrid of an emission and a pulse theory.[29] Since he believed that the theory of a corporeal emanation and that of an impulse propagated through an intervening medium were equally pressed with difficulties, he assumed "that light is sometimes produced in each way, both through corporeal effluvia, and through a continuous impulse; and it will be better to attribute some of its effects to the latter, others to the former."[30] His objection to an emission theory is a rather standard one – namely, how is it possible for the flame from a tiny lamp to fill an immense space with light without being immediately consumed? His objection to a pulse theory shows how little wave motion was understood at this time, for in essence he could not understand how it was possible for a bare impulse or motion to be reflected and refracted.

Barrow's own eclectic theory draws heavily on Descartes:

> I assume that every luminous body, as such, is a certain mass of corpuscles of a smallness and fineness almost beyond the imagination; however, each of them, when struck by a very violent motion, goes off somewhere in a straight line (according to that law of nature sufficiently received and investigated). Moreover, it is surrounded by a medium, also fluid (namely, whose parts are not connected by any bonds and are freely set in motion in every direction), formed out of very fine bodies, but thicker and more solid with respect to the first. It therefore has channels and pores capable of admitting the finer corpuscles. However, by running into the thicker corpuscles the progress of many of those [finer] corpuscles, which dwell in the luminous surface or rush from it, is impeded, so that being blocked and repelled, they necessarily retreat inwards. Thus the said mass (with other corpuscles of the same nature having also flowed into it from elsewhere) is to some extent restrained within its own boundaries, and it is not totally dissipated by expanding into the air at once. Meanwhile, very many of them having found a path through the said channels continue their course in a straight line setting in motion and pushing before themselves the matter in there which does not resist so strongly. Other corpuscles, similarly proceeding from the luminous body, follow their paths forwards, and all together they produce a long stream of light flowing out in an unbent series. In fact, some of them by chance strike the thicker corpuscles of the medium with such a strong blow that sometimes they are also forced to yield, and acting together with them, propel the adjacent bodies in a straight line, which in a similar way produces a force in the next ones in succession, and so continuously. Thus all together, a series of such corpuscles indefinitely extended moves and endeavors [*connitatur*] forwards; such a propagation of light, produced in either way, is ordinarily called a *ray*.[31]

The theory of matter is Cartesian: The fine matter in luminous bodies, which is the cause of light, is Descartes's first matter; and the larger corpuscles, which form the fluid ether and surround luminous bodies, comprise Descartes's second matter.[32] According to Barrow (at least insofar as I can understand his terse explanation) light is, in the first place, propagated by means of the particles of first matter that leave the luminous body and travel through the pores of the ether formed of second matter. In the pores of the ether these particles encounter other particles of the first matter, which they hit, press forward, and set in motion; at the same time they similarly set some of the second matter in motion. Yet Barrow's theory differs fundamentally from Descartes's. Whereas for Descartes light consists in the endeavor to motion (*conatus*) of the particles of the second matter alone, for Barrow light consists of an actual motion and an endeavor (pressing) of the first and second matter together.

The six hypotheses are introduced next, and each of them is elucidated by Barrow's theory of light: (1) Light propagates in straight lines in a uniform medium (I:8). (2) Every point on the surface of a luminous or illuminated body emits rays in all directions (I:9). (3) A light ray incident along the perpendicular to the surface of any medium either proceeds in a straight line or returns along its incident path (I:10). (4) A reflected or refracted ray always lies in the plane perpendicular to the reflecting or refracting medium (I:11). (5) The angle of incidence always equals the angle of reflection (II:2). (6) The ratio of the sine of the angle of incidence to the sine of the angle of refraction is constant for any pair of media (II:4). Barrow briefly justifies these hypotheses by appeal to simple experiment and observation and explains them by his theory of light, especially the concept of physical ray.

Barrow took the concept of physical ray – a three-dimensional ray whose front is always normal to its sides – and his mathematical model of refraction directly from Hobbes's *Tractatus opticus,* which was published in 1644 by Mersenne in his *Universae geometriae, mixtaque synopsis.*[33] Hobbes's explanation of refraction was known by virtually every optical writer of the seventeenth century, though it was not widely recognized as his. It was generally attributed to Emanuel Maignan and, somewhat less often, to Barrow, from whose works it became known.[34] Hobbes developed his model as part of a true pulse theory of light. The model was, however, independent of a theory of the nature of light: Maignan embedded it within an effluvium theory and Barrow within his hybrid effluvium–impulse theory.

The essential feature of the physical light ray is that its front is always perpendicular to its sides. From its obvious analogy to a wave front I call it a "ray front," and we can conceive of the physical rays as infinitesimal portions of wave fronts. Physically, according to Barrow, a light ray is a long, thin, right cylinder or prism, which is the path traced out by the shape of the emitted or propelled light particles (I:8). Mathematically, these rays can be considered, just as in classical geometrical optics, to be straight lines that are parallel to the axis or sides of the cylinders or

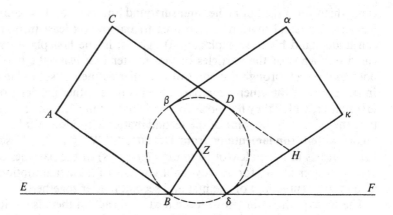

Figure 2

prisms. Barrow's derivation of the law of reflection with his physical rays should help to illuminate that concept (II:2).

Rectangle *ABCD* (Figure 2) is a cross section of a physical light ray in the plane of incidence. The ray front *AC* is successively propagated to *BD* in the direction of the mathematical rays *AB* and *CD*, to which it is perpendicular. When end *B* of the ray front strikes the reflecting plane *EF*, it attempts to rebound toward *A* along its previous path. Since the other end *D* simultaneously tends to continue forward along *CDH*, the ray front begins to rotate about the midpoint *Z* and describes a portion of the circle *BβDδ*. End *B* will have described the arc *Bβ* and end *D* the arc *Dδ* when end *D* strikes the reflecting surface at *δ* and rebounds toward *κ*. The ray front now moves off along the tangents *βα* and *δκ* and describes the physical ray *αβδκ*. The law of reflection follows directly: Angle *ZBδ* is equal to angle *ZδB*, and if the right angles *ABZ* and *κδZ* are added to them, then angle *ABF* is equal to angle *κδE* or angle *ABE* to angle *κδF*. An essential feature of the ray fronts is that they are conceived of as rigid bodies inalterable in length and shape. Barrow's proof of the law of reflection generalized Hobbes's original theory, which treated only refraction, but Hobbes (in an unpublished manuscript) and Maignan had independently developed quite similar derivations.[35] Features of their derivations that might trouble the modern reader would not have particularly disturbed their contemporaries, since the principles of rigid-body motion were barely developed at this time.

Although Barrow's derivation of the sine law of refraction (II:4) was based on Hobbes's, it was simpler and more elegant. It is assumed that light moves more slowly in optically denser media such as glass than in air. Moreover, because the model requires that the ray fronts always be normal to the rays and of constant length, the rays are again assumed to

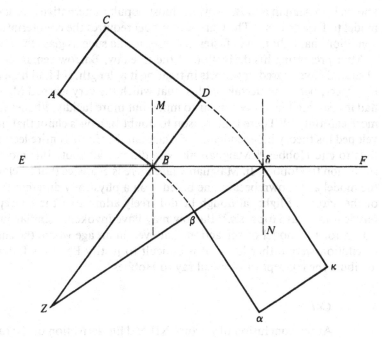

Figure 3

rotate during refraction. The physical ray *ABCD* (Figure 3), whose ray front is successively propagated from *AC* to *BD*, is incident from a rarer medium upon the refracting surface *EF*. Since point *B* of the ray enters the denser medium first, it moves with a slower velocity before end *D*, which remains in the less dense medium. During refraction the two parts of the ray progress with unequal velocity, and the ray traverses the path *BD*$\delta\beta$, the portion of the sector described by rotating *DB* about the point *Z*, which is on the extension of *DB* such that $ZD/ZB = v_i/v_r$, where v_i and v_r are the velocities of light in the upper and lower media, respectively. The unequal progress of the two portions of the ray continues until *D* enters the denser medium at δ, when the entire ray once more moves rectilinearly, since it is again in a uniform medium. The ray will move in the direction $\beta\alpha$, $\delta\kappa$, the perpendiculars to the ray front $\beta\delta$. The sine law of refraction follows directly from the model, for by applying the trigonometric law of sines in triangle $Z\delta B$, there results $Z\delta/ZB = ZD/ZB = \sin ZB\delta/\sin Z\delta B = \sin ZBE/\sin Z\delta B = \sin i/\sin r$, where $\widehat{ABM} = \widehat{ZBE} = i$, the angle of incidence, and $\widehat{N\delta\kappa} = \widehat{Z\delta B} = r$, the angle of refraction. Thus, $ZD/ZB = v_i/v_r = \sin i/\sin r$ is a constant independent of the angle of incidence. This result – namely, that light moves more slowly in an optically denser medium where it is bent toward the perpendicular – is probably what made the Hobbesian model, as transformed and made known by Maignan and Barrow, the most popular alternative mechanical model to Descartes's.[36] The Cartesian model requires the counterintuitive condition that light moves faster in denser media such as glass than in air.

After presenting his derivation of the sine law, Barrow remarked that "I should have spared my efforts in treating it at length, if I had happened to inspect, before undertaking this, that which the very learned Maignan had investigated in a way similar to mine, but more lucidly, I believe, and more carefully."[37] There is no reason to doubt Barrow's claim that he developed his theory independently of Maignan, but in this entire lecture he fails to cite Hobbes.[38] Maignan likewise did not attribute this model of refraction to Hobbes. In Maignan's case there is evidence that he claimed the model as his own because he based it on a physically different theory of the nature of light, although he did freely admit that the concept of physical ray was Hobbes's.[39] Barrow may have invoked a similar justification for the model of refraction, but even in an age where the canons of citation were rather loose, it is difficult to justify Barrow's failure to attribute the concept of physical ray to Hobbes.

Color

At the conclusion of Lecture XII and his derivation of the radius of the primary rainbow, Barrow remarked that "since colors happened to be mentioned, what if I were to guess a few things about them, although it is contrary to my practice and rule?"[40] Barrow's conjectures on the origin of colors, like his remarks on the nature of light, are very eclectic and terse and consequently difficult to understand fully. He adopted a modification theory of color, that is, the idea that colors arise from some modification or alteration of pure white light, such as a mixture with shadow or a compression by refraction or reflection.[41] In the new theory of color that Newton was then developing, white light was, on the contrary, a mixture of all the simple spectral colors, which could be separated from one another but in no way be modified.

Although Barrow's remarks on color were prompted by the rainbow, he does not discuss the colors generated by refraction but rather considers those produced by luminous or illuminated bodies. He first explains the nature of white and black bodies:

> *White* is that which pours around a great quantity of uniformly dense light; such, in general, are bodies punctured with pores farther apart, especially those that have many little surfaces facing every way. . . .
> *Black* is that which pours out the least or very little light; such for the most part are very transparent bodies, and also those which have abundant pores and little holes absorbing light.[42]

White and black bodies therefore differ in the quantity of light they emit. From Aristotle to Newton colors were essentially ranged on a scale of luminosity from white to black, and following this approach Barrow invoked changes in the density of light and the intermixture of shadow to explain the nature of red and blue. "*Red* is that which pours all around condensed and unusually compressed light, but broken and interrupted by shadowy interstices."[43] These bright bodies can, according to Barrow, be imagined to consist of particles in the shape of concave burning mirrors or spherical burning glasses that focus the light falling on them. Blue, on the contrary, is closer to black. These bodies emit "a rare light or a light excited by a rather sluggish force" and consist of a mixture of white and black particles.[44] At this point Barrow essentially gives up:

> *Green* is very closely related to blue. Let wiser men explore the difference; I dare not guess.
>
> The other colors emerge from these variously mixed and tempered. . . . But let this much suffice for exploring things beyond our comprehension and delivering ourselves to the derision of the more exacting critics of a physics of causes.[45]

Barrow's brief forays away from mathematical physics into natural philosophy to explain light and color are neither particularly insightful nor original. His "guesses" on color were offered "by way of digression," and it is evident that Barrow did not intend that they be taken any more seriously than that.[46] Nonetheless, in 1675 the anonymous reviewer in the *Journal des sçavans* (undoubtedly the Cartesian Pierre Sylvain Regis) took them quite seriously and devoted most of his review to these few paragraphs of the *Lectiones XVIII*. He especially approved of Barrow's account of red and blue, since "the new philosophers after explaining black and white so well, do not ordinarily speak of other colors except in general terms and by very remote conjectures." He concluded by observing that Barrow "is not of the Cartesian opinion, which we will perhaps explain elsewhere."[47] Fifteen years later in his *Cours entier de philosophie* Regis reinterpreted Barrow's views in terms of the rotating balls of the Cartesian ether.[48] Since Regis's comments on Barrow's ideas about color are the only ones that I have found in this period, it hardly seems credible to conclude that Newton allowed Barrow to become a subject of ridicule.

Kepler and the principles of optical imagery

As a preliminary to his own theory of image formation and vision, Kepler devoted the third chapter of his *Ad Vitellionem paralipomena* to replacing the ancient cathetus rule of image location with a new, more general one.[49] According to this ancient rule, the image is located at the intersection of the reflected or refracted ray and the cathetus (produced when necessary), which is the perpendicular drawn from the object

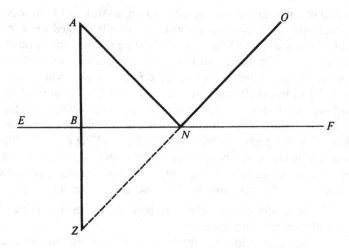

Figure 4

to the reflecting or refracting surface.[50] Since the cathetus is simply the axis of the optical system, the image will therefore always be located on the axis. The principle, which was first formulated for reflection by Euclid in his *Catoptrics,* is strictly true for plane mirrors alone. In Figure 4 the object *A* emits ray *AN*, which is reflected to the eye *O* by the mirror *EF*, and the image *Z* is located at the intersection of the reflected ray produced and the perpendicular or cathetus *AB* drawn from the object to the mirror. By analogy the rule was extended to curved surfaces: In the concave spherical mirror *ENB* (Figure 5) the image *Z* of the object *A* seen by the eye *O* is located at the intersection of the cathetus *AB* produced and the reflected ray *NZO*. When the object is shifted to *A'* and the cathetus *A'B'* becomes parallel to the reflected ray, the image will lie at infinity; and when it is shifted still farther to *A"*, the image will be at *Z".*[51] Qualitatively, at least, this theory appears to agree with the modern one, with point *A'* corresponding to the focal point and the images *Z* and *Z"* to real and virtual images, respectively (although ancient and medieval opticians did not distinguish between the two sorts of image). Let us, however, examine this case more carefully in order to define the range of validity of the image locations predicted by the cathetus rule. In the first place, insofar as the rule was supposed to locate the (tangential) image for rays lying in a single plane of incidence – and it was – it is erroneous. According to the modern theory of Kepler and Barrow, the image will not lie on the axis or cathetus; in particular, the images *Z* and *Z"* will be shifted away from the axes *AC* and *A"C* along *ON* toward point *N* and will lie on the caustic. For nearly perpendicularly incident or paraxial rays, however, as both Kepler and Barrow noted, the cathetus rule scarcely

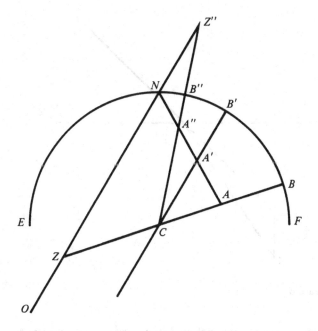

Figure 5

differs from the modern one. If now we consider rays lying out of the initial plane of incidence (which was not the intention of ancient opticians), the cathetus rule correctly predicts the location of the sagittal image on the axis, as Kepler first observed. Thus, the range of validity of the cathetus rule is greater than generally recognized, but it was only with Newton's investigation of astigmatism that the ancient and modern theories could be properly reconciled. The cathetus rule was extended to refraction at plane and curved surfaces by medieval opticians. For example, in Figure 6 the point *A* in a denser medium below the refracting surface *EF* emits ray *AN*, which is refracted at *N* to the eye at *O*; and the image *Z* is at the intersection of the refracted ray produced *ONZ* and the perpendicular *AB* to the refracting surface.[52]

Despite the capacity of the cathetus rule to describe the position of the image correctly in certain circumstances, the concept of image employed differs fundamentally from a Keplerian image, which is formed by a pencil of rays arriving from each point of the object. In the ancient and medieval approach an image of a point is determined by a single ray leaving or entering the eye, so that its optical geometry is founded on a visual pyramid that has its base on the object and vertex in the eye and is composed of single rays extending from each point of the object to a corresponding point of the eye.[53] It is equally characteristic of medieval

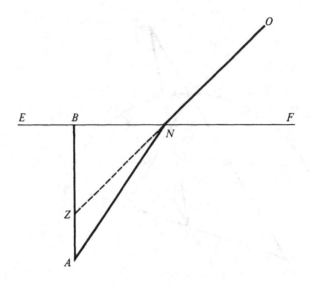

Figure 6

optics that the eye is always included in an analysis of images, for without an eye to receive the form or species of an object, there can be no vision or images. Thus, the concept of real images as pictures painted on a screen by pencils of light rays independent of the eye is a foreign one.

Kepler's treatment of images in Chapter III, "On the Foundation of Catoptrics and the Place of the Image," is fundamentally a psychological one. He adopts the traditional definition of the image as an error of the faculties in not seeing an object in its true place and size.[54] After beginning with an extended critique of the cathetus rule "of Euclid, Witelo, and Alhazen," Kepler explains judgments of distance. Distances that are not too great are determined by means of a "geometry of the triangle" (*trianguli Geometria*) *ZOP* in Figure 7a that utilizes the distance between our two eyes *O* and *P* and the angle of convergence of the axes of the eyes (Proposition VIII); this is known as convergence. We also learn to make similar judgments of (smaller) distances with a single eye from experience with two eyes, but now we use a "distance-measuring triangle" (*triangulum distantiae mensorium*) *ZOP* in Figure 7b whose base *OP* is the opening of the pupil (Proposition IX). Kepler explains other cues, such as light density, for judging distance, before he presents his new principle of image location. The eye imagines objects to be in the place from which the reflected or refracted rays come, since it is unaware of any changes in the direction of the rays before they enter the eye. Consequently, Kepler explains in Proposition XVII, "the true place of the image is that point where the visual rays *from each eye* produced through

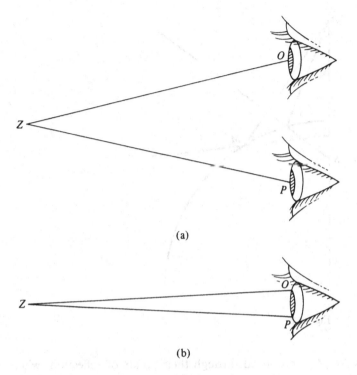

(a)

(b)

Figure 7

their points of refraction or reflection come together, according to Prop. VIII of this third chapter."[55] Although Kepler frequently asserts that his discussion of vision with two eyes applies equally to the opening of one eye, it is apparent from the formulation of this rule and his citation of Proposition VIII that he is thinking in terms of binocular vision and convergence; this will be even more evident from his applications of this rule. Turning to an analysis of the cathetus rule, Kepler first shows that when the planes of incidence for each eye do not coincide (which is ordinarily the case), the image will lie on the intersection of the two planes, which will also be perpendicular to the surface. In particular, since all surfaces that are perpendicular to a sphere pass through its center, the image will lie on a line passing through the center, just as the cathetus rule demands.[56] By considering the formation of an image in three dimensions, Kepler has essentially identified the location of the sagittal image. In Proposition XVIII, he now demonstrates that if the two eyes are in the same plane of incidence (which is *contra naturam*), the image will not lie on the perpendicular, and the cathetus rule cannot therefore be universally true. The image of point *A* (Figure 8) reflected from the convex mirror *BNM* to the eyes *O, P* will be at the intersection *Z* of the reflected

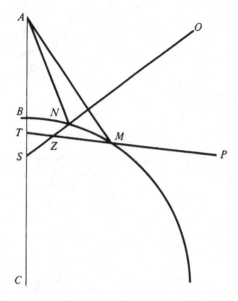

Figure 8

rays *NO, MP* produced through their points of reflection, whereas according to the cathetus rule it will lie on the axis at *S* or *T*, depending on which eye views it. At the conclusion of the proposition, Kepler notes that with one eye this rule breaks down, because the arc *MN* is then too small to be sensible.[57] Kepler did not attempt to reconcile the two fundamentally different locations for the image depending on the plane in which the eye(s) is located. He afterward ignores the (sagittal) image on the axis, for he will be concerned with the (tangential) image in a single plane of incidence, the traditional problem of geometrical optics. Newton would later return to consider image formation in three dimensions and observe that in fact both can be seen simultaneously, so that there are two images in two locations.

Kepler has not clearly distinguished between the perceived and geometrical images. Thus far he has been concerned with the location of the perceived image by convergence and accommodation, and in this sense the point *Z* where he locates the image is not the vertex of a pencil of rays, but the vertex of the "distance-measuring triangle" made by two rays. However, the cathetus rule, which this chapter is devoted to refuting, is equally a rule of geometrical optics. Consequently, it must be replaced by a rule that defines a geometrical image and applies to only one eye, or rather one pencil of rays. In this sense a single pencil of rays for each eye must be considered, because in modern (i.e., Keplerian) geometrical

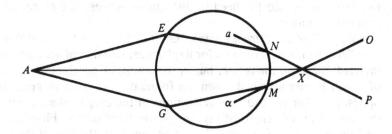

Figure 9

optics an image is defined as the place where a pencil of rays from a point of the object is reassembled. If we consider the formation of a geometrical image in Kepler's example, the two rays originate from the same point, but instead of being brought together again at a single point of one image they enter different eyes and contribute to two images. When pencils are considered with two eyes, two different pencils of rays with axes *NO* and *MP* entering each eye will form two images.[58]

From his application of his rule of image location in Chapter V of his *Paralipomena* we can see even more plainly that Kepler intended it to apply to the perceived location of the image. He begins Chapter V by examining the place of the image of a point seen through a sphere filled with water. He locates the image of point *A* (Figure 9) seen through the sphere *ENM* with two eyes *O* and *P* at *X*, the intersection of rays *AENP* and *AGMO* (Proposition I). However, with one eye he puts it on the sphere at *M* and *N* (Proposition VI), for with one eye the two rays are so close together that they effectively become one, and the sphere is the first place intersected by the ray at which the eye can locate the image.[59] Yet later in this chapter Kepler describes the formation of geometrical images in the water-filled sphere by the intersections of pencils of rays, without regard to any eye, and locates them at various places on the axis, depending on the nature of the incident rays.[60] In this example, then, Kepler has by the rule of Chapter III located the perceived image on the axis with two eyes and on the sphere with one, whereas by the geometrical methods of Chapter V he places the geometrical image on the axis. In the earlier refutation of the cathetus rule by the example of the image in a convex mirror, he had not yet introduced the concept of geometrical image. Barrow in his *Lectiones XVIII* would attack Kepler's failure ("otherwise rarely careless"), even in the case of the sphere (Figure 9), to consider two pencils and admit the existence of two images *a*, *α* with two eyes off the axis.[61] Barrow would reinterpret and restrict Kepler's rule to small pencils entering one eye and convert it into a fruitful rule of geometrical optics. In the process, however, he would all but abandon

Kepler's insight into the need to distinguish between the perceived and geometrical images.

Chapter V of the *Paralipomena*, "On the Manner in Which Vision Occurs," has attained its renown for Kepler's explanation of vision by a real inverted image on the retina, but in the course of developing his theory of vision Kepler also laid down the foundation of modern geometrical optics. In order to study the refractions in the eye, Kepler investigated the refractions of pencils of rays in a water-filled sphere. First he examined the intersection of parallel incident rays with the axis of the sphere after refraction and demonstrated that the closer the incident rays are to the axis the farther their intersection will be from the sphere (Proposition IX). Then he showed that all rays making an angle of less than 10° with the axis intersect the axis very near the extreme limit of intersections, that is, at the focal point (Proposition XV). Kepler correctly determined the sphere's focal point to be equal to its radius by calculating the paths of the refracted rays with the approximation that the angles of incidence and refraction are proportional (Proposition XIV), since he did not know the true law of refraction. "I have," he lamented, "despaired of defining geometrically the precise point where the extreme intersection occurs. I beseech you, reader, to help me here."[62] Kepler then proceeded to describe in a purely qualitative way the points of intersection with the axis for rays from objects at a finite distance from the sphere. All of this was preliminary to his introduction of the concept of a "picture" or real image cast on a screen (as opposed to "images" perceived by the eye) that would serve as his model for the retinal image.[63] We cannot dwell on his account of the retinal image other than to note that he here presents a thorough qualitative description of the caustic and its intersection with the axis, or the image point, where most of the rays come together.[64]

Even this incomplete sketch should indicate the extraordinary nature of Kepler's achievement in attaining such fundamental and sophisticated results with such a simple – almost trivial – model of refraction in an aqueous sphere. Though Kepler had been unable to reach any general rules for defining the image or focal points, his concepts and methods would serve as the foundation for future research in geometrical optics. Barrow, in particular, would pursue Kepler's approach to optical imagery and synthesize and extend his ideas on the location of the image, caustics, and image points as limits, all of which are based on the intersection of reflected and refracted rays.

Shortly after Galileo published his *Sidereus nuncius* announcing his telescopic discoveries in 1610, Kepler published his *Dioptrice,* in which he applied the concepts of the *Paralipomena* to explaining in a brief and lucid manner the operation of telescopes and lenses. Kepler succeeded in finding the focal point for a single spherical surface and for a thin equiconvex lens.[65] Still unaware of the sine law of refraction, he assumed that up to an angle of incidence of 30° from air into glass the deviation is approximately equal to one-third the angle of incidence (which corresponds to an index of refraction of $\frac{3}{2}$).[66] He was unable to determine either the focal point for any other lenses or the image of a point at a finite distance for any surface whatever. In the *Dioptrice* Kepler was concerned solely with the geometrical image. His definition of the location of an image was now confined to its direction alone (namely, in the direction of the rays entering the eye), and he did not use intersecting rays to find its position. We need not be concerned with his fundamental contribution to telescopic optics, but henceforth the study of geometrical optics would be closely related to the design of optical instruments, especially the telescope.

Optical imagery from Kepler to James Gregory

Kepler's work was rather rapidly assimilated. His theory of vision was advocated and advanced by Christoph Scheiner's *Oculus hoc est: Fundamentum opticum* (1619) and Descartes's *La dioptrique* (1637), though neither gave due credit to Kepler. Barrow simply adopted Kepler's theory of vision as a starting point and made no contributions to it. Descartes's *La dioptrique* also contained his discovery of the sine law of refraction, which at last provided the science of dioptric imagery with an exact law to replace small-angle approximations. Descartes devoted himself principally to systems with perfect imagery, that is, systems that bring rays to a focus at a single point. He discovered the class of curves with this property, "Cartesian ovals," which under certain conditions reduce to the conic sections. With the goal – vain, as it turned out – of avoiding the spherical aberration of ordinary lenses, Descartes described how to grind lenses in the form of conics and succeeded in directing much of the effort in seventeenth-century practical optics toward nonspherical lenses. James Gregory likewise devoted much of his *Optica promota* to reflection and refraction in conoids, although in the isolation of Aberdeen he seemed unaware of Descartes's *La dioptrique*. Gregory's most important contribution, the Gregorian reflecting telescope, was designed with both paraboloidal and ellipsoidal mirrors.[67] Barrow (and also Huygens) stood apart from this trend and only briefly treated reflection and refraction in conics, since it had been treated thoroughly by others and did not seem especially useful.[68]

Cavalieri took the first significant step beyond Kepler in analyzing the focal properties of spherical surfaces and lenses in his *Exercitationes geometricae sex.* He begins the section "On the Foci of Lenses" by observing that whereas conic sections have precise foci, spherical lenses, as Kepler showed, possess a focus only to a very close approximation.[69] He then succeeded in deriving a general rule for the focal point for all varieties of lenses, by assuming, as Kepler did, thin lenses and the small-angle approximation, but without now mentioning the newly discovered sine law of refraction. Using these approximations, he found for the focus f of a lens with index of refraction $\frac{3}{2}$, $(\rho_1 \pm \rho_2)/\rho_1 = 2\rho_2/f$, where $\rho_{1,2}$ represent the radii of the two surfaces and the plus or minus sign is taken depending on whether the surfaces face in the opposite or the same direction, respectively.[70] Cavalieri derived these results by means of single paraxial rays – not pencils – very much like a modern elementary textbook.[71]

Just a few years later, in 1653, and independently of Cavalieri, Huygens launched a more general attack on the problem of optical imagery in his *Dioptrica*. He determined the image point not just for parallel rays, but for an object at any distance on the axis for plane and spherical surfaces and thick lenses. For thin lenses he generalized Cavalieri's rule for the focal point for an arbitrary index of refraction n to $(\rho_1 \pm \rho_2)/\rho_1 = \rho_2/(n-1)f$, and he found for the image point s', $(s-f)/s = s/(s+s')$;[72] together these yield the familiar formula $1/s + 1/s' = 1/f = (n-1)(1/\rho_1 \pm 1/\rho_2)$, where the object and image distances s and s' are measured from the lens. Huygens's approach to geometrical optics shows his deep understanding of Kepler's methods and concepts. He employed pencils of rays and defined the "points of concourse and dispersion" (image points for real and imaginary images) as the limiting point of intersections with the axis of rays infinitely close to the axis.[73]

Shifting our attention from the geometrical to the perceived image, we see that in *La dioptrique* Descartes's position on this problem was very much like Kepler's. He endorsed Kepler's rejection of the cathetus rule for catoptrics and considered the location of the image to be mainly a psychophysiological problem involving a number of mechanisms, including accommodation and convergence.[74] In general, he believed that "all the means that we have for knowing distance are very uncertain."[75] However, in the ray diagrams that Descartes uses to illustrate the principle that objects seen by reflection and refraction are judged to be in the direction in which the rays enter the eye, he draws all the images at the place from which the pencils that enter the eye diverge.[76] I do not believe that we can infer much about Descartes's ideas on vision with one eye from a ray diagram, since there is little alternative to drawing the image at the place where the pencils converge. I do believe, though, that ray diagrams, with their pencils of rays extending from object point to image point, did serve to teach that in geometrical optics a single pencil must be traced through an optical system to its convergence at one image point.

James Gregory asserted that the place of the image seen with one eye is judged to be at the point of intersection of a pencil of reflected or refracted rays, and thus he unambiguously identified the perceived and geometrical images. In Proposition 29 of his *Optica promota*, Gregory extended the validity of Kepler's principle of distance judgment for a single

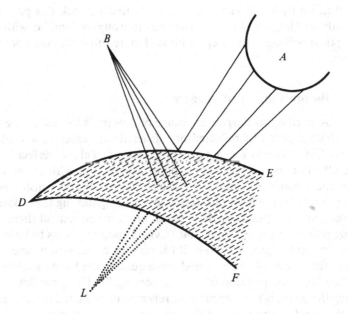

Figure 10

eye, his "distance-measuring triangle," to all distances. He explained that from our sensations of the alterations made in our eye to form a distinct picture on the retina, we are able to estimate from experience the distance of an object. Proposition 36 – given the positions of a surface, a visible point, and an eye, to determine the place of the image – applies this principle to vision by reflection and refraction. Gregory's solution to this problem (Figure 10) for a mirror *DEF* is a straightforward transformation of Kepler's rule for two eyes:

> From the points of the pupil [*A*], draw through the points of reflection all the lines of reflection, in whose concourse *L* (provided they concur) will be the apparent place of the image of the point *B*. If, however, they do not concur in one point, no distinct and fixed place of the image of the visible point *B* will exist.[77]

Gregory's diagram is similar to Kepler's (Figure 8), but now a pencil of rays diverges from the image point and falls upon the pupil of a single eye. Although Gregory transformed Kepler's rule to one of geometrical optics, he did not adopt Kepler's concept of an image point as the limit of intersecting rays, for in this case he considered the image "indistinct" and "indeterminate." Rather, Gregory restricted himself to perfect imagery and could not exploit Kepler's rule to obtain further results. That task was taken up by Barrow. It is quite likely that he took this path independently of Gregory, since Barrow was thoroughly familiar with Kepler's optical writings, and Kepler himself stated that his rule applied to one eye.

Barrow's theory of imagery

According to Barrow an image of a point is located at the place from which a pencil of reflected or refracted rays entering a single eye diverges. Since Barrow was not generally concerned about perfect images, where all the rays diverge from a point exactly, but with refraction at planes and spheres, he adopted Kepler's approach and considered the image point to be the limit of intersections of neighboring rays. Barrow's achievement was to convert these ideas into a mathematical theory. His starting point was the principle that "a visible point appears to be located on that ray which proceeds from it (directly or by inflection) and passes through the center of the eye, and consequently the location of objects is judged from the position of rays so passing."[78] Thus, the first step in locating the image is to determine the reflected or refracted ray that passes from the visible point through the center of the eye when the position of each is given; Barrow calls this ray the "principal ray."[79] This step generates a class of difficult mathematical problems such as Alhazen's and the anaclastic problem. For his next step, to determine where on the principal ray the image of the point is located, he finds the intersections of the principal ray with those rays that are infinitely close to it and enter the pupil of the eye. With exhaustive rigor he then demonstrates that the closer the rays are to this principal ray the nearer their intersection falls to a limit point, which is his strict definition of the image point. To illustrate his approach let us jump ahead (see Figure 13) to his determination of the image of a point A refracted at a plane EF: The refracted ray ONZ passes through the center of the pupil COD, and Barrow demonstrates that Z is the image point (through which ray OKZ alone passes) by showing that all other rays of the pencil that enter the eye intersect that ray very close to Z but above or below it at points V and X. He finds the image point for oblique pencils by limit-increment arguments in the various cases of spherical surfaces, but for plane surfaces he is able to avoid them with a clever insight; and for paraxial rays such arguments are unnecessary. In his second "epistola" to the reader Barrow boasts, quite legitimately, that he is the first one in print to have erected the foundations of dioptrics on the sine law of refraction.[80] Without the exact law of refraction, results can be derived only for paraxial rays, as had been done up to his time.

Barrow's very conception of the problem of image location is in terms of rays entering the eye and does not at all distinguish between the geometrical and perceived images. In the central Lectures IV–XIII he consistently

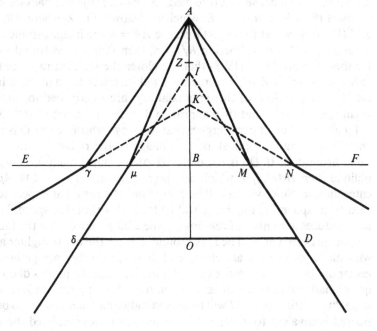

Figure 11

applies this approach to determining images in reflection and refraction at planes and spheres. He attacks each of these problems in a methodical way, beginning with the simplest case of parallel incident rays with the eye on the axis and proceeding through parallel rays with the eye off the axis (i.e., oblique pencils) to pencils of rays at a finite distance with the eye on and finally off the axis; when the eye is on the axis, he always determines the focal and image points for the limiting case of paraxial rays. The essence of his mathematical demonstrations is often hidden within a geometrical formulation and manipulation of proportions, which are so foreign to us today. My aim will be to epitomize his demonstrations, so that the key steps and their physical import can be grasped.

Refraction at a plane surface

Barrow readily dispatches plane mirrors, which form perfect images (III:13–17), and the trivial case of the image for parallel rays falling on a plane refracting surface (IV:2–3). He seriously sets out on his program by finding the image Z (Figure 11) of a radiant point A viewed across a refracting surface EF; the eye O is located in the rarer medium and is on the axis ABO. By means of a construction to determine the refraction of any ray incident from point A, Barrow establishes that all rays refracted from a dense to a rare medium, when projected backward, will intersect the axis at points I, K, which are below point Z, where $AB/ZB = R/I$ (IV:5). If we let the object distance $AB = s$, the image distance $ZB = s'$, and the index of refraction $R/I = n$, then Barrow has found $s = ns'$. He then demonstrates (IV:8) that the closer the incident rays are to the axis the closer to Z their refractions will intersect the axis, and indeed that the intersections of these paraxial rays are so crowded together in a small space near Z that they appear to flow from that point (IV:15–19).

Finally (IV:20), Barrow argues that to an eye whose center O is placed anywhere along the optical axis the image of the radiant point A is located around Z. If $D\delta$ is the diameter of the pupil, then only rays less oblique then $IMD, I\mu\delta$, which are the refractions of rays $AM, A\mu$, can enter the eye. Since vision will be caused only by rays that appear to proceed from space ZI, the image will be located within that space. Barrow now adduces a number of reasons – some a bit strained – for the fact that the image is seen at Z: The rays around Z enter the eye straighter and so with more force; they are closer and denser; and they are gathered together again in the eye more easily.[81] Finally, since the pupil's diameter is quite small and the eye's distance from the refracting surface is relatively large, "the entire space ZI will be exceedingly small and deserve to be considered equivalent to a point."[82] Barrow has now completed the mathematical transformation of Kepler's principle for locating the perceived image by applying it to the infinitesimally small region of intersecting rays entering a single eye, and not limiting it, as Gregory did, to perfect images. This, however, is how Kepler conceived of the geometrical image, and Barrow has fruitfully combined the two approaches.

The results are even more fruitful when Barrow turns to finding points on the caustic or the image point for an obliquely incident pencil of rays. Barrow, however, formulates the problem in a different way, namely, to find the place of the image to an eye viewing the radiant point off the axis. This nearly medieval conception of the role of the eye affects his solution to the problem. Rather than determining the image point for an arbitrary pencil of rays emitted from the object, Barrow assumes the position of the eye (and object) to be given and searches for the ray emitted from the object that enters the center of the eye, and then he determines the image point. This approach leads him to begin with what is known as the anaclastic problem (Figure 12): Given the radiant point A and a point X through which a refracted ray is to pass, find the incident ray AN.[83] Barrow considers the point X to be on the same side of the refracting plane EF by projecting the refracted ray $N\beta$ backward, because he is ultimately interested in locating the image that will be seen on the same side of the plane as the object. As a preliminary, Barrow deduces from the sine law of refraction (IV:10) that if $KN\beta$ is the refracted ray of any incident ray AN, and a point Y is chosen such that $YB/AB = I/(I^2 - R^2)^{1/2}$, then it

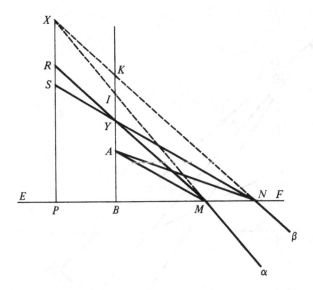

Figure 12

follows that $XP/SN = (I^2 - R^2)^{1/2}/R$. To solve the anaclastic problem (V:12) – given the points A and X – all one need do is insert through the point Y the given line $SN = XP \times R/(I^2 - R^2)^{1/2}$ subtending the angle ABF. Barrow presents a construction (V:7) involving the intersection of a circle and rectangular hyperbola to solve this "neusis," or "verging," problem of inserting a line of given length in a right angle so that it passes through a given point.[84] One feature of posing the anaclastic problem with both points on the same side of the refracting plane is that in general two points N, M satisfy the solution, or $SN = RM$. When the points N, M or the neighboring refracted rays XN, XM coincide, then X will be the image point. Barrow then observed (V:9) that in this case the line SN is a minimum, which occurs when $BN^3 = BP \times BY^2$. Thus, the image point X for any refracted ray $KN\beta$ that passes through the center of the eye is readily found (V:15–16): Take the point Y according to the condition specified in IV:10, find the distance BP as determined in V:9, and through point P drop to the refracting surface the perpendicular PX, which will intersect ray KN at X (which image point Barrow henceforth designates by Z).[85]

It is now a straightforward matter for Barrow to argue (V:17–21), just as he had in the preceding case for the eye on the axis, that for an eye whose center O (Figure 13) is on the refracted ray NK and whose pupil COD is perpendicular to it, Z is the place of the image of point A. A ray more oblique than KNO will intersect it beyond Z at V, while a less

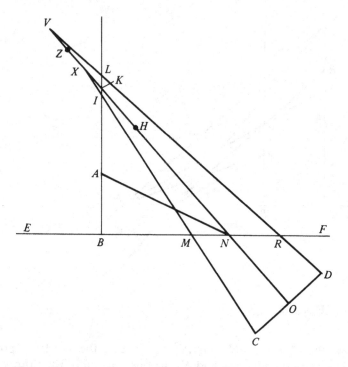

Figure 13

oblique one will intersect it on the near side at *X*; and since the intersections around *Z* are especially dense, the eye will judge the image to be around *Z*. According to the cathetus rule "of Alhazen and most of the multitude of opticians after him," the image is at *K*, the intersection of the perpendicular and the principal ray; but Barrow notes that only one ray passes through that point unless the perpendicular passes through the eye.[86] Hobbes's view that the image is located at *H*, which is as far from the refracting point *N* as the object *A*, is even more erroneous according to Barrow, since no rays at all intersect within angle *ABF*.[87] In his criticism of contemporary theories of optical imagery, which he considered "extremely deficient, not to say thoroughly faulty," he failed to mention Kepler's earlier critique.[88] Most of the experiments reported in the *Lectiones XVIII* are drawn from others, but Barrow introduced one of his own here – indeed, a "not inelegant" one – that "clearly destroys the doctrine of Alhazen and his followers."[89] An object *HG* (Figure 14) is placed over a surface of water *RS* so that it is adjacent to a vertical string or rod *GBF* with a weight attached at *F*. When these are viewed by the obliquely placed eye *O*, the point *G* is seen by reflection at γ on the perpendicular, while the point *F* is seen by refraction on the near side of the perpendicular

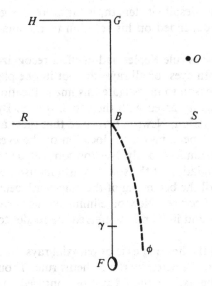

Figure 14

at ϕ and not on it, as the cathetus rule predicts. The principal virtue of this frequently cited experiment is that the perpendicular to the refracting surface is found optically by the law of reflection, for it is otherwise difficult to determine the perpendicular precisely.[90]

The existence of two images

We can gain additional insight into Barrow's method, as well as the cathetus rule, by briefly turning to Newton's *Optical Lectures,* in which he fully adopts Barrow's approach and further develops his solution.[91] Newton observes that the point Z (Figure 13) is the image point solely for rays in the plane of incidence ABN, for there is another image point for rays out of this plane: The refracted rays of all those incident rays that lie on the surface of the cone with axis AB, vertex A, and half-angle BAN will intersect the principal ray NZ in point K on the perpendicular and produce a second image point there. Newton has extended the concept of pencil to three dimensions and thereby disposed of Barrow's argument against the cathetus rule that only one ray passes through point K. Having identified the existence of two image points, he then defines a single point in the interval KX as the place of the image, which turns out to be equivalent to the modern circle of least confusion.[92] With the sagittal image point K being trivially defined by the sine law (namely, $NK/NA = n$), Newton produces a limit-increment derivation of the tangential image point Z that, despite its length, is far more straightforward than Barrow's, which depended on his solution to the anaclastic problem.

In his criticism of the cathetus rule Kepler had in effect recognized the sagittal image point when both eyes, or all rays, are not in one plane of incidence. He did not have reason to investigate this image location further, in part precisely because it agreed with the cathetus rule. By pursuing Barrow's line of investigation, Newton perceived that these are not two locations for the image depending on the location of the eyes, but that they are two images seen simultaneously. Newton completed the erection of the mathematical foundation of the study of astigmatism, which, however, was to languish until the beginning of the nineteenth century.[93] When he revised his *Optical Lectures,* Newton eliminated his derivation of the tangential image point and in its place referred the reader to Barrow's *Lectiones XVIII.*[94]

Barrow understood that in the limiting case of paraxial rays the image fell on the axis in the position predicted by the cathetus rule. Though in this case the image fell on the axis, it should not be confused with the sagittal image; if we return to Figure 1, the paraxial image, which Barrow considered to be the limit of the tangential image Z, is at S, whereas Newton's newly discovered sagittal image – and that of the cathetus rule – is at K. Barrow considered it sufficiently important to distinguish between the paraxial and tangential images that he gave them distinct names. Near the end of his *Lectiones XVIII,* where he derives the images of straight lines, he indicated that

> the image of any point is in a certain manner double (which has
> frequently been insinuated in the preceding): the one is simple,
> absolute, and principal [*simplex, absoluta, principalis*], namely,
> that which dwells in the straight line that is perpendicular to the
> inflecting surface and passes through both the radiant point
> and the center of the eye... but the other is relative, mutable,
> and less important [*relata, mutabilis, ac minus praecipua*],
> which is such with respect to an eye arbitrarily placed off the
> perpendicular to the inflecting surface.[95]

Barrow's terminology for the paraxial and tangential images – absolute and relative or mutable – I believe, offers an insight into his formulation of geometrical optics and concept of oblique pencils and shows the fundamental role of the eye in his thought. His principle of image location, that the image was located at the point from which a single narrow pencil of rays diverged and entered one eye, provided him the means to develop a mathematical theory of optical imagery by investigating the intersection of infinitely close rays. One can, of course, define an image mathematically without introducing an eye. Huygens took precisely this approach

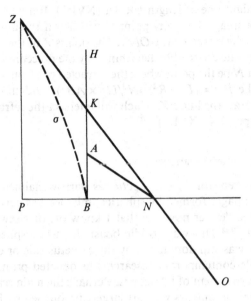

Figure 15

and substantially advanced geometrical optics. Yet he did not develop the imagery of oblique pencils, one of the most fruitful concepts of geometrical optics. I conjecture that it was Barrow's thinking in terms of the eye that enabled him to develop this concept. The geometrical image, as formulated by Kepler, is determined by the intersections of a pencil of parallel rays with the fixed axis of the optical system. On one occasion, in investigating the retinal image, Kepler – with his customary insight – did consider the image formed by an oblique pencil, but the *Paralipomena* and especially the *Dioptrice* were devoted to the study of images on the axis of the system.[96] Barrow's "optical axis," the principal ray that passed through the center of the eye, was, on the contrary, an axis for a pencil that moved with the eye. The intersections of neighboring rays with this oblique and mobile axis allowed him to conceive of an image anywhere off the axis of the optical system – whence we can understand the origin of his definition of our tangential image as *relative* and *mutable*.

In Lecture XVI Barrow extended his solutions for the image of a point to those of a vertical line viewed across a refracting plane: "In a nearly similar way the image of any magnitude can be considered to be double; one in fact absolute (which I at least designate by this name), which is, as it were, made up of the absolute images of the individual points standing in it... but indeed the other relative."[97] If a line *AB* (Figure 15) in air is perpendicular to the refracting surface *PBN*, and an eye located in water views the line from directly above point *B*, the absolute image *BH* of the line will be a straight line of length $n \times AB$ (XVI:7). But when the eye is at O, the relative image Z of any point A will lie off the line AB away from the eye O on the refracted ray ONK. The points Z of the curve $B\sigma Z$ on which the images lie are readily determined by the preceding construction (V:15–16). Let N be the point where the refracted ray from point A cuts the surface; take $BP = (I^2 - R^2)BN^3/(I^2 \times AB^2)$, and through point P parallel to AB draw the line PZ, which will intersect the refracted ray ONK in the image point Z (XVI:7).

Reflection at a spherical surface

In the prefatory "epistola" to his *Lectiones* Barrow claimed that his investigation of the images formed by spherical mirrors was new and hitherto had been treated "either nowhere that I know of, or elsewhere for the most part falsely."[98] This was no idle boast, for when spherical mirrors were studied, it was still dominated by the cathetus rule or erroneous alternatives, while contemporary research was devoted primarily to refraction for the improvement of telescopes. Perhaps the main reason that the imagery of spherical mirrors was left largely untouched by Kepler's followers was that in contrast to refraction, where the exact sine law of refraction had been discovered only recently, the imagery of spherical mirrors had been at least approximately worked out by ancient and medieval scholars by the cathetus rule and the long-known law of reflection. Moreover, much of the ancient and medieval investigation of mirrors was devoted to images off the axis, and no one before Barrow had developed a mathematical theory of oblique pencils to supplant the cathetus rule. The modern theory of image formation in spherical mirrors is consequently essentially Barrow's creation. He attacks the problem in a thorough, if tedious, way: first finding the image point for pencils of parallel rays incident upon concave and convex mirrors along and oblique to the axis (Lectures VI and VII), then for pencils from a near point for convex mirrors with the eye on and off the axis (Lectures VIII and IX), and finally applying the preceding to concave mirrors (Lecture X).

Barrow derives the focal points for concave and convex mirrors simultaneously, for they are identical. In Figure 16 if ray MNP is incident on the mirror NB and reflected into GNK, for a convex mirror the reflected ray GN of MN must be produced backward through NK to intersect the axis, whereas for a concave mirror the reflected ray NK of PN passes directly through the axis. Thus, Barrow's mathematical approach makes no distinction between real and virtual images. From the law of reflection it is easy for him to deduce (VI:5) that if NQ is dropped perpendicular to the axis, and the radius BC is bisected at point Z, then $CZ/CK = CQ/CN$, or (since angle $NCB = i$, the angle of incidence) $CZ = CK \cos i$. It is then apparent that Z is the limit, or focal point, beyond which no

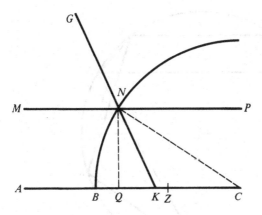

Figure 16

reflected ray intersects the axis (VI:7). Just as for refraction at a plane, Barrow demonstrates that incident rays closest to the axis intersect it nearer to the point Z, where the intersections are densest, so that Z must be considered to be the place of the image of a distant point to an eye with its center on the axis (VI:15–18). Barrow's conclusion that the focal point of a mirror bisects the radius had already been discovered in antiquity by means of the burning mirror, but the rest of his results are new.

Having found the focal point, Barrow next turns to locating the (tangential) image point when the eye is placed off the axis. First, however, he must solve a special case of Alhazen's problem of finding the reflected ray that passes through the eye when the incident rays are parallel (VII: 5–9). To determine the image point (VI:11–14) he begins by showing that for the infinitely close parallel incident rays SR, PN (Figure 17) $3NR = \pi\sigma$. The infinitesimal arcs can be replaced by their chords, so that $\frac{1}{3} = \widehat{NR}/\widehat{\pi\sigma} < NR/\pi\sigma = RX/X\pi < RX/X\sigma$, and hence $R\sigma < 4RX$. Similarly, he shows for the same intercept NX on the less oblique ray that $NX < N\pi/4$. Hence, in the reflected ray, where $NZ = N\pi/4$, Z will be the image point for an eye placed on the chief ray $N\pi$, since only one ray passes through that point and the intersection of neighboring rays is densest around that point (VII:15–18). Since $N\pi = -2\rho\cos i$, where $-\rho$ is the radius of a convex sphere, if we let the image distance NZ be s', Barrow has correctly located the tangential image point, or the point where the reflected ray is tangent to the caustic, to be $s' = -(\rho/2)\cos i$.

To locate the image point for objects at a finite distance Barrow now treats convex and concave mirrors separately. He begins with a pencil of rays from the object A incident along the axis upon the convex mirror NB (Figure 18) and finds (VIII:7), for the location of the paraxial image point Z, $AC/AB = CZ/BZ$. If we let the object distance $AB = s$, the

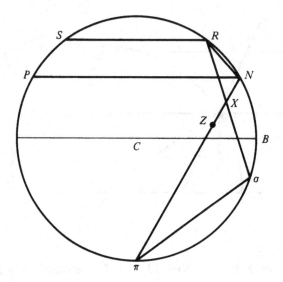

Figure 17

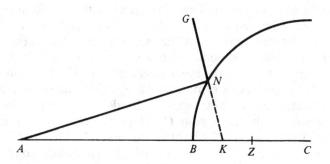

Figure 18

image distance $BZ = s'$, then Barrow's equation becomes $(s + \rho)/s = -(\rho + s')/s'$, or $1/s + 1/s' = -2/\rho$.

Barrow now undertakes the most general problem of reflection at a sphere, namely, determining the tangential image point when the object is at a finite distance. Since the usual preliminary problem confronting Barrow of drawing a reflected ray through a given point is in this case the renowned Alhazen's problem, I will later treat it separately. In Figure 19 two infinitely close rays AR, AN are incident from point A upon the convex mirror $RN\sigma$, where they are reflected at N and R along NG and RH, and when they are projected backward their point of intersection

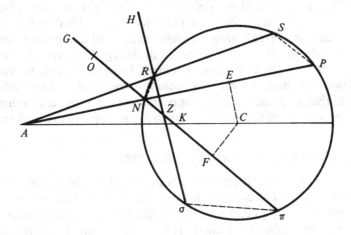

Figure 19

or image point Z is to be found. To prepare for his derivation (IX:10–12) Barrow shows that for the arcs cut off by the incident and reflected rays $\widehat{PS} + 2\widehat{NR} = \widehat{\pi\sigma}$, and also that if the chords NP and $N\pi$ are bisected at E and F, then $AE = (PA + NA)/2$. Since the arcs are indefinitely small, they can be replaced by their chords and AR can be taken equal to AN, so that $PS/NR = PA/AN$, or $(PA + 2AN)/AN = (PS + 2NR)/NR = \pi\sigma/NR = \pi Z/NZ$.[99] From the result $\pi Z/NZ = (PA + 2AN)/AN$, it readily follows that $FN/NZ = (AP + 3AN)/2AN$, or $FZ/NZ = (AP + AN)/2AN = AE/AN$ (IX:12). If we once again designate the object distance AN by s and the image distance NZ by s', and recall that $N\pi = -2\rho\cos i$, then Barrow's determination of the tangential image point or the catacaustic condition, $FZ/NZ = AC/AN$, becomes $1/s + 1/s' = -2/\rho\cos i$. He concludes by summarizing the method that he has so frequently invoked:

> Hence it is gathered that point Z is the very place around which the image of point [A] is located with respect to an eye placed, as at O, on the reflected ray $GN\pi$. For... as it can now be considered like a certain and fixed rule or law, that that image dwells where the reflections of rays closer to the principal incident ray (that is, whose reflected ray, passing through the center of the eye, acts as the optical axis) intersect that principal reflected ray.[100]

To complete his account of reflection at spherical mirrors, in Lecture X Barrow rapidly retraces the preceding two lectures on convex mirrors and shows how to apply those results to concave ones. The equations for the image's location are intrinsically the same, so that we need not follow him. Since a concave mirror yields both real and virtual images, Barrow discusses the conditions for the appearance of each. He does not use the terms "real" or "virtual" image, or equivalents, and distinguishes the two solely by whether the rays converge to or diverge from the axis after reflection. Barrow does not restrict himself to paraxial rays, so that his conditions are somewhat different from the modern ones. The condition, though, that a virtual image will appear when the object is at a distance from the mirror less than its focal length is equivalent to the modern one.

Images of lines and the cathetus rule

To appreciate Barrow's method in treating reflection at spherical surfaces, let us briefly compare his account of images in a convex mirror with those of his predecessors. We should first recall that Kepler illustrated his case against the cathetus rule with this example and found two different locations for the image, depending on the plane of the eyes; but his solution to the problem was only qualitative. Barrow was the first one to apply Kepler's theory to a mathematical derivation of the image point in all cases of reflection at a sphere. According to the cathetus rule (Figure 19), the image to the eye O will always be at the intersection of the reflected ray and the cathetus, or at the point K. Barrow correctly placed his absolute or paraxial image on the perpendicular only when the eye was on the perpendicular, whereas for oblique reflections he found that his relative or tangential image was always off the perpendicular (on the caustic). The contrast between Barrow's theory and the ancient one appears even more clearly in his analysis of the image of an infinitely long straight line BAS (Figure 20) perpendicular to the convex mirror BM. This problem was treated in most works on catoptrics that adopted the cathetus rule, since its solution was so simple: The image must always lie on the straight line, which is the cathetus itself, from B to L, where L, the image of the infinitely far point S, is defined by the condition that $CL = LD$.[101] For Barrow the appearance of the image is more complicated, and he considers two cases (XVI:11). When the eye is on the axis, the absolute image X of any point A of the line is determined by the previously derived rule (VIII:7) that $AC/AB = CX/BX$; and the image of the infinite line BAS will be the line BZ, where $Z = -\rho/2$. When the eye O is off the axis, the relative image of point A will lie on the ray OMK at point α, where α is above point X and off the axis toward the eye O; and similarly the image of the point S will be at σ. Thus, the image of the straight line will lie along the curve $B\alpha\tau\sigma$, where the points α are determined by the rule (IX:12) that $F\alpha/\alpha M = AE/EM$, and E and F bisect the portion of the reflected rays MP and $M\pi$ intercepted by the mirror.[102] Barrow then declares the nearly universally adopted cathetus rule to be shattered, for it is "gratuitously assumed and contrary to reason."[103] In

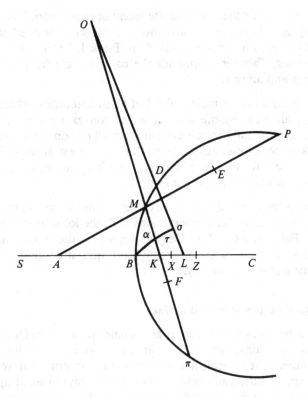

Figure 20

arguing against "a very erudite recent optical author, following in the ancients' footsteps," who claims to have experimentally verified "a hundred times" that the image is a straight line, Barrow concedes that it is exceedingly difficult to distinguish between his curve and their straight line (XVI:14). In addition to the difficulty (if not impossibility) of exactly determining the perpendicular, he notes that the points of his curve are removed from the perpendicular by such a small distance that the difference of the two images would not be apparent even to an acute observer. Rather, for experimental confirmation he appeals to his earlier experiment on refraction (V:22) in which the image of a plumb line immersed in water appears removed from the perpendicular (which is defined optically by reflection), since that experiment is based on identical principles. The significant difference with the cathetus rule in this case is not so much in the quantitative prediction of the image's position as it is in the conceptual foundation of determining images by the use of pencils. The differences are far more striking when the object is perpendicular to the axis and manifestly curved.[104]

Tacquet is most likely the "very erudite recent optical author" that Barrow was replying to here, for on two other occasions he criticized him by name for adhering to the cathetus rule.[105] In Book I, Proposition XXII of his *Catoptrica*, Tacquet propounds the cathetus rule for plane and convex mirrors and affirms:

> This theorem is the most fruitful of all of catoptrics, whereby nearly all the phenomena of plane and convex mirrors are demonstrated, as will become evident from all of book two and book three. Consequently, its truth is in turn extraordinarily established: for it cannot be false, since it agrees wonderfully with all phenomena without exception.[106]

Tacquet restricts the applicability of the cathetus rule to concave mirrors, because of certain anomalous observations with the location of blurred images (the "Barrovian case").[107] Despite Kepler's earlier attack on the cathetus rule and Barrow's more thorough critique, it continued to be taught into the eighteenth century.[108]

Refraction at a spherical surface

When he takes up refraction at a single spherical surface, Barrow returns to the mainstream of contemporary optics; but still his interests remain purely theoretical, and he shows no concern whatever for telescope design. His derivations are carried out for rays incident upon a denser convex surface but are easily extended to other cases. He devotes Lecture XI to parallel rays and shows (XI:2) that the focal point $f = \rho n/(n-1)$. This result was not new. Kepler had already derived it in his *Dioptrice* by assuming paraxial rays and $n = \frac{3}{2}$, but Barrow, in his now familiar way, rigorously shows that the focal point is the limit point beyond which no rays cross the axis and that the intersecting rays are densest about that point.

Barrow breaks new ground when he determines the image point off the axis for parallel rays incident upon a sphere and then applies this result to determining the radius of the primary rainbow. In Figure 21 let the indefinitely close rays MN and QR be refracted along NZ and RZ, and their place of intersection Z must be found (XII:7). Let CE and CF be perpendicular to the incident rays and CG and CI to the refracted rays. The infinitesimal arcs NR, $\pi\sigma$ can be replaced by their chords, so that $NR/\pi\sigma = NZ/\pi Z$ and therefore $NR/\frac{1}{2}(NR + \pi\sigma) = NZ/\frac{1}{2}(N\pi + 2\pi Z) = NZ/GZ$. It can also be shown (XII:6) that $NR/\frac{1}{2}(NR + \pi\sigma) = (NR/CN) \times (CN/\frac{1}{2}[NR + \pi\sigma]) = (EF/EN)(NG/GI) = (CE/CG)(NG/EN)$.[109] Equating the two conclusions we find, for the image point Z, $NZ/GZ = (NG/NE)(CE/CG)$. If we denote the angles of incidence and refraction ENC and GNC by i and r, NZ by s', and CN by ρ, then $CG/NG =$

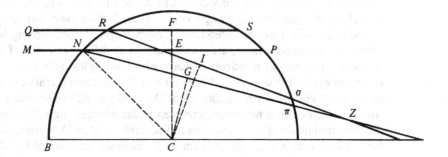

Figure 21

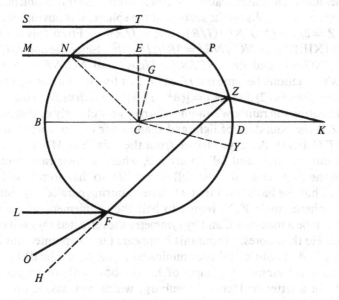

Figure 22

$\tan r$, $CE/NE = \tan i$, and $GZ = s' - \rho \cos r$. The tangential image point s', or the diacaustic condition, is then readily put in the form familiar since the work of L'Hospital, $s' = \rho n \cos^2 r/(n \cos r - \cos i)$.

Descartes's investigation of the rainbow was the point of departure for all subsequent research on the rainbow in the seventeenth century. In *Les meteores*, Discours VIII, he announced that the primary bow is seen when in a spherical rain drop (Figure 22) the inclination of the emergent ray *FO* with the incident ray *MN*, after two refractions at *N* and *F* and an internal reflection at *Z*, is a maximum. In his unpublished papers from about 1600, Thomas Harriot went on to deduce from this condition the rule that the tangent of the angle of incidence is then twice the tangent of the angle of refraction. Descartes, however, ponderously compiled a table with twenty-seven separate calculations at different angles of incidence on a spherical raindrop to learn at what angles light rays could enter our eyes: "I found that after one reflection and two refractions very many more could be seen under an angle of 41° to 42° than under any smaller one, and that none could be seen under a larger angle."[110] One of the principal goals of investigations of the rainbow after Descartes, therefore, was to derive a general rule for its appearance, as Harriot had done earlier, and to demonstrate that the inclination of the incident and refracted rays is then a maximum.

Barrow recognized that his determination of the image point for parallel rays incident on a sphere readily yielded a solution to this problem. When the image point Z falls on the surface of the sphere, it is obvious that then $NZ/GZ = 2/1 = (NG/NE)(I/R)$, where $I/R = n$. From this it directly follows (XII:11) that $NC^2/NE^2 = 3R^2/(I^2 - R^2)$. Since the angles $CNE = i$ and $CNG = r$, and thus $NC/NE = 1/\cos i$ and $NG/NE = \cos r/\cos i$, Barrow's equations become $\cos r/\cos i = 2/n$ (or, in Harriot's equivalent form, $\tan i/\tan r = 2$) and $\cos i = [(n^2-1)/3]^{1/2}$, which provide the grounds for calculating the rainbow's radius. Barrow now cleverly establishes that point Z is the boundary of light and shadow for rays refracted in quadrant BT (XII:13): All rays farther from the axis than MN intersect ray NZ within the circle and fall on arc ZD, whereas closer rays intersect it beyond the circle and must also fall on arc ZD, so that no rays will reach arc ZT. Thus, he has shown that when neighboring parallel rays intersect on the sphere, angle ZCD (equal to half the supplement of the deviation)[111] will be a maximum, and by symmetry the refracted rays will emerge parallel. He then notes, "From this it appears (as was pointed out by the distinguished M. Sluse and communicated to me by a friend) Descartes could have determined the angle of his rainbow without preparing tables."[112] In a letter to Henry Oldenburg, which was passed on to Barrow by his "friend" Collins, Sluse related his discovery that the arc ZY ($= 2$ arc ZD) illuminated by the incident rays is a measure of the radius of the primary rainbow.[113] Armed with this insight, Barrow readily demonstrated that angle ZCY, which he just showed was a maximum, is equal to angle LFO, the radius of the primary bow or the angle made by the refracted and incident rays. Since Z is the image point, the rays will be concentrated there and the region beyond will be in darkness.

By exploiting his concept of the image point, Barrow was able to formulate such a simple solution to the radius of the rainbow and was the first to establish that the bow appears when indefinitely close rays coincide at the point of reflection.[114] His approach cleverly avoids the mathematical problem of finding the maximum inclination, as both Huygens

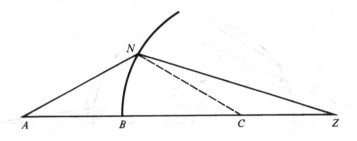

Figure 23

and Sluse had to do in their earlier solutions, which were then unknown to Barrow.[115] Barrow left the determination of the radius of the secondary bow and the breadth of the colored band of both bows untouched, although, as we have seen, he did speculate about the physical cause of the appearance of the colors. In his *Optical Lectures* Newton adopted Barrow's method not only to solve the radius of the secondary bow, but also to generalize it brilliantly to bows of all orders.[116] Newton – to be sure, altogether independently of Barrow – also explained the breadth of the bow by means of the different refrangibility of rays of different color. Other than for his crucial influence upon Newton, Barrow's researches on the rainbow have been virtually ignored from his day to our own. Newton too lost priority for his solution of the nth-order rainbow to Edmond Halley and Jakob Hermann, who published their solutions in 1702 and 1704.[117] In his *Opticks* (1704) Newton set forth a rule for the radius of any-order bow without demonstration, so that his Barrovian derivation became known only with the posthumous publication of his *Optical Lectures* in 1728.[118]

Near object points

For an object point at a finite distance Barrow, as usual, first determines (XIII:15) the image point or limit Z (Figure 23) for paraxial rays. In the case of rays from point A in the rarer medium falling upon the convex surface BN and refracted along NZ, he finds $I/R = (AC/AB)(BZ/CZ)$. If we set $AB = s$, $CB = \rho$, and $NZ = s'$, this becomes the familiar formula $n/s' + 1/s = (n - 1)/\rho$. In an extended discussion (XIII:3–12) of the conditions under which real and virtual images are formed, he introduces a reference length that is equivalent to determining whether the object distance is greater or less than the focal length. These results are applied to image formation in lenses in the following lecture.

The high point of this lecture, and perhaps of the entire work, is Barrow's impressive derivation of the tangential image point for rays incident

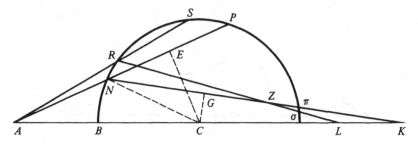

Figure 24

away from the axis (XIII:24). His limit-increment proof is quite similar to the simpler case of parallel incident rays, and some of the steps in
the two proofs are identical. Two indefinitely close incident rays *ANP*
and *ARS* (Figure 24) are refracted by the sphere *BNR* into $N\pi$ and $R\sigma$,
and their intersection *Z* must be found. Erect *CE* and *CG* perpendicular
to the incident and reflected rays *NP* and $N\pi$. Since the arcs *PS* and
NR are infinitesimally small, $AP/AN = PS/NR$, or $\frac{1}{2}(PS+NR)/NR =
AE/AN$. Similarly, it was found in the case of parallel incident rays
that $NZ/ZG = NR/\frac{1}{2}(NR+\pi\sigma)$ and, multiplying this by the preceding,
there results $(AE/AN)(NZ/ZG) = \frac{1}{2}(PS+NR)/\frac{1}{2}(NR+\pi\sigma)$. It can,
however, be shown that $\frac{1}{2}(PS+NR)/\frac{1}{2}(NR+\pi\sigma) = [CN/\frac{1}{2}(NR+\pi\sigma)]/
[\frac{1}{2}(PS+NR)/CN] = (NG/NE)(CE/CG)$. Whence it follows that $NZ/
ZG = (I/R)(NG/NE)(AN/AE)$, where $CE/CG = I/R = n$. If we adopt
the notation of the case for parallel incident rays, while setting $AN = s$
and $NZ = s'$, then $NZ/ZG = s'/(s' - \rho\cos r)$, $NG/NE = \cos r/\cos i$, and
$AN/AE = s/(s + \rho\cos i)$; and we find for the tangential image point s',
or the diacaustic condition,

$$1 = \rho\cos^2 i/s(n\cos r - \cos i) + \rho n\cos^2 r/s'(n\cos r - \cos i).$$

In his *Optical Lectures* Newton once again adopted the Barrovian approach, and he gave an equivalent limit-increment proof of the diacaustic
condition.[119] Newton's demonstration is more elegant and far less pro-
lix than Barrow's. Indeed, Barrow thought enough of Newton's construction to publish it immediately following his own with a note that it was
"communicated by a friend and discovered by him by another method
and elegantly demonstrated."[120] In turn, Newton in his *Lectures* advised
his audience that "for more about these matters see Dr. Barrow's *Lectures*," which was in fact more comprehensive.[121] As with the other Barrovian problems that he pursued, Newton here too made a significant advance by again recognizing that there were two image points, *Z* and *K*.[122]
    Newton did not publish his *Optical Lectures* with its applications of
Barrow's mathematical theory of imagery and its citations of the *Lectiones*

XVIII, but he did manage to include Barrow's principle of image location in the *Opticks.*[123] Geometrical optics was not treated in the *Opticks*
other than eight axioms at the beginning, which, Newton claimed, epit-
omized "the summ of what hath hitherto been treated of in Opticks."[124]
Despite his grandiose claim, he did do a remarkable job of compressing
elementary geometrical optics into nine pages. Axiom VI contained rules
for finding the image points for reflection and refraction at plane and
spherical surfaces, including the first statement of "Newton's formula."
In the next axiom Newton explained that if the rays converge after reflection or refraction, they will form a real image or "make a picture" on a
screen; and he applies this to explain the retinal image. The final axiom
is Barrow's principle of image location for a single eye: "An Object seen
by Reflexion or Refraction, appears in that place from whence the Rays
after their last Reflexion or Refraction diverge in falling on the Specta-
tor's Eye."[125] Newton justifies this geometrical principle of image location, just as Barrow did, by claiming that vision of an image is no different than that of the object itself: "For these Rays do make the same
Picture in the bottom of the Eyes as if they had come from the Object
really placed at [the image]...and all Vision is made according to the
place and shape of that Picture."[126] Moreover, like Barrow, he nowhere
discusses psychological or physiological processes of distance perception.
Since Newton does not give any citations in his axioms, he left this principle unattributed. There is, in any case, no evidence that he had given
any thought to the author of such an obvious principle, which he now
put in the public domain as axiomatic. When more than thirty years later
Robert Smith attacked Barrow's principle of image location, which was
"universally received," he all but quoted Newton's Axiom VIII.[127]

Lenses

Lecture XIII successfully concludes the central problem of Barrow's *Lectiones XVIII,* namely, to define the image of a point after any
reflection or refraction in plane and spherical surfaces. In the remainder
of his lectures Barrow applies these results, with varied success, to images
at multiple surfaces and images of extended objects. He explains (XIV:1)
that the conclusions of the preceding lectures can be applied to finding
the image for any number of surfaces simply by treating the image resulting from the preceding surface as if it were a real object with respect
to the following surface. Barrow devotes Lecture XIV to refraction at
two successive plane or spherical surfaces, that is, lenses "chosen for the
sake of common practice and prepared especially to diminish the work
of those who come upon these things."[128]

After the passage of more than half a century, Barrow was the first
to publish a complete solution to the problem bequeathed by Kepler,

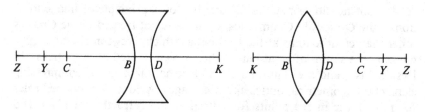

Figure 25

namely, to find the focal and image points for all species of lens. Hitherto, the only exact solution for paraxial rays available was that of Cavalieri, who had gotten as far as finding the focal point for thin lenses. Barrow's solution, however, was not the triumph one might expect, for it
was excessively complex and cumbersome. In what amounted to a long
table, he laid down without demonstration a separate set of equations for
each species of lens, except where symmetry allowed him to combine certain sorts. Thus, his treatment of focal lengths required four cases, which
combined, for example, plano–convex and plano–concave, and convex-
concave and concave–convex lenses. Yet the last case, in turn, required
four separate subcases depending on the ratio of the radii of the two surfaces. Matters became even more complex for the equations describing
the image point when the object point was at a finite distance; this required nine cases, with some having as many as six subcases, depending
on such factors as the relative position of the object and focal point. One
can only speculate as to the form Barrow's lens equations took before he
revised this lecture.[129] Although Barrow did not derive his equations, they
do follow in a straightforward way – as he claimed – from the preceding
results for single surfaces. I will illustrate his method with the example
(his Cases V and VI) of the focal length of a biconvex or biconcave lens,
whose two surfaces, despite the diagrams (Figure 25), need not be of
the same radius. If parallel rays are incident from the left upon the first
surface with vertex B and center C, the focal point Z is determined by
XI:2 to be $I/R = BZ/CZ$ or, in terms of directly measurable quantities,
$(I - R)/I = BC/BZ$.[130] Now point Z is considered the object point for the
second surface with vertex D and center K, so that the focal point Y
is found from XIII:15 to be $R/I = (DY/DZ)(KZ/KY)$, or $DK/DY =$
$[(I/R)KZ - DZ]/DZ$.[131] The focal point Y is specified solely in terms of
the lens's shape and index of refraction, since $DZ = BZ \pm BD$ and $KZ =$
$DZ + DK$, where BD is the thickness of the lens and DK the radius of its
second surface, while BZ was determined at the first surface.

   Barrow's solution is complete and valid, but it certainly could not be
considered particularly elegant or instructive. The brevity of his presentation, which could scarcely serve to introduce the uninitiated – namely,

undergraduates – to the properties of lenses, shows once again that Barrow was concerned principally with the mathematical theory of image formation and not with its applications. At the conclusion of this lecture Barrow included a more general construction that was "communicated by a friend" – Newton – which he aptly described as an "elegant and convenient method for geometrically representing the image for any case."[132] Newton later included his construction, though applied to only one surface, in his *Optical Lectures;* consequently, in its application to lenses it was known solely through Barrow's *Lectures.*[133] As I noted earlier, Huygens had investigated imagery in lenses more thoroughly and earlier than either Barrow or Newton.[134]

A flurry of works in the early 1690s, nearly a quarter of a century after the publication of the *Lectiones XVIII,* superseded Barrow's determination of the image point of a thick lens with both generalizations and simplifications for thin lenses. In 1692 William Molyneux, who was familiar with Barrow's *Lectiones XVIII,* published his *Dioptrica nova,* which was a practical treatise on lenses and telescopes. He independently arrived at Huygens's rule for images in thin lenses, though in a slightly different form and stated less generally.[135] In the following year Jean Picard's posthumous writings on dioptrics also contained a similar rule for thin lenses as well as a series of equations for thick lenses.[136] Picard had read and admired the *Lectiones XVIII* shortly after it appeared.[137]

An inherent limitation of the geometrical approach adopted by Barrow and his contemporaries is that all line segments must be positive. This requirement frequently necessitates that a single problem be inconveniently broken up into separate cases with distinct solutions. Edmond Halley in his landmark paper, "An Instance of the Excellence of the Modern Algebra, in the Resolution of the Problem of Finding the Foci of Optick Glasses Universally," which appeared in the *Philosophical Transactions* for 1693, overcame this drawback and derived a single equation that yielded the image point for all varieties of thick lenses. He observed of his formula for a biconvex lens that "to bring this to the other Cases, as of...Concave Glasses, the Rule is ever composed of the same terms, only changing the signs of + and −."[138] By the early years of the eighteenth century a variety of methods for determining images in lenses were widely available, and the problem was no longer in the province of advanced mathematical physics.

Special problems

Images of lines

Whereas most treatises on geometrical optics in the second half of the seventeenth century thoroughly discussed the magnification and

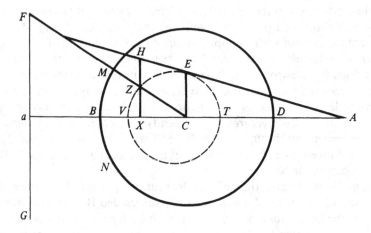

Figure 26

orientation of extended objects, or lines, Barrow treated these subjects perfunctorily. As was his wont, he attempted to restrict himself to mathematically tractable aspects of these problems in the last four lectures. Earlier I presented his solutions in Lecture XVI to the image of a straight line perpendicular to the surface of a plane refracting medium and of a convex mirror. Though these problems are of some significance for theoretical optics, they are of little interest for applied optics, which primarily treats lines perpendicular to the axis.

Barrow concludes with two stunning lectures in which he demonstrates that the absolute (paraxial) image of a straight line perpendicular to the axis seen by either reflection or refraction from a spherical surface is always a conic section. As a preliminary he first proves a general property of conics (XVII:1, 2): If in any right triangle ACE (Figure 26) two sides AC and AE are extended, and through any point X on the side AC the line XH is drawn parallel to CE, and if in the angle CXH line CZ is drawn equal to XH, then the point Z lies on a conic section. Barrow now presents a construction (XVII:6) to find the locus of the image points of the straight line FaG perpendicular to the axis aC of mirror MBN, and he then shows that the points lie on the ellipse VZT with one focus at the center C of the mirror. Perpendicular to Ca, draw through the mirror's center the line $CE = BD/4$; make $CA = Ca$; from any point F on the object join FC, and its image Z will lie on this line; and from the image point Z draw ZX perpendicular to aCA. Earlier (VIII:7) Barrow found that the (absolute) image in a convex mirror of a point a or A at a finite distance on the axis when the eye is on the axis is $aC/aB = CV/BV$. Since the mirror is circular, by imagining the eye and axis continuously to change direction or rotate about C, he is able to consider the absolute image of

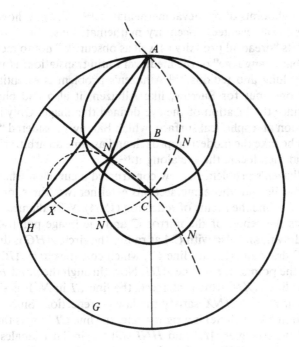

Figure 27

the line *FaG* to be composed of the locus of (absolute) images *V* (now *Z*) of the points *F* on that line, or that $FC/FM = CZ/MZ$. From this it is straightforward for Barrow to show that $XH = CZ$, so that according to the preceding theorem the image *Z* must lie on an ellipse with one focus at the center of the mirror *C*, provided that *Ca* is greater than *CE*. The demonstration for reflection from the concave part of the mirror is similar. He shows more generally that according to whether the object is located before the focus, at it, or on its other side, its absolute image will be an ellipse, parabola, or hyperbola, respectively. When the eye is at the center of the circle, the absolute and relative images will be identical and precisely defined; for other situations of the eye Barrow provides a qualitative description of the relative image. In the final lecture he extends these results to the image of a line refracted at a single spherical surface.[139]

Alhazen's problem

Alhazen did, in fact, solve the problem that since the seventeenth century has been named after him, namely, given the position of an eye *A* and radiant point *X* to find the point of reflection *N* on the surface of a spherical mirror *NBN* (Figure 27). His solution is rightly considered to be one of the highpoints of medieval mathematics.[140] It did not, however, satisfy the tastes of seventeenth-century mathematicians. Barrow complained about its "dreadful prolixity as well as obscurity," not to mention "the rude barbaric language" of the medieval Latin translation; Huygens found it "very long and tedious."[141] Alhazen's problem is essentially a mathematical one, but for Barrow, like Alhazen, it also had physical significance, since the location of the eye defined the image. Only in the case of refraction at a spherical surface, which Barrow considered "most intricate," did he take the modern approach and assume an arbitrarily refracted ray and then locate the eye along it.[142]

To solve Alhazen's problem Barrow constructed a curve on which the object and image lie and whose intersections N (either two, three, or four) with the mirror define the points of reflection (IX:4). With diameter CA, which connects the center of the mirror C and the image point A, the circle AIC is drawn; and also with CA as radius the circle AHG is drawn. From center C draw an arbitrary line CI, which cuts the circle AIC in I, and then join the points A, I by line AIH. Now through the point H and object X draw line HXN, which intersects the line CI in N. It is simple to establish that rays AN, NX satisfy the law of reflection: Since angle CIA (inscribed in a semicircle) is a right angle, the line CI bisects the line AH in I, and the triangles ANI and HNI will be similar triangles with the angle of incidence INX equal to the angle of reflection INA.

Barrow's solution to Alhazen's problem appeared at the beginning of a competition between Huygens and Sluse that thoroughly examined the problem and produced a series of elegant solutions. Huygens had invited Sluse to attack Alhazen's problem as early as 1657 and sent his first solution to Oldenburg in June 1669, a few months before Barrow's *Lectiones XVIII* appeared.[143] Neither Huygens nor Sluse seemed to be particularly impressed by Barrow's solution. Their contest culminated in Huygens's justly famous "la bonne construction" that determined the reflected points for all cases by the intersection of a hyperbola with the mirror.[144] Barrow's contribution was subsequently largely forgotten. No doubt, an additional cause for its neglect was Barrow's failure to derive an equation for his curve, or even to recognize that it was a cubic. It was more than a century before Kaestner derived the equation for it in polar coordinates.[145] Lohne has examined Barrow's curve and shown its usefulness for studying the various solutions of Alhazen's problem, so that I need not pursue it here.[146]

Reception of the *Lectiones XVIII*

Contemporary response

The audience for Barrow's *Lectiones XVIII* was a small one, restricted to the mathematically astute. Indeed, "the booksellers have lost

so much by... the Optic & Geometric Lectures of Dr. Barrow" and other
mathematical treatises that – Collins lamented – "it is no easy task to
persuade booksellers to undertake any thing but toys that are mathematical."[147] Almost as soon as the *Lectiones* came off the press in early
November 1669 Oldenburg and Collins were proudly distributing it, and
the immediate response was very favorable. James Gregory writing from
St. Andrew's in January was enthusiastic: "Mr Barrow, in his Optics,
sheweth himself a most subtle geometer, so that I think him superior to
any that ever I looked upon.... I esteem that author more than ye can
easily imagine."[148] From Paris the English diplomat Francis Vernon was
able to relay to Oldenburg the response of the astronomers of the Royal
Academy of Sciences. Auzout, he wrote in July, was "extreamly satisfied," and six weeks later he was able to give a fuller report: "Sigre Cassini
extreamely admires Mr Barrows Optiques, wch I sent him & soe did Monsieur Picart who had Read it before. & all that have seen it say it a very
pretty piece, & full of Curiosity."[149] No doubt its immediate attraction to
the astronomers was that for the first time they possessed a set of for-
mulas, even if not in the most convenient form, that fully described the
image points in lenses.

Sluse's reaction to Barrow's *Lectiones XVIII* was a bit more restrained,
since he had worked on many of the same problems as Barrow and was
anxious to lay down some priority claims. After some polite comments on
the book's "worthiness" and "geometrical acuity" (which applied equally
to some other works that Oldenburg had sent along with the *Lectiones
XVIII*) Sluse got down to the business at hand: "A few problems occur
in the *Lectiones* upon which I had once had some thoughts, the solution
of which, with your leave, I here transcribe for communication to the
famous Collins, if they seem to merit that."[150] He then proceeded to give
his own solutions to the anaclastic problem and a special case of Alhazen's
problem, while alluding to his general solution of it.[151] A few weeks later
Sluse had gotten as far in the *Lectiones* as Barrow's determination of the
radius of the rainbow with its citation and a criticism of him, and he set
out his own solution in greater detail than in his earlier letter to Oldenburg. He concluded by expressing his "delight" at finding that Barrow
had determined the aplanatic points of a sphere in the same way as he
had done "many years before."[152]

Huygens had far more at stake than Sluse, who had in fact solved only
a series of special problems. For more than fifteen years Huygens had
systematically investigated the focal properties of refracting surfaces and
had designed, constructed, and used telescopes; and he had a large unpublished treatise, his *Dioptrica,* beside him. Although Oldenburg had
known about the *Dioptrica* for a number of years, he became particularly concerned about its publication and protecting Huygens's priority in early 1669, when Barrow's lectures were being readied for publication.[153] In March he twice warned Huygens that he would "be anticipated

by a certain other person who, to my knowledge, is working strenuously at that subject, and is a very capable man."[154] Huygens, ever reluctant to publish, complained about the pressure of other matters, but was sufficiently concerned to take two steps to protect his priority. In June he sent Oldenburg his first solution to Alhazen's problem and in September thirteen anagrams announcing various discoveries, including three on optics.[155]

Huygens had a copy of Barrow's *Lectiones XVIII* for about two months before acknowledging its receipt in January 1670, which was sufficient time for him to recognize how much of his work had been preempted. His letter to Oldenburg, with its faint praise for Barrow, does not disguise his disappointment:

> I received Mr. Barrow's treatise on dioptrics, which equally displays the knowledge and ingenuity of its author; but although he appears to have exhausted this whole subject you will some day see that what I have written about it is still quite different. The difficulty he found in Alhazen's problem concerning the point of reflection must without doubt have made the solution of it that I sent to you very pleasing to him. . . . As for the position of the image, I dare to say that he is wide of the mark, and the difficulty that he himself raises at the end ought to have warned him. Please let me know what your Fellows think about it.[156]

"The opinion here," Oldenburg replied, "is that he has done well, but that there is still much more to be said; that is what we expect from you."[157] Barrow, as was evident from his own assessment of his work with respect to Huygens, would no doubt have accepted Oldenburg's judgment.[158] Barrow's recognition of his limited achievement and failure to treat applied optics, together with his loss of interest in the mathematical sciences, probably lay behind his request to Collins that the review in the *Philosophical Transactions* be only "a short and simple account of their subject" with no "commendation or discommendation" – a request that was honored.[159] Regis's review in the *Journal des sçavans* was equally uninformative, since he could not appreciate Barrow's mathematics and devoted most of it to the remarks on color.[160] I will return to Huygens's thoughts on Barrow's principle of image location after we briefly touch upon Huygens's investigation of caustics.

Caustics and image points

As a consequence of his development of the wave theory of light, Huygens was able to describe caustic curves by means of the theory of involutes and evolutes that he had published earlier in his *Horologium oscillatorum* (1673). The problem confronting Huygens was to describe the nature of wave fronts after refraction or reflection, for in general they are not spherical waves. Having discovered, in the first place, that the rays are always normal to the wave front and then that they are tangent to the caustic, he was able to demonstrate that the wave front is the involute of the caustic and, conversely, that the caustic is the evolute of the wave front.[161] Huygens set forth this result in his *Traité de la lumière,* which he wrote in 1678, read to the Royal Academy of Sciences in 1679, but published only in 1690.[162] For reflection of a plane wave (or parallel incident rays) at a spherical mirror, he found that the caustic was an epicycloid, but he was unable to determine the curve for the equivalent case of refraction. He did recall, however, that Barrow had determined the points on this curve, "though for another purpose" (they were, of course, Barrow's image points, but Huygens rejected his principle of image location), and referred the reader to the twelfth of the *Lectiones XVIII.*[163] This reference helped to preserve a place for Barrow in the historiography of caustics.

Tschirnhaus in 1682 was the first in print with a determination of the catacaustic, but Philippe de La Hire quickly recognized that his solution was erroneous.[164] In 1690 Tschirnhaus did publish a correct solution without, however, citing Barrow and Huygens.[165] This was followed in 1692 and 1693 by a series of papers by the Bernoulli brothers, Jakob and Johann, that culminated in Jakob's publication in 1693 of the general solution (though without demonstration) for the diacaustic curve.[166] Unlike Huygens, Jakob Bernoulli did not reject Barrow's concept of image. He had carefully studied Barrow's optical and geometrical lectures in his period of mathematical self-education in the 1680s.[167] In a scholium to his solution he observed that

> all those things that Barrow so painstakingly constructed for determining the place of the image of a radiant point by reflection or refraction at the surface of a circle are only very special corollaries of our general relation of caustics and diacaustics to evolutes, since by the very name *image* nothing else enters than the concourse of the reflected or refracted rays.[168]

Thus, Barrow's investigation of the place of the image for oblique pencils of rays entered the mainstream of optical theory: The image point was nothing other than the tangent of the reflected or refracted ray to the caustic. David Gregory, as is evident from his *Catoptricae et dioptricae sphaericae elementa* (1695), was thoroughly familiar with Barrow's contributions in his *Lectiones XVIII.* In his notebook, as he was working his way through the latest Continental papers on caustics a few years after they had appeared in the *Acta eruditorum,* Gregory assigned Barrow – and the English – a more significant place in the history of caustics:

> Amongst the Germans and others named by Bernoulli in the *Acta* for May 1692 it is asserted that Tschirnhaus was the first

to have considered caustics, when nevertheless those (catacaustics as well as diacaustics) were first considered by Barrow. For these are the same as the place of the image with respect to an eye variously placed off the axis of radiation.[169]

Montucla greatly admired Barrow's investigation of caustic points, and in his *Histoire des mathématiques* expressed surprise that Barrow, "with a decided taste for everything which approached pure and sublime geometry," did not investigate the caustic curve:

> Dr. Barrow very nearly touched upon the discovery of caustics, because these curves are nothing other than the succession of all the images of the same point seen by reflection or refraction of all the different places that the eye can occupy.... It could very well be that it was the inspection of this passage of Barrow's *Optical Lectures* that caused Mr. Tschirnhaus to take up this consideration.[170]

Montucla's assessment of Barrow's contribution was repeated often in eighteenth- and early-nineteenth-century histories and encyclopedias; but his conjecture on his influence on Tschirnhaus, who had a dubious reputation, is most likely erroneous.[171] Although Tschirnhaus was probably familiar with Barrow's *Lectiones XVIII*, it is more likely that he had appropriated Huygens's work. Huygens claimed that he showed Tschirnhaus parts of the manuscript of his *Traité* in 1678 and also that his book had appeared before Tschirnhaus published his second article.[172]

The publication in 1696 of L'Hospital's *Analyse des infiniments petits*, with its clear and direct demonstrations of the caustic conditions using the differential calculus, put the theory of caustics on solid ground and made it an integral part of eighteenth-century mathematics.[173] Until the beginning of the nineteenth century, when a new approach based on the theory of curves was adopted, further research on caustics was essentially a mathematical problem of discovering particular caustic curves. For Barrow and Newton, we should recall, the determination of caustic points was actually the determination of image points for reflected and refracted oblique pencils or the study of astigmatism, which is more fundamental to geometrical optics. Their pioneering research would have suffered nearly total neglect had not Robert Smith incorporated it into his *Compleat System of Opticks* in 1738. Smith extended their investigation to any number of surfaces and clearly related the determination of the tangential image point to the study of caustics.[174] Although he confined his own research to the tangential image point, he gave a clear account of Newton's investigation of both image points in the appended "Remarks."[175] After Newton's *Opticks* Smith's work was the most important optical treatise of the eighteenth century, and its many annotated translations, particularly Kaestner's German in 1755, but also the Dutch

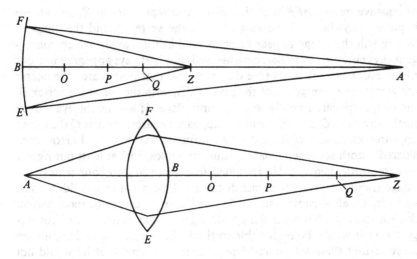

Figure 28

in 1755 and the two French in 1767, contributed to its broad influence. Thomas Young, who in 1801 made the first contribution to the study of astigmatism after Smith by recognizing that the image points were in fact lines, was led back to Newton's research through Smith's *Compleat System*.[176] Since Young failed to cite Barrow's *Lectiones XVIII* here, Barrow's contributions to the study of astigmatism remained largely unrecognized until Culmann called attention to it in 1904.[177]

The principle of image location and the "Barrovian case"

By the turn of the century Barrow's *Lectiones XVIII* and his principle of image location would most likely have been largely forgotten, except for the occasional historical citation, had he not concluded the *Lectiones* with a phenomenon that seemed to contradict that principle. This phenomenon – which has since become known as the "Barrovian case" – was widely discussed throughout the eighteenth century; indeed, it achieved some notoriety after Berkeley included it in his attack on mathematical theories of vision in his *Essay towards a New Theory of Vision* (1709). "Before I quit," to use Berkeley's translation of Barrow, "the fair and ingenuous dealing that I owe both to you and to truth obligeth me to acquaint you with a certain untoward difficulty, which seems directly opposite to the doctrine I have been hitherto inculcating, [or] at least admits of no solution from it."[178] The phenomenon involves the judgment of the distance of blurred images, which are produced by observing a real image before the rays converge to form an image behind the head. The object *A* (Figure 28) is placed on the axis of a convex lens or concave mirror *BEF* such that it forms a real image at *Z*, and an eye is placed anywhere on the axis between the vertex *B* and the image.[179] Where will the image appear to be located in this case in which convergent rays enter the eye? Barrow observed that it always appeared in front of his eye, nearer than its true distance, and that it appeared the nearer the greater the divergence of the rays. Thus, with the eye at the vertex *B*, the image appeared nearly in its natural place *A*; and as the eye moved farther away to *O* and *P*, the image appeared nearer, until at *Q* the image appeared extremely close and very confused. "All which," Barrow considered, "doth seem repugnant to our principles, [or] at least not rightly to agree with them."[180] He reasoned that according to "our tenets" the image should appear "at a vast distance off, so great as should in some sort surpass all sensible distance."[181] We judge objects to be more distant the less the rays that enter the eye diverge; so that in the case of convergent rays it would be reasonable to think that objects would seem even more distant than with parallel rays. Barrow vowed that he would not abandon his principle because of this "odd and particular" case. "For in the present case," he concluded, "something peculiar lies hid, which being involved in the subtilty of nature will, perhaps, hardly be discovered till such time as the manner of vision is more perfectly made known.... I shall, therefore, leave this knot to be untied by you, wishing you may have better success in it than I have had."[182]

Barrow's reasoning – that convergent rays should appear even more distant than parallel rays – seems to invoke a new principle, rather than, as he claims, to conflict with his principle of image location, which was formulated not for blurred images but solely for distinct images formed by divergent or parallel rays. D'Alembert, who rejected Barrow's principle on other grounds, stated this point quite clearly:

> ...it seems to me that it little weakens the principle. For when rays converge in entering the eye, vision must be confused and the principle can extend only to rays that diverge in entering the eye, because first these rays are the only ones that can produce distinct vision and consequently a clear view of the image; and second because convergent rays are reunited behind the eye, where one certainly cannot ascribe an image, of which moreover one has only confused vision.[183]

In his *Dioptrice,* Kepler had already observed that distinct vision is impossible with rays converging toward the eye, since "the eye is made to see either remote or near things distinctly."[184] Though Kepler does not consider the position of images in the *Dioptrice,* it is clear that his principle of image location, for either one or two eyes, would break down. James Gregory, in the same proposition of his *Optica promota* in which he converted Kepler's principle of image location to a single eye, explicitly rejected the idea that convergent rays could be assigned a location. After considering image location for divergent and parallel rays in the first three corollaries, he explains in the fourth that "if the rays of one point converge to another point behind the eye, no place can be assigned to this point, except (if anyone wishes) at the intersection of the rays behind the eye."[185] In the following Proposition 30 he cites Kepler's argument that distinct vision is impossible with convergent rays, "which arise artificially and not naturally."[186]

Barrow's principle of image location and the challenge of the "Barrovian case" stimulated debate on the place of image almost as soon as the *Lectiones XVIII* appeared. Huygens, as we saw, objected to his principle of image location upon first receiving the book.[187] After a visit to England, Leibniz wrote to Oldenburg from Paris in April 1673 that "I brought with me Barrow's *Lectiones opticae*; at the end of the book the very learned author reports a phenomenon which he says he cannot explain. . . . Yet Huygens and Mariotte say they possess its solution."[188] At just about this time Huygens prepared an outline for a revision of his *Dioptrica,* to which he intended to add a chapter on image location. His very brief notes for the chapter succinctly present the grounds for his criticism of Barrow: "One cannot at all judge distance with a single eye."[189] Huygens never wrote this chapter, but his ideas emerge from a number of later documents.[190] From his earliest research Huygens clearly distinguished between the geometrical and perceived place of the image.[191] He completely rejected the possibility of judging location with one eye without prior knowledge of an object's true size and distance, a position that put him at odds with virtually everyone who considered the problem in the seventeenth century.[192] Even judgments with two eyes, he argued, were effective up to distances of only about fifteen or twenty feet without motion of the head or prior knowledge of the object's true size.[193] Although he frequently referred to the "Barrovian case," Huygens never offered an explanation for the phenomenon.

In his notes, Huygens had correctly identified William Molyneux as a supporter of Barrow's principle of image location. In Book I, Proposition XXXI of his *Dioptrica nova* (1692) Molyneux set forth the "Barrovian case" and concluded by maintaining "the resolution of the same admirable Author [Barrow], of not quitting the *evident* Doctrine... for determining the *Locus Objecti,* on the account of being pressed by *one Difficulty*."[194] Elsewhere in this proposition, "Concerning the Apparent Place of Objects Seen through Convex-Glasses," he consistently located distinct images at their image points, just as Barrow did. Yet, unlike Barrow, he did not simply identify the geometrical and perceived images, for he recognized that the Estimate we make of the *Distance* of Objects (especially when so far removed, that the Interval between our two Eyes, bears no sensible Proportion thereto; or when look'd upon with one Eye only) is rather the Act of our *Judgment,* than of *Sense*; and acquired by *Exercise* and a Faculty of *comparing,* rather than *Natural.* For *Distance* of it self, is not to be perceived; for 'tis a Line (or a Length) presented to our Eye with its End towards us, which must therefore be only a *Point,* and that is *Invisible.*[195]

Retracing ground covered earlier by Kepler, Descartes, and James Gregory, he distinguished between the perception of near and distant objects, and to explain the former he introduced the processes of convergence for vision with two eyes and accommodation with one eye. It is not clear to me why, with his understanding of distance perception, he still placed the image at its geometrical location independent of distance, unless he implicitly assumed that if the image can be drawn on a ray diagram its distance is always near.

Berkeley's *Essay towards a New Theory of Vision* contained a broad attack on contemporary theories of vision, of which his discussion of the "Barrovian case" and judgments of distance were only one aspect. He drew much of his understanding of contemporary optics from Molyneux's *Dioptrica nova,* and it is most likely that Molyneux's discussion of the "Barrovian case" initially stimulated his interest in it.[196] The first aim of the *Theory of Vision* was to refute geometrical theories of vision and the concept of distance judgment by means of a natural triangle. Berkeley argued that the lines and angles of those triangles are a fiction and are neither perceived nor thought of by those not knowledgeable in optics; they are "only an hypothesis framed by the mathematicians, and by them introduced into optics, that they might treat of that science in a geometrical way."[197] As an alternative explanation, Berkeley held that from experience we learn to judge distance from the sensations accompanying vision: (i) the disposition of the optical axes, (ii) the confusion attending an image, since near objects are seen confusedly, and (iii) the eye strain to avoid confused vision.[198] After setting forth Barrow's own account of his case in both the original Latin and a translation, Berkeley proceeded to explain the phenomenon by means of the amount of confusion. As the eye moves away from the vertex *B* (Figure 28) toward *Z*, the confusion increases and consequently the image will seem nearer – just as Barrow observed – since by experience we associate greater confusion with nearer objects. Berkeley's neat explanation requires that only the quantity of confusion be judged, regardless of whether it arises from convergent or divergent rays.[199] In this way he was able to account for the embarrassing failure of Barrow's assumption that convergency is perceived directly and implies exceedingly great distances.

Smith in his *Compleat System* commended Berkeley for being the first to refute Barrow's principle of image location. Nonetheless, he believed that his "highly entertaining and useful" work would have been improved by "experiments with glasses, or even geometrical inferences from glasses," for then he would have seen the inadequacy of his own new principles.[200] Though Smith had the highest regard for Barrow's contributions to mathematical optics and continued his (and Newton's) research on astigmatism, he judged his ideas on the place of the image to be a failure. Accordingly, Smith defined an image strictly geometrically, without any regard to the eye, as the place where pencils of rays from points of the object reconverge.[201] In the appended "Remarks" he presented a thorough review of contemporary ideas on image location, including an experimental refutation of both Barrow's and Berkeley's principles.[202] Against Barrow, he observed, for example, that if a concave lens is inserted close to the eye, the apparent distance of the image does not vary, even though the geometric place of the image has changed.[203] Against Berkeley, he objected that since confused vision begins for most people at a distance of one to two feet, the image should always appear to be less distant than that. Yet the apparent distance of the image is observed to change from the true distance to the naked eye to successively smaller ones as the eye is moved backward from the vertex toward the image point.[204] Smith's own explanation for the judgment of apparent distance was based on the apparent magnitude of the image, so that "an object seen by refraction or reflection, appears at the same distance from the eye, as it usually does from the naked eye, when it appears of the same magnitude as in the glasses."[205] Smith's principle was as vulnerable as Barrow's and Berkeley's, and we need not follow its fate.[206]

It would be a mistake to believe that Berkeley and Smith had succeeded in at once discrediting Barrow's principle. In 1740 Krafft published a paper, "On the Place of the Image of a Radiant Point in a Curved Mirror," in which he attempted to clarify the problem of image location. He surveyed the field from the cathetus principle of Alhazen, Witello, and Tacquet through the contributions of Kepler and Barrow, but was unaware of later work from the British Isles. Krafft concluded that Barrow had developed a legitimate principle to replace the older ones and lamented the neglect of his writings, a view that reflects the state of science in eastern Europe as much as Krafft's opinion.[207] Most of his paper was devoted to the mathematics of imagery and not to its psychological aspects. Thus, he rederived Barrow's result for the image of a straight line in a spherical mirror by means of L'Hospital's analytic approach, but he did not discuss the "Barrovian case."[208] Montucla, writing his *Histoire des mathématiques*

in Paris at midcentury, was far more *au courant;* and he took the opportunity in his discussion of Barrow's *Lectiones XVIII* to give a general review of the problem of image location and explanations of the "Barrovian
case." After considering the views of Berkeley and Smith (he learned of
the former from the latter) and admitting problems with Barrow's principle, he still supported Barrow's principle as the most satisfactory yet
proposed.[209]

Pierre Bouguer in his posthumous *Traité d'optique* adopted Barrow's
principle of image location but added an interesting twist to its interpretation.[210] He combined it with the theory of caustics, but, following Newton, he noted that there are in fact two images. His lucid discussion of
the two image points is, to my knowledge, unique in the eighteenth century. In certain circumstances, he observed, it can be difficult for us to
distinguish the two images and locate them. He concluded his account
with the remark, "This is the solution of a difficulty which greatly occupied the Reverend Father Tacquet, Barrow, and Smith, as well as several
other authors, and which Mr. Newton himself regarded as forming a very
thorny problem, even if it were not really insoluble."[211] Since Bouguer
does not explain exactly how knowledge of two image points can resolve
the difficulty, I can only conjecture what he had in mind. It seems to me
that he is confusing two fundamentally different sorts of blurred or confused images. Newton's "very thorny problem" was defining a sensible
location for the two image points arising from divergent rays (or, in modern terms, defining the circle of least confusion), whereas Tacquet's, Barrow's, and Smith's difficulty was the altogether different one of locating
a single blurred image arising from convergent rays.

As we approach the latter part of the eighteenth century, more than a
century after the publication of the *Lectiones XVIII,* we no longer find any
active supporters of Barrow's principle. Priestley in his *History* (1771)
based his account of Barrow's principle and the ensuing controversies
over it upon Montucla's, but he did not follow him in endorsing that
principle.[212] However, after describing the problems with Berkeley's and
Smith's alternative explanations, Priestley leaves the impression that Barrow's principle still led the field. Three years later, Joseph Harris in his
Treatise of Optics clearly distinguished between geometrical optics and
a theory of vision. He defined image in Book I in strictly geometrical
terms; and at the beginning of Book II on vision, he stressed that geometrical optics – or, as he calls it, "abstract theories of images" – "has no
necessary connection with vision, and may be perfectly understood by a
blind man."[213] Only for "moderate distances" (one or two feet with one
eye, and six or nine feet with two) does he allow that the perceived image
will be at the place of intersection of the rays.[214] In the article "Bild" or
"image" in Gehler's *Physikalisches Wörterbuch* (1787) we find the now familiar story largely drawn from Montucla, although Kepler is here assigned

a role in first rejecting the cathetus rule and proposing an alternative. The article concludes by observing that so many unexplained concepts are involved in judging distance that "we cannot reduce our judgment of the place of these images to any simple and definite principle."[215]

Conclusion

Barrow's principle of image location was, well into the eighteenth century, widely known and generally attributed to him. Benjamin Robins, who admired Barrow's *Lectiones XVIII,* attributed it, not unjustifiably, to James Gregory; and Smith, though he associated the principle with Barrow, did not ascribe it to him or anyone else, but simply noted that it was "universally received."[216] I was unable to find anyone who assigned that principle to Kepler. Though eighteenth-century scholars did not for the most part sufficiently appreciate the magnitude and subtlety of Kepler's contributions to geometrical optics, they were on this point of attribution correct. Kepler's principle of image location, from which Gregory's and Barrow's derived, was intended as a perceptual principle and not one to locate the geometrical image. By the end of the eighteenth century scientists returned to Kepler's position and rejected Barrow's identification of the perceived and geometrical image. No doubt, the reason that Gregory's formulation was largely ignored and Barrow's principle was taken so seriously so long was that Barrow developed it into a powerful mathematical–physical method. By means of his concept of oblique pencils he made pioneering contributions to the study of astigmatism, caustics, and the images of extended objects. This achievement was widely admired and cited by proponents of Barrow's principle, and even by many of those who did not support it, but it did not lead to the establishment of a major school of research in geometrical optics. Only Newton and Smith carried on Barrow's research on the refraction of oblique pencils. Although Smith rejected Barrow's principle for locating the perceived image, he recognized more completely than anyone else the significance of his contributions to geometrical optics. He observed that by means of his concept of narrow oblique pencils and the "law of refraction then newly discovered," Dr. Barrow was able "to handle the subject of dioptricks and catoptricks in a more extensive manner than any writer had then done."[217]

Notes

This research was made possible by support from the John Simon Guggenheim Memorial Foundation, the National Science Foundation, and the Rowland Foundation. I thank them for their generous assistance. A. G. Bennett kindly made many useful suggestions.

1. Isaac Barrow, *Lectiones XVIII, Cantabrigiae in scholis publicis habitae; in quibus opticorum phaenomenonωn genuinae rationes investigantur, ac exponuntur* (London, 1669); henceforth cited as *Lectiones XVIII*. Despite the notice on the title page that "Some Geometrical Lectures Are Adjoined" (*Annexae sunt lectiones aliquot geometricae*), his *Lectiones geometricae: In quibus (praesertim) generalia curvarum linearum symptomata declarantur* were published separately in 1670 and then reissued together with the *Lectiones XVIII* in 1672 and 1674. The *Lectiones XVIII* is reprinted in *Scriptores Optici; or, a Collection of Tracts Relating to Optics,* ed. Francis Maseres and Charles Babbage (London, 1823); and in Vol. 2 of *The Mathematical Works of Isaac Barrow, D. D.,* ed. William Whewell, 2 vols. (Cambridge, 1860); henceforth cited as Barrow, *Mathematical Works*. All of my page references are to the Whewell edition, and I refer to the lectures and articles by roman and arabic numerals, respectively. *Isaac Barrow's Optical Lectures ('Lectiones XVIII'),* trans. H. C. Fay, ed. A. G. Bennett and D. F. Edgar (London, 1987), appeared too late for me to use.

2. I have used Berkeley's translation, with a few omitted phrases restored, from his *Essay towards a New Theory of Vision* (Dublin, 1709), §29, in *The Works of George Berkeley, Bishop of Cloyne,* ed. A. A. Luce and T. E. Jessop, 9 vols. (London, 1948–57), 1:179; *Lectiones XVIII,* XVIII:13, p. 152. A second edition appeared later in 1709, and two more in 1732.

3. Barrow to Collins, Easter Eve 1669, *Correspondence of Scientific Men of the Seventeenth Century,* ed. Stephen P. Rigaud, 2 vols. (Oxford, 1841), 2:70–1.

4. Huygens composed the first part of his *Dioptrica* in 1653 and continued to add to it until a few years before his death in 1695. A version of it was finally published in 1703 in his *Opuscula postuma,* which was edited by Burchard de Volder and Bernard Fullenius. The complete manuscript is in Vol. 13 of the *Oeuvres complètes de Christiaan Huygens,* publiées par la société hollandaise des sciences, 22 vols. (The Hague, 1888–1950); hereafter cited as Huygens, *Oeuvres*. The introduction and notes to this volume, to which I am indebted, provide a very useful guide to geometrical optics in the seventeenth century.

5. This mistaken view can be found in Vasco Ronchi, *Optics: The Science of Vision,* trans. Edward Rosen (New York, 1957), pp. 43–5, especially the fanciful Figure 9, which contains two pencils of rays. It has, however, been most thoroughly expounded by Colin M. Turbayne in his "Berkeley and Ronchi on optics," in *Proceedings of the XIIth International Congress of Philosophy* (Florence, 1961), 12:453–60; and *The Myth of Metaphor,* rev. ed. (Columbia, S.C., 1971), Chap. VI. The source of their error is that they fail to distinguish between Kepler's "distance-measuring triangle" used by a single eye to locate the perceived image – especially in the process of accommodation – and his pencils of rays to locate the geometrical image. I explain Kepler's views below. Readers who are concerned with historical veracity must take Turbayne's caveat about his discussion of Kepler's rules of vision seriously: "Although Kepler stated all these rules, in some of them I summarize what he meant rather than state what he actually said" (ibid., p. 143). Citing Turbayne, I adopted this erroneous interpretation in my edition of *The Optical Papers of Isaac Newton,* 1 vol. to date (Cambridge, 1984–), 1:216, n. 19; hereafter cited as Newton, *Optical Papers*.

6. *Lectiones XVIII,* I:5, p. 14; see also III:16.

7. The plane of incidence is the plane that contains the axis and incident ray. It is a fundamental law of optics that the reflected and refracted rays also lie in this plane.

8. Jakob Bernoulli, "Curvae dia-causticae, earum relatio ad evolutas, aliaque nova his affinia," *Acta eruditorum* (June 1693):244–9.

9. These points are in fact focal lines, as Thomas Young recognized in the nineteenth century; see note 176 below. The tangential focus or image point is also known as the "primary" one.

10. The sagittal focus or image point is also known as the "secondary" one.

11. For paraxial rays, therefore, the sines of incidence and refraction can be replaced by the angles themselves and the cosines by unity.

12. In a letter written in mid-1669 to Edmund Baldroe (or Boldero), the vice-chancellor, Collins referred to Barrow's "Treatise of Optics, which he prepared to deliver in to the former Vice-Chancellor, as his anniversary lectures, according to the laudable constitution or injunction laid upon your mathematic professor" (Rigaud, ed., *Correspondence*, 1:137–8). Barrow had already agreed to the publication of his optical lectures by the end of 1668, for on 30 December 1668 Collins wrote James Gregory that "Mr Barrow doth hereafter intend a Treatise of Opticks for the Presse but is not yet ready" (*James Gregory Tercentenary Memorial Volume*, ed. Herbert Westren Turnbull [London, 1939], p. 55).

13. *Lectiones XVIII*, "Epistola ad lectorem," pp. 5–6.

14. Barrow to Collins, 23 February 1668/9, Rigaud, ed., *Correspondence*, 2:67–8.

15. Barrow to Collins, 20 July 1669 and 20 August 1669, Royal Society Ms. 81, No. 1; see also the letter of 31 July. This manuscript collection contains the original documents that were published in the *Commercium epistolicum D. Johannis Collins et aliorum de analysi promota: Iussu societatis regiae in lucem editium* (London, 1712).

16. Rigaud, ed., *Correspondence*, 2:71.

17. The dates 1667 and 1668 are supported by a note by John Ellis (or Ellys), a close friend of Newton's at Cambridge, in his manuscript extract of the first two of Barrow's *Lectiones XVIII*. At the head of the extract he entered, "Praelectiones quaedam ejusdem Mr Isaaci Barrow in scholis publicis Cantabrig. habitae. Annis 1667 & 1668" (Gonville and Caius College Library, Ms. 629/578); see also Montague Rhodes James, *A Descriptive Catalogue of the Manuscripts in the Library of Gonville and Caius College*, 2 vols. (Cambridge, 1907–8). The title page of Barrow's posthumously published *Lectiones mathematicae XXIII* (London, 1683) states that they were delivered at Cambridge in 1664, 1665, and 1666.

18. On 1 November 1669, Oldenburg wrote Huygens that "Mr. Barrow's book on optics has come from the press" (*The Correspondence of Henry Oldenburg*, ed. A. Rupert Hall and Marie Boas Hall, 13 vols. [Madison, Wis., and London, 1965–86], 6:304–5); hereafter cited as Oldenburg, *Correspondence*. On 11 November he sent a copy to Huygens; ibid., p. 312.

19. For Newton's early optical research see Newton, *Optical Papers*, 1:1–15.

20. *The Mathematical Papers of Isaac Newton*, ed. D. T. Whiteside, 8 vols. (Cambridge, 1967–81), 1:573–4; hereafter cited as Newton, *Math Papers*.

21. Newton, *Optical Papers*, 1:47, 281. There are two versions of Newton's *Optical Lectures* extant, both in Latin, an early one (Cambridge University Library [CUL] Ms. Add. 4002) and a greatly expanded and reorganized revision (CUL Dd.9.67). Newton did not publish either one. The later version was published posthumously in 1729 as the *Lectiones opticae*. Volume 1 of his *Optical Papers* contains a critical edition and translation of both versions.

22. See, e.g., Frank E. Manuel, *A Portrait of Isaac Newton* (Cambridge, Mass., 1968), p. 96; and Louis Trenchard More, *Isaac Newton: A Biography* (New York, 1934; rpt. New York, 1962), p. 81. For a general assessment of this question, see I. Bernard Cohen, *Franklin and Newton*, Memoirs of the American Philosophical Society, 43 (Philadelphia, 1956), pp. 49–53.

23. See Richard S. Westfall, *Never at Rest: A Biography of Newton* (Cambridge, 1980), Chap. 7.

24. Collins to Francis Vernon (and Richard Towneley?), 26 December 1671, Newton, *Math Papers*, 3:23; a Latin translation of this letter was included in the *Commercium epistolicum*, pp. 27–8.

25. See the section on Barrow's theory of color.

26. *Lectiones XVIII,* "Epistola; in qua...scopus breviter exponitur," pp. 7–8. For the different terms for the hypotheses, see II:1, 11 and III:1; and Cohen, *Franklin and Newton,* pp. 580–1, 583.

27. *Lectiones XVIII,* "Epistola; in qua...scopus breviter exponitur," p. 8. See J. A. Lohne, "Die eigenartige Einfluss Witelos auf die Entwicklung der Dioptrik," *Archive for History of Exact Sciences,* 4 (1968):414–26.

28. *Lectiones XVIII,* "Epistola; in qua...scopus breviter exponitur," and I:7. This attitude toward physical models is reminiscent of Newton's; see Robert Kargon, "Newton, Barrow, and the Hypothetical Physics," *Centaurus,* 11 (1965):45–56.

29. I have presented Barrow's theory of light in greater detail and put it into the broader context of seventeenth-century optics, especially its relation to Thomas Hobbes's theory, in "Kinematic Optics: A Study of the Wave Theory of Light in the Seventeenth Century," *Archive for History of Exact Sciences,* 11 (1973):134–266.

30. *Lectiones XVIII,* I:6, p. 15.

31. Ibid., I:7, pp. 16–17.

32. Barrow's first two parenthetical remarks in this passage also clearly indicate a Cartesian influence. The first remark is a reference to Descartes's law of inertia, i.e., his first two laws of motion; *Principia philosophiae,* Pt. II, §§37, 39. In the second remark Barrow adopts the Cartesian definition of a fluid; *Principia philosophiae,* Pt. II, §54. See William Whewell, "Barrow and His Academical Times as Illustrated in His Latin Works," in *The Theological Works of Isaac Barrow,* ed. Alexander Napier, 9 vols. (Cambridge, 1859), 9:i–lv, esp. v–ix for Barrow and Descartes's mechanical philosophy.

33. Marin Mersenne, *Universae geometriae, mixtaque, synopsis et bini refractionum demonstratum tractatus* (Paris, 1644), pp. 567–89. Hobbes's work first received the title *Tractatus opticus* when it was reprinted in *Thomae Hobbes Malmesburiensis opera philosophica quae latine scripsit omnia,* ed. William Molesworth, 5 vols. (London, 1839), 5:215–48.

34. Emanuel Maignan, *Perspectiva horaria sive de horographia gnomonica tum theoretica, tum practica libri quatuor* (Rome, 1648). The derivation of the sine law of refraction is in Bk. IV, Prop. 34, pp. 634–6.

35. Shapiro, "Kinematic Optics," pp. 175–8.

36. To all those who adopted, presented, or studied the Hobbesian model and are cited in ibid., pp. 186–7, there should be added Nicolas Hartsoeker, *Essay de dioptrique* (Paris, 1694), pp. 21–3.

37. *Lectiones XVIII,* II:11, p. 35.

38. Since Maignan's *Perspectiva* was not in Barrow's library, it is quite possible that he had only recently come across a copy. Barrow refers to Hobbes's *Tractatus opticus,* though not by title, in III:8.

39. Shapiro, "Kinematic Optics," pp. 163, 182–4.

40. *Lectiones XVIII,* XII:16, p. 107.

41. For a good discussion of modification theories of color see Michel Blay, *La conceptualisation Newtonienne des phénomènes de la couleur* (Paris, 1983), Pt. I.

42. *Lectiones XVIII,* XII:17, p. 107.

43. Ibid., p. 108.

44. Ibid.

45. Ibid.

46. About four years earlier in his *Mathematical Lectures,* Barrow explained that, unlike mathematics, disciplines such as physics are incapable of providing rigorous demonstrations. Among his examples was that of color: "It is hard to imagine distinctly, and define exactly what is *Colour* in *Physics*...because we have seldom any Notion of these Things.... Whence it falls out, that there are almost as many different Conceptions and Explications of such Things, as there are Authors and Interpreters. Consequently

it proves to be difficult to determine or demonstrate anything true about ambiguous and undefined matters of this sort" (Isaac Barrow, *The Usefulness of Mathematical Learning Explained and Demonstrated: Being Mathematical Lectures Read in the Publick Schools at the University of Cambridge,* trans. John Kirkby [London, 1734; rpt. London, 1970], IV:1, p. 54). This is a translation of the *Lectiones mathematicae XXIII*; in Barrow, *Mathematical Works,* 1:67. I have added the last sentence, which the translator omitted. Barrow's views on the philosophical status of explanations of color evidently did not change in the intervening four years.

47. Review of Barrow's *Lectiones opticae & geometricae,* in *Journal des sçavans* (18 November 1675):268–71, esp. 271.

48. Pierre Sylvain Regis, *Cours entier de philosophie, ou systeme general selon les principes de M. Descartes, contenant la logique, la metaphysique, la physique, et la morale,* new ed., 3 vols. (Amsterdam, 1691), Bk. VIII, Pt. II, Chap. XXI, 3:205–7; this edition is based on that of Paris, 1690.

49. Johannes Kepler, *Ad Vitellionem paralipomena, quibus astronomiae pars optica traditur* (Frankfurt, 1604), which is in Vol. 2 of Johannes Kepler, *Gesammelte Werke,* ed. Walther von Dyck and Max Caspar, 18 vols. to date (Munich, 1937–). I have found the annotated translation of the first five chapters by Catherine Chevalley to be very helpful; Johannes Kepler, *Les fondements de l'optique moderne: "Paralipomènes a Vitellion" (1604)* (Paris, 1980). Part of Chap. V has been translated into English by Alistair C. Crombie, "Kepler: De modo visionis," in *Mélanges Alexandre Koyrè,* Vol. 1, *L'aventure de la science* (Paris, 1964), pp. 135–72. The secondary literature is devoted principally to Kepler's theory of vision, with his geometrical optics being treated only in passing. See A. C. Crombie, "The Mechanistic Hypothesis and the Scientific Study of Vision: Some Optical Ideas as a Background to the Invention of the Microscope," in *Historical Aspects of Microscopy,* ed. S. Bradbury and G. L'E. Turner (Cambridge, 1967), pp. 3–112; Huldrych M. Koelbing, "Kepler und die physiologische Optik: Sein Beitrag und seine Wirkung," in *Internationales Kepler-Symposium, Weil der Stadt, 1971,* ed. Fritz Krafft, Karl Meyer, and Bernhard Sticker (Hildesheim, 1973), pp. 229–45; David C. Lindberg, *Theories of Vision from al-Kindi to Kepler* (Chicago, 1976); Gérard Simon, "On the Theory of Visual Perception of Kepler and Descartes: Reflections on the Role of Mechanism in the Birth of Modern Science," *Vistas in Astronomy,* 18 (1975):825–32; and Stephen Mory Straker, "Kepler's Optics: A Study in the Development of Seventeenth-Century Natural Philosophy" (Ph.D. dissertation, Indiana University, 1970).

50. On the cathetus rule, see Albert Lejeune, *Recherches sur la catoptrique grecque d'après les sources antiques et médiévales,* Académie royale de Belgiques, Classe de lettres et des sciences morales et politiques, Mémoires, Vol. 52, fasc. 2 (Brussels, 1957); Colin M. Turbayne, "Grosseteste and an Ancient Optical Principle," *Isis,* 50 (1959):467–72; and his *Myth,* pp. 148–58.

51. Lejeune, *Catoptrique grecque,* pp. 74–80. Alhazen's formulation of the cathetus rule in his *Optica,* Bk. V, §8, is straightforward: "The image in any mirror is seen at the intersection of the perpendicular of incidence and the line of reflection" (*Opticae thesaurus. Alhazeni arabis libri septem, nunc primùm editi . . . Item Vitellonis thuringopoloni libri X,* ed. Friedrich Risner [Basil, 1572; rpt. New York and London, 1972], p. [129]. Witelo's formulation is quite similar, except that he has replaced Alhazen's term "perpendicular of incidence" with "cathetus of incidence"; *Optica,* Bk. V, §37, ibid., p. $_2$207.

52. Alhazen, *Optica,* Bk. VII, §18; and Witelo, *Optica,* Bk. X, 15, *Opticae thesaurus,* pp. 253, $_2$416.

53. See Lindberg, *Theories of Vision.*

54. Kepler, *Paralipomena,* Chap. III, Def. I, *Werke,* 2:64. Compare this definition with Barrow's (note 6 above), in which an image, far from being a mistake, is considered simply as an object in a different place.

55. Ibid., Prop. XVII, p. 72; italics added.

56. Ibid., pp. 72-3.

57. Barrow, on the contrary, would base his rule for the location of the image with one eye on the requirement that the arc *MN* be very small.

58. Kepler introduced the term "pencil" (*penicillum*) in his *Dioptrice seu demonstratio eorum quae visui & visibilibus propter conspicilla non ita pridem inventa accidunt* (Augsburg, 1611; rpt. Cambridge, 1962), §XLV, p. 17; but he thoroughly applied the concept in his *Paralipomena*. For the development of Kepler's concept of pencil see Straker, "Kepler's Optics."

59. We can now appreciate why Kepler emphasizes binocular vision. Judgments of distance with two eyes is prior to that with one, for we learn to judge distance with one eye from our experience with two; *Paralipomena*, Chap. III, Prop. VIII and its "note." Moreover, distance judgments with one eye have a more limited range than do those with two; the base of the "distance-measuring triangle," or the breadth of the pupil, is so small that at appreciable distances the two rays effectively become one, and the rule is no longer applicable.

60. See especially Chap. V, Prop. XVII (*Werke*, 2:173-4), where Kepler describes the intersection of the rays with the axis, or the geometrical place of the image, but places the image perceived by a single eye on the sphere.

61. *Lectiones XVIII*, XV, App., p. 133.

62. Kepler, *Paralipomena*, Chap. V, Prop. XV (*Werke*, 2:172).

63. Ibid., Chap. V, Def., p. 174.

64. See Chap. V, Props. XIX, XX, XXIII, XXIV, and especially the frequently reproduced figure of intersecting rays generating the caustic in Prop. XIX, ibid., pp. 175-9. In Prop. XXIII Kepler even considers an image point on the caustic in describing the image formed on the retina when objects off the eye's axis cast oblique pencils through the aperture of the pupil.

65. Kepler, *Dioptrice*, §§XXXIV, XXXV, XXXIX, pp. 10-12, 14-15.

66. Ibid., §IIX, pp. 3-4.

67. James Gregory, *Optica promota, seu abdita radiorum reflexorum & refractorum mysteria, geometrice enucleata* (London, 1663), p. 93.

68. *Lectiones XVIII*, XIII:30, p. 119.

69. Bonaventura Cavalieri, *Exercitationes geometricae sex* (Bologna, 1647), p. 458. The section "De perspicillorum focis" (pp. 458-95), consisting of Props. VII-XIX of "Exercitatio sexta: De quibusdam propositionibus miscellaneis," contains his optical theorems. Cavalieri had treated reflection earlier in his *Lo specchio ustorio, overo trattato delle settioni coniche, et alcuni loro mirabili effetti intorno al lume, caldo, freddo, suono, e moto ancora* (Bologna, 1632), which was in Barrow's library.

70. Cavalieri, *Exercitationes sex*, p. 462.

71. See, e.g., the lucid elementary derivation of the focus of a plano-convex lens in Prop. VIII, ibid., pp. 462-4.

72. For the focal points see *Dioptrica*, Pt. I, Bk. I, Props. XIV-XVII, and for the image points Prop. XX, Huygens, *Oeuvres*, 13, i:80-93, 98-109.

73. Huygens, *Oeuvres*, Prop. III, pp. 16-19.

74. René Descartes, *La dioptrique*, Discours VI, in *Oeuvres de Descartes*, ed. Charles Adam and Paul Tannery, 13 vols. (Paris, 1897-1913), 6:144. Descartes's *Le dioptrique* was appended, together with *Les meteores* and *La geometrie*, to his *Discours de la methode pour bien conduire sa raison, & chercher la verité dans les sciences* (Leyden, 1637).

75. Ibid.

76. Ibid., p. 143. Note that the figures in the popular English translation by Paul J. Olscamp, *Discourse on Method, Optics, Geometry, and Meteorology* (Indianapolis, Ind., 1965), p. 109, are not from the first French edition.

77. Gregory, *Optica promota,* pp. 46–7.

78. *Lectiones XVIII,* III:1, p. 36. "Inflection" is Barrow's term encompassing both reflection and refraction; ibid., I:11, pp. 21–2.

79. Ibid., IX:13, p. 88 (quoted in note 100 below); see also III:17, V:21, VI:18. Barrow calls the line through the center of the eye along which the principal ray passes the "optical axis" (*axis Opticus*) to distinguish it from the fixed "axis" of the optical system that passes through the radiant point and is perpendicular to the reflecting or refracting surface; V:21, VI:17.

80. By restricting his claim to first publication, Barrow was undoubtedly attempting – successfully, as it turned out – to avoid a priority dispute with Huygens.

81. See also VI:17.

82. *Lectiones XVIII,* IV:20, p. 52. Huygens had already solved this problem for paraxial rays, but he did not proceed to the more difficult problem of an oblique pencil; Huygens, *Dioptrica,* Pt. I, Bk. I, Props. IV–VII, *Oeuvres,* 13, i:18–27.

83. This was a popular problem in the seventeenth century. Newton gave his own solution in the first version of his *Optical Lectures,* but he abandoned it in the revised version (Pt. I, Prop. 6) and referred the reader to Barrow's *Lectiones XVIII*; see Newton, *Optical Papers,* 1:226–9, 350–1; Newton, *Math Papers,* 3:450–3 (esp. note 6, where Barrow's solution is treated and his preceding analysis is reconstructed); and J. A. Lohne, "Fermat, Newton, Leibniz und das anaklastische Problem," *Nordisk Matematisk Tidskrift,* 14 (1966):5–25. In 1601 Thomas Harriot found a solution (unpublished in the seventeenth century) that was similar to Newton's; see J. A. Lohne, "Dokumente zur Revalidierung von Thomas Harriot als Algebraiker," *Archive for History of Exact Sciences,* 3 (1966):185–205. Lohne mistakenly attributes Harriot's solution to William Lower; personal communication, Whiteside to Shapiro, 23 March 1982. For Sluse's solution, see note 151 below.

84. Barrow repeats this solution in his *Lectiones geometricae,* VI:3, *Mathematical Works,* 2:210.

85. From Newton's equivalent analytical solution, $AN/ZN = (BA^2/BK^2)(NK/NA)$, in the first version of his *Optical Lectures* (Newton, *Optical Papers,* 1:216–21), we can put Barrow's solution into a recognizable form. If we let angle $BAN = i$ and angle $BKN = r$, then $BA = (\cot i)BN$, $BK = (\cot r)BN$, and $NK/NA = n$, and his solution becomes $AN/ZN = \cos^2 i/n\cos^2 r$. See also Newton, *Math Papers,* 3:460, n. 20.

86. *Lectiones XVIII,* V:21, p. 62.

87. Thomas Hobbes, *Elementorum philosophiae sectio secunda: De homine* (London, 1658), Chap. VII, §§1, 2, *Opera,* 2:59–60. Barrow suggests that Hobbes fell into this error because in reflection at a plane mirror the object and image are equidistant from the plane. He also believes – "if I am not mistaken" – that Euclid, Alhazen, and Stevin made a similar error in catoptrics; *Lectiones XVIII,* V:21, p. 63. I could not find such a view in Euclid and Alhazen, but Stevin did hold that the image and object are equidistant from a spherical mirror; see his *L'Optique,* Bk. II, "Des catoptriques ou des reflexions," Props. V and VII, in *Les oeuvres mathématiques de Simon Stevin de Bruges,* trans. Albert Girard, 6 vols. in 1 (Leyden, 1634), 5:569, 570.

88. *Lectiones XVIII,* V:21, p. 63.

89. Ibid., V:22, pp. 63–4.

90. On the difficulty of determining the perpendicular see note 103 below. Barrow's figure and the curve $B\phi$ are carefully redrawn to scale in Barrow, *Optical Lectures,* p. 85. Barrow's tangential image points lying on the curve and the corresponding sagittal image points of the cathetus rule lying on the axis are clearly indicated; the circle of least confusion lies between the axis and Barrow's curve. Jean-Etienne Montucla and, following him, Joseph Priestley and others misunderstood Barrow's experiment and claimed that the image of the string is not contiguous with that part of the string above the water; Jean-Etienne Montucla, *Histoire des mathématiques,* 2 vols. (Paris, 1758),

2:598; Joseph Priestley, *The History and Present State of Discoveries Relating to Vision, Light, and Colours* (London, 1772; rpt. Millwood, N.Y., 1978), p. 203.

91. Newton, *Optical Lectures,* the scholium to Prop. 3 of the first version and to Pt. I, Prop. 8 in the second, *Optical Papers,* 1:214–21, 352–5.

92. See A. G. Bennett, "Some Unfamiliar British Contributions to Geometrical Optics," *Transactions of the International Ophthalmic Optical Congress, 1961* (London, [1962]), pp. 274–91, esp. pp. 281–2.

93. For Newton's solution for the two image points at a spherical surface see note 122 below. P. Culmann first recognized the historical significance of Barrow's and Newton's investigations of astigmatism; see his "Historische Notizen. B. Ueber den Astigmatismus," in *Die Bilderzeugung in optischen Instrumenten vom Standpunkte der geometrischen Optik,* ed. Moritz von Rohr (Berlin, 1904), pp. 199–205; or the translation by R. Kanthack, *Geometrical Investigation of the Formation of Images in Optical Instruments* (London, 1920), pp. 201–9. See also Bennett, "British Contributions"; and note 176 below.

94. Newton, *Optical Lectures,* Pt. I, Prop. 8; Newton, *Optical Papers,* 1:352–5.

95. *Lectiones XVIII,* XVI:2, p. 136. Although *relata* should more literally be translated as "removed" or "withdrawn," I have chosen "relative" and treated it as a technical term. Fay similarly chose "related"; Barrow, *Optical Lectures,* p. 196.

96. See note 64 above.

97. *Lectiones XVIII,* XVI:2, pp. 136–7.

98. Ibid., p. 8.

99. In IX:14 Barrow apologizes for the lack of rigor – but not of certainty – in these approximations, which he could eliminate only at the cost of great tedium.

100. *Lectiones XVIII,* IX:13, p. 88.

101. This condition, $CL = LD = \rho/2 \cos i$, immediately yields the correct location of the sagittal image point L as previously noted; Barrow, to be sure, never considered the sagittal image. For a description of the image of a straight line perpendicular to a convex mirror according to the cathetus rule, see Andreas Tacquet, *Catoptrica tribus libris exposita,* Bk. III, "De speculis curvis," Prop. III, *Opera mathematica* (Antwerp, 1669), p. $_2$246, and Prop. XXII, pp. $_2$256–7, for a concave mirror.

102. In XVI:12 Barrow gives a clever construction to determine the points of bisection F and E and thus α. Barrow's figure with the curve $B\alpha\tau\sigma$ is accurately redrawn in Barrow, *Optical Lectures,* p. 201.

103. *Lectiones XVIII,* XVI:14, p. 140.

104. Barrow's investigations of the images of straight lines were widely admired in the eighteenth century; see notes 139 and 208 below.

105. *Lectiones XVIII,* XVI:18, XVIII:13, pp. 142, 153.

106. Tacquet, *Opera,* p. $_2$223.

107. Ibid., Bk. III, Props. XXIX and XXX, p. 259. Tacquet concluded Prop. XXX by observing, "Therefore Alhazen, Witello, and other opticians following them err in considering that just as in plane and convex mirrors so in concave ones the image never appears outside the intersection of the reflected ray with the cathetus of incidence."

108. In an annotation to Jacques Rohault, *System of Natural Philosophy, Illustrated with D^r. Samuel Clarke's Notes Taken Mostly out of Sir Isaac Newton's Philosophy,* trans. John Clarke, 3d ed., 2 vols. (London, 1735), which was used as a textbook at the University of Cambridge, Samuel Clarke chastises Tacquet for restricting the validity of the cathetus rule; Pt. I, chap. 34, 1:278. Yet Clarke was familiar with Barrow's *Lectiones XVIII,* for he translated his description of the "Barrovian case" in Pt. I, Chap. 33, 1:260–1. Christian Wolff, following Kepler's critique in the *Paralipomena,* still considered the cathetus rule to be valid for two eyes, provided that they were not in the same plane of incidence (Kepler's Prop. XVIII, note 56 above). See his *Elementa matheseos universae,* new ed., 5 vols. (Halle and Magdeburg, 1733–42), "Elementa

catoptricae," §§41–2, 151–2, 3 (1735):143, 171–2; I have used the facsimile reprint in Christian Wolff, *Gesammelte Werke*, ed. J. Ecole et al., Ser. II, Vol. 31 (Hildesheim, 1968).

109. For brevity I have assumed that the normals *CG* and *CI* coincide when the refracted rays *R*σ and *N*π are infinitely close, although Barrow does not make this simplification.

110. René Descartes, *Les meteores*, Discours VIII, *Oeuvres*, 6:336. For Harriot, see J. A. Lohne, "Regenbogen und Brechzahl," *Sudhoffs Archiv für Geschichte der Medizin und der Naturwissenschaften*, 49 (1965):401–15; and "Thomas Harriot als Mathematiker," *Centaurus*, 11 (1965):19–45. On the history of the rainbow, see Carl B. Boyer, *The Rainbow: From Myth to Mathematics* (New York and London, 1959), which, however, makes no mention at all of Barrow.

111. The angle of inclination $ZCD = \frac{1}{2}$ angle $ZCY = 2r - i$, whereas in modern optics the deviation is defined as $\pi - 2(2r - i)$. Consequently, Barrow's angle is a maximum rather than a minimum. Since Barrow will show that angle *ZCY* measures the bow's radius, the radius may be readily calculated using the two equations of XII:11. Barrow does not himself calculate it.

112. *Lectiones XVIII*, XII:14, p. 106.

113. Sluse to Oldenburg, 24 November 1667 [N.S.], Oldenburg, *Correspondence*, 3:596, 598.

114. Robert Hooke *assumed* that the rays coincide at the point of reflection in his *Micrographia* (London, 1665; rpt. Brussels, 1966), pp. 60–2, and esp. Fig. 3, Scheme VI; Barrow owned this work.

115. See Huygens, *Oeuvres*, 13, i:146–53; Sluse to Oldenburg, 20 August 1670 [N.S.], Oldenburg, *Correspondence*, 7:115–19; and also Newton, *Math Papers*, 3:502, n. 47. Newton had a copy, in Collins's hand, of this letter from Sluse to Oldenburg (as well as a number of other letters of Sluse), Cambridge University Library, Ms. Add. 3971, ff. 3–6, 13–21.

116. Newton, *Optical Lectures*, Pt. I, Props. 35, 36; Newton, *Optical Papers*, 1:418–25.

117. Edmond Halley, "De iride, sive de arcu coelesti, dissertatio geometrica, qua methodo directâ iridis utriusque diameter, data ratione refractionis, obtinetur...," *Philosophical Transactions*, 22 (1700–1):714–25. Jakob Hermann, "Méthode géométrique & générale de déterminer le diamétre de l'arc-en-ciel, quelque hypothèse de la refraction qu'on suppose dans l'eau, ou dans toute autre liqueur transparante...," *Nouvelles de la republique des lettres*, 32 (1704):658–71.

118. Isaac Newton, *Opticks: Or, a Treatise of the Reflexions, Refractions, Inflexions and Colours of Light* (London, 1704; rpt. Brussels, 1966), Bk. I, Pt. II, Prop. IX, pp. 126–9; hereafter cited as Newton, *Opticks*.

119. Newton, *Optical Lectures*, Pt. I, Prop. 32; Newton, *Optical Papers*, 1:410–15.

120. *Lectiones XVIII*, XIII:26, p. 117.

121. Newton, *Optical Papers*, 1:415.

122. Ibid.

123. Newton cites Barrow by name three times (ibid., pp. 349, 351, 415), in addition to the reference (note 21 above) in his inaugural lecture.

124. Newton, *Opticks*, Bk. I, Ax. VIII, p. 12.

125. Ibid., p. 11. Newton illustrates this with virtual images in a plane mirror (his Figure 9), a prism (Figure 2), and a convex lens (Figure 10). The axiom clearly applies to the location of real images as well, and his Figure 3 shows pencils of rays diverging from a real image formed by a convex lens and entering the eye, just as for the virtual image in Figure 9. It is not clear to me whether Newton actually intended to restrict this axiom to virtual images; see Turbayne, *Myth*, pp. 147–8.

126. Newton, *Opticks*, p. 12.

127. Robert Smith, *A Compleat System of Opticks, in Four Books, viz. a Popular, a Mathematical, a Mechanical, and a Philosophical Treatise. To Which Are Added Remarks upon the Whole*, 2 vols. (Cambridge, 1738), "Remarks," §209, 2:33.

128. *Lectiones XVIII*, XII:31, p. 119.

129. See the passage quoted in note 14 above.

130. If $BZ = f'$, $BC = \rho_1$, and $I/R = n$, so that $CZ = f' - \rho_1$, then Barrow's formula yields $f' = \rho_1 n/(n-1)$.

131. If we let $BD = d$, $DY = f$, and $DK = \rho_2$, so that (for the biconvex lens) $DZ = f' - d$ and $KZ = f' - d + \rho_2$, where f' was found in the preceding note, we get $f = [n\rho_1\rho_2/(n-1) - \rho_2 d]/[n(\rho_1 + \rho_2) - d(n-1)]$. This yields for the focal point in the limiting case of a thin lens (which was not considered by Barrow) $f = n\rho_1\rho_2/(\rho_1 + \rho_2)$.

132. *Lectiones XVIII*, p. 127.

133. Newton, *Optical Lectures*, Pt. I, Prop. 29; Newton, *Optical Papers*, 1:400-5.

134. See note 72 above.

135. See William Molyneux, *Dioptrica Nova: A Treatise of Dioptricks in Two Parts* (London, 1692), Pt. I, Props. V, VIII, XIV, XV, pp. 42, 48, 63, 66. He cites Barrow in the "Admonition to the Reader," p. [ix] and elsewhere; but see especially the passage cited at note 194 below. See also J. G. Simms, *William Molyneux of Dublin, 1656-1698*, ed. P. H. Kelly (Blackrock, Ireland, 1982).

136. Jean Picard, "Fragmens de dioptrique," in *Divers ouvrages de mathematique et de physique: Par messieurs de l'academie royale des sciences* (Paris, 1693), pp. 375-412, esp. pp. 383, 392-4. See also Michel Blay, "Recherches sur les travaux de dioptrique et de catoptrique de Jean Picard à l'académie royale des sciences," in *Jean Picard et les débuts de l'astronomie de précision au XVIIᵉ siècle: Actes du colloque du tricentenaire*, ed. Guy Picolet (Paris, 1987), pp. 329-44.

137. See note 149 below.

138. *Philosophical Transactions*, 17 (1693):960-9, esp. 963.

139. Barrow's derivations of the images of lines were frequently cited. See, e.g., Robert Smith, "Remarks," §436, *Compleat System*, 2:72; Montucla, *Histoire* 2:599; Priestley, *History*, p. 204; Joseph Harris, *A Treatise of Optics: Containing Elements of the Science; in Two Books* (London, 1775), p. 30; and also at note 208 below.

   In the nineteenth century Airy and Joseph Petzval proved that the image of a plane object formed by reflection or refraction in a single spherical surface is a curved surface (when spherical aberration and astigmatism are eliminated) and that in the paraxial region this surface approximates a sphere whose radius is independent of the distance of the object. Bennett and Edgar have shown that these results, as well as others, may be deduced from Barrow's constructions, though he himself did not derive them; see A. G. Bennett, "The True Founder of Point-Focal Lens Theory: George Biddell Airy," *Optician*, 150 (1965):395-8, 422-5; and Barrow, *Optical Lectures*, "Appendix A. Barrow's Treatment of Image Curvature," pp. 228-32. They show that in Barrow's construction for refraction the approximating sphere is defined by the region between the perpendicular through the focus *CE* (or the semilatus rectum) and the nearer vertex. For reflection, *CE* is by construction one-half of the diameter of the reflecting sphere.

140. Alhazen, *Optica*, Bk. V, 32-9; Risner, *Opticae thesaurus*, pp. 142-51. Of the large literature on Alhazen's problem the following are particularly useful: Paul Bode, "Die Alhazensche Spiegel-Aufgabe in ihrer historischen Entwicklung nebst einer analytischen Lösung des verallgemeinerten Problems," *Jahresbericht des physikalischen Vereins zu Frankfurt am Main* (1891-92):63-107; J. A. Lohne, "Alhazens Spiegelproblem," *Nordisk Matematisk Tidskrift*, 18 (1970):5-35; and A. I. Sabra, "Ibn al-Haytham's Lemmas for Solving 'Alhazen's Problem,'" *Archive for History of Exact Sciences*, 26 (1982):299-324.

141. *Lectiones XVIII*, IX:5, pp. 84-5; Huygens, *Oeuvres*, 20:330.

142. *Lectiones XVIII*, XIII:23, p. 116.

143. Huygens set forth Alhazen's problem for Sluse on 13 August 1657 [N.S.]; Huygens, *Oeuvres*, 2:45. He went on to observe, "For me nothing memorable appears in the

entire work of Alhazen except this one thing, and I am always amazed that he could construct it without the aid of algebra." On 26 June 1669 [N.S.], he sent Oldenburg his solution to Alhazen's problem; Huygens, *Oeuvres,* 6:459–62, and plate facing p. 462.

144. The correspondence of Huygens and Sluse with their solutions was published as "Excerpta ex epistolis non-nullis, ultrò citróque ab Illustrissimis viris, Slusio & Hugenio, ad editorem scriptis, de famigrato Alhazeni problemate circa punctum reflexionis in speculis cavis aut convexis," *Philosophical Transactions,* 8 (1673):6119–26, 6140–6.

145. Abraham Gotthelf Kaestner, "Problematis Alhazeni analysis trigonometrica," *Novi commentarii societatis regiae scientiarum Gottingensis,* 7 (1776):92–141, esp. 101–2.

146. Barrow, Huygens, and Sluse also independently discovered the aplanatic points for refraction at a sphere. These are a unique pair of points possessing the property that all rays emitted from one of the points after refraction pass exactly through the other, that is, there is no spherical aberration. See *Lectiones XVIII,* XI:9, XIII:14, XIV, pp. 97–8, 112, 126–7; Huygens, *Dioptrica,* Pt. I, Bk. I, Prop. XII; Huygens, *Oeuvres,* 13, i:64–7; and Sluse to Oldenburg, 20 August 1670 [N.S.], Oldenburg, *Correspondence,* 7:117, 119. By 7 September 1657 [N.S.], Huygens had found the aplanatic points, as he then wrote Sluse, and had also seen a solution by Roberval; Huygens, *Oeuvres,* 2:55.

147. Collins to Thomas Baker, 10 February 1676/7, Rigaud, ed., *Correspondence,* 2:14; see also p. 22.

148. Gregory to Collins, 29 January 1670, ibid., p. 190.

149. Vernon to Oldenburg, 19 July and 4 September 1670 [N.S.], Oldenburg, *Correspondence,* 7:62, 140.

150. Sluse to Oldenburg, 25 July 1670 [N.S.], Oldenburg, *Correspondence,* 7:74, 78.

151. Sluse's (incomplete) solution to the anaclastic problem was, in fact, similar to Newton's, though no doubt the two were independent; see Newton, *Math Papers,* 3:451, n. 5. Newton had a copy of this letter; see note 115 above.

152. Sluse to Oldenburg, 20 August 1670 [N.S.], Oldenburg, *Correspondence,* 7:115–19; see note 146 above.

153. In replying to an inquiry by Oldenburg in the fall of 1665, Spinoza wrote that the *Dioptrica* could be expected after Huygens worked out one more point on correcting for spherical aberration; Oldenburg, *Correspondence* 2:497–500, 540–2. On Huygens's close scientific relations with Oldenburg and the Royal Society, see M. B. Hall, "Huygens' Scientific Contacts with England," in *Studies on Christiaan Huygens: Invited Papers from the Symposium on the Life and Works of Christiaan Huygens, Amsterdam, 22–25 August 1979,* ed. H. J. M. Bos, M. J. S. Rudwick, H. A. M. Snelders, and R. P. W. Visser (Lisse, 1980), pp. 66–82.

154. Oldenburg to Huygens, 8 March 1668/9, Oldenburg, *Correspondence,* 5:436, 437. Again, on 29 March he warned that "if you delay still longer the publication of your *Dioptrica* the same mishap may occur with that, as has occurred with motion, for there is a very skillful man here who is preparing a treatise on the same subject; it will be out of the ordinary" (ibid., pp. 464, 466).

155. On 6 February 1669 [N.S.], Huygens had already sent Oldenburg one cipher on lenses, after having almost lost priority on the collision problem; ibid., pp. 361, 362. He sent the other thirteen on 4 September 1669 [N.S.]; ibid., 6:213–18. On Alhazen's problem see note 143 above.

156. Huygens to Oldenburg, 22 January 1670 [N.S.], ibid., pp. 424, 425; I have slightly altered the Halls's translation.

157. Oldenburg to Huygens, 31 January 1669/70, ibid., pp. 458, 459.

158. Quoted in note 3 above.

159. Barrow to Collins, 23 April 1670, Rigaud, ed., *Correspondence,* 2:74; and "An Accompt of Some Books. I. *Lectiones 18...,*" *Philosophical Transactions,* 6 (1671): 2258–9.

160. See note 47 above.

161. See Shapiro, "Kinematic Optics," pp. 232–6. Huygens made this discovery about 1676 or 1677; see Huygens, *Oeuvres,* 19:420–3.

162. Christiaan Huygens, *Traité de lumière: Où sont expliquées les causes de ce qui luy arrive dans la reflexion, & dans la refraction* (Leyden, 1690), in *Oeuvres,* 19:534–7. Silvanus P. Thompson in his popular English version has mistranslated Huygens's "faite par l'Evolution" as "formed as the evolute" instead of the correct "the involute"; *Treatise on Light* (London, 1912; rpt. New York, 1962), pp. 124–6. Huygens introduced the terms "evoluta" and "descripta ex evolutione" for the evolute and involute in his *Horologium oscillatorium sive de motu pendulorum,* Pt. III, Defs. III, IV, *Oeuvres,* 18:188–9; in French the terms adopted are "développée" and "développante," respectively.

163. Huygens, *Oeuvres,* 19:536.

164. Ehrenfried Walther Tschirnhaus, "Nouvelles découvertes dans les mathematiques proposees à messieurs de l'academie royale des sciences, par Mr. de Tschirnhaus," *Journal des sçavans* (8 June 1682):210–13; and his more widely cited paper, "Inventa nova, exhibita Parisiis societati regiae scientarum à D. T.," *Acta eruditorum* (November 1682):364–5. In addition to La Hire, Cassini and Mariotte were referees at the Royal Academy for Tschirnhaus's paper, which did not contain a demonstration; see La Hire, "Examen de la courbe formée par les rayons réfléchis dans un quart de cercle," *Mémoires de l'académie royale des sciences: Depuis 1666 jusqu'à 1699,* 9 (1730):294–310, which was first published in his *Mémoires de mathématique et de physique* (Paris, 1694); and also Fontenelle's eloges of Tschirnhaus and La Hire, *Histoire de l'académie royale des sciences* (1709):114–24, esp. 115–16, and (1718 [Amsterdam ed.]):95–112, esp. 104–5.

165. Tschirnhaus, "Methodus curvas determinandi, quae formantur a radiis reflexis, quorum incidentes ut paralleli considerantur, per D. T.," *Acta eruditorum* (February 1690): 68–73; and also "Curva geometrica, quae seipsam sui evolutione describit, aliasque insignes proprietates obtinet, inventa a D.T.," ibid. (April 1690):169–72. In the latter paper (p. 170), Tschirnhaus, I believe, first introduced the term "caustic curve" (*curva caustica*), which he quite naturally drew from the burning (*caustica*) mirror, sphere, and point.

166. In his classic 1693 paper Jakob Bernoulli introduced the terms "catacaustic" and "diacaustic"; see note 8 above. For Johann's paper see *Acta eruditorum* (January 1692): 30–5; and for Jakob's earlier ones, ibid. (March 1692):110–16 and (May 1692):207–13.

167. J. E. Hofmann, "Jakob Bernoulli," in *Dictionary of Scientific Biography,* ed. Charles Gillespie, 16 vols. (New York, 1970–80), 2:47. See, e.g., Bernoulli's notes from about 1680 on Barrow's explanation of reflection in the *Lectiones XVIII,* in *Die Werke von Jakob Bernoulli,* Herausgegeben von der Naturforschenden Gesellschaft in Basel, 2 vols. to date (Basel, 1969–), 1:317.

168. Bernoulli, "Curvae dia-causticae," p. 246.

169. "A Germanis alijsque nominatim a Bernoullio in Actis Mensis Maij 1692, perhibetur Tschirnhausum primum considerasse Causticas; cum tamen illae (tum CataCausticae quam Diacausticae) fuerint consideratae a Barrovio, Nam hae eandem sunt cum locis imaginum respectu oculi extra axem radiationis variè constituti" (Gregory notebook "E," Christ Church Library Ms. 346, p. 11).

170. Montucla, *Histoire,* 2:599.

171. See, e.g., Priestley, *History,* pp. 204; and Johann Samuel Traugott Gehler, *Physikalisches Wörterbuch,* 4 vols. (Leipzig, 1787–91), "Bild," 1:355. On Tschirnhaus's reputation see J. E. Hofmann, "Tschirnhaus," *Dictionary of Scientific Biography,* 13: 479–81.

172. See Huygens's letter to Leibniz, 9 October 1690 [N.S.] and his memorandum on Tschirnhaus from 7 April 1691 [N.S.], *Oeuvres,* 9:498–9, 511–15.

173. Guillaume François Antoine de L'Hospital, *Analyse des infiniment petits, pour l'intelligence des lignes courbes,* 2d ed. (Paris, 1715), Pt. I, Secs. VI, VII, pp. 104-30.

174. Smith, *Compleat System,* Bk. II, Ch. IX, "Determination of Focuses of Rays Falling with Any Degrees of Obliquity upon Any Number of Reflecting and Refracting Surfaces of any Sort, and also of the Properties of Causticks," 1:160-81.

175. Ibid., "Remarks upon Chapter 9," §§490-4, 2:81-2.

176. Young not only recognized that the image points were lines, but also described the changing cross section of an incident pencil of cylindrical rays in the region of the two image points *Z* and *K* (Figure 24 above); "On the mechanism of the eye," *Philosophical Transactions,* 91 (1801):23-88, esp. Prop. IV., Schol. 4, p. 30. He remarked here that the work of "Sir Isaac Newton, and extended by Dr. Smith... appears, however, to have been too little noticed." Young omitted his "dioptrical propositions" in the reprint of this paper in his *Course of Lectures on Natural Philosophy and the Mathematical Arts,* 2 vols. (London, 1807); but in the "Mathematical Elements of Natural Philosophy" appended to his *Lectures,* he added a new series of theorems extending the study of astigmatism still further (2:73-6). For nearly two centuries after Barrow's pioneering research the study of astigmatism was the sole province of Englishmen – indeed, Cambridge men – and even its name is due to another master of Trinity, William Whewell. See Marius H. E. Tscherning's excellent annotated translation, *Oeuvres ophtalmologiques de Thomas Young* (Copenhagen, 1894); and also John R. Levene, "Sir George Biddell Airy, F.R.S. (1801-92) and the Discovery and Correction of Astigmatism," *Notes and Records of the Royal Society of London,* 21 (1966):180-99; and *Clinical Refraction and Visual Science* (London and Boston, 1977).

177. Young cited Barrow earlier in this paper in Prop. I, "Mechanism of Vision," p. 27. For Culmann see note 93 above.

178. Berkeley, *Theory of Vision,* §29; Berkeley, *Works,* 1:179; *Lectiones XVIII,* XVIII:13, p. 152.

179. I have added rays to Barrow's figure to make it easier to comprehend the phenomenon.

180. Berkeley, *Works,* 1:180; *Lectiones XVIII,* p. 153.

181. Berkeley, *Works,* 1:179; *Lectiones XVIII,* p. 152.

182. Berkeley, *Works,* 1:181; *Lectiones XVIII,* p. 153.

183. Jean le Rond d'Alembert, "Doutes sur différentes questions d'optique," in *Opuscules mathématiques, ou mémoires sur différens sujets de géométrie, de méchanique, d'optique, d'astronomie &c,* 8 vols. (Paris, 1761-80), 1 (1761):277-8.

184. Kepler, *Dioptrice,* §LXV, p. 28; on blurred images see also §LXXI, pp. 30-1.

185. Gregory, *Optica promota,* Prop. 29, Corol. 4, p. 41; see also Prop. 45, pp. 60-4, on blurred images.

186. Ibid., Prop. 30, p. 41.

187. Quoted in note 156 above.

188. Leibniz to Oldenburg, 16 April 1673, Oldenburg, *Correspondence,* 9:595-6, 599. I have been unable to find Mariotte's views on the place of the image.

189. Huygens, *Oeuvres,* 13, ii:745. Another note simply reads "Embarras de Barrow."

190. Fullenius (see note 4 above) sought Huygens's thoughts on image location and the "Barrovian case" on 10 August 1683 [N.S.], and Huygens expounded upon them on 12 December 1683 [N.S.] and 31 August 1684 [N.S.]; Huygens, *Oeuvres,* 8:443-51, 474-8, 533-6. In various notes from about 1692 on image location – composed for his *Dioptrica* and from his reading of Molyneux's *Dioptrica Nova* – Huygens criticized Barrow (as well as Kepler, Descartes, and Molyneux); *Oeuvres,* 13, ii:771, 775-6, 779-80, 830-1.

191. At the conclusion of Prop. XXVI of his *Dioptrica,* Pt. I, Bk. I, on the construction of the eye and formation of the retinal image, Huygens vowed that he would not treat the psychology of vision, "which I judge too obscure for any mortals to be able to examine" (*Oeuvres,* 13, i:135).

192. Ibid., 8:477; 13, ii:775–6.

193. In his first letter to Fullenius, Huygens estimated that distance to be twelve or fifteen feet, and in the second fifteen or twenty; ibid., 8:477, 535.

194. Molyneux, *Dioptrica Nova*, Pt. I, Prop. XXXI, p. 119.

195. Ibid., p. 113.

196. See the "Editor's Introduction," Berkeley, *Works*, 1:146, 156–7.

197. Berkeley, *Theory of Vision*, §14, *Works*, 1:173.

198. Ibid., §§16–27, pp. 174–7.

199. Ibid., §§32–6, pp. 181–4, esp. Berkeley's diagrams in §35.

200. Smith, "Remarks," §217, *Compleat System*, 2:36. From Smith's quotations of the *Theory of Vision*, it is evident that he used the 1732 edition.

201. Smith, *Compleat System*, Bk. I, Ch. II, §25, 1:7.

202. Ibid., "Remarks," §§197–248, 2:31–42.

203. Ibid., §§213–14, p. 35.

204. Ibid., §§217–29, pp. 36–8.

205. Ibid., Bk. I., Ch. V, §139, 1:51.

206. See, e.g., Montucla, *Histoire*, 2:600–2. Insofar as I can determine there is still no generally accepted explanation for the "Barrovian case."

207. Georg Wolffgang Krafft, "De loco imaginis puncti radiantis in speculum curvilineum, dissertatio catoptrica," *Commentarii academiae scientiarum imperialis Petropolitanae*, 12 (1740):243–60, esp. 249–50.

208. Ibid., p. 252.

209. Montucla, *Histoire*, 2:602. In the second edition Montucla began this section with the observation that one should not find it surprising that despite the efforts of "a number of the greatest mathematicians" the problem of image location was still unresolved, since it raises psychological questions that are not at all susceptible to mathematical treatment; *Histoire des mathématiques*, new ed., ed. Jérôme de la Lande, 4 vols. (Paris, 1799–1802; rpt. Paris, 1960), 3:435.

210. Pierre Bouguer, *Traité d'optique sur la gradation de la lumiere* (Paris, 1760), Bk. II, Sec. I, §§I–III, pp. 98–104; I have used the translation by W. E. Knowles Middleton, *Pierre Bouguer's Optical Treatise on the Gradation of Light* (Toronto, 1961), pp. 76–81. Bouguer treats Barrow's principle as common knowledge, although he was familiar with Barrow's *Lectiones XVIII*. According to Middleton (p. ix) Bouguer probably wrote Bk. II a few months before he died in 1758.

211. Bouguer, *Traité*, p. 104; Middleton, *Bouguer's Treatise*, pp. 80–1. I have slightly altered Middleton's translation. Bouguer cites the Scholium to Prop. 8 of Newton's *Optical Lectures* here.

212. Priestley, *History*, pp. 202–5, 688–94.

213. Harris, *Treatise of Optics*, pp. 8, 89, 180; for his knowledge of Barrow see note 139 above.

214. Ibid., pp. 178–9. Kaestner, likewise, could not support Barrow's principle in his "De obiecti, in speculo sphaerico visi, magnitudine apparente," *Novi commentarii societatis regiae scientiarum Gottingensis*, 8 (1777):96–123.

215. Gehler, *Physikalisches Wörterbuch*, 1:357. For yet another account in this genre see John Robison's article "Optics," *Encyclopaedia Britannica*, 3d ed., 18 vols. (Edinburgh, 1797), 13:327–31.

216. Benjamin Robins, *Remarks on Mr. Euler's 'Treatise of Motion,' Dr. Smith's 'Compleat System of Opticks,' and Dr. Jurin's 'Essay upon Distinct and Indistinct Vision'* (London, 1739), §9, in *Mathematical Tracts of the Late Benjamin Robins, Esq.*, ed. James Wilson, 2 vols. (London, 1761), 2:225–8. See §58, p. 258, for his appreciation of Barrow's contributions, and also Wilson's preface, 1:xii–xiii. For Smith see note 127 above.

217. Smith, "Remarks," §490, *Compleat System*, 2:81.

3

Barrow's mathematics: between ancients and moderns

MICHAEL S. MAHONEY

Introduction

Isaac Barrow's earliest mathematical publication, an edition of Euclid's *Elements* in a symbolic shorthand inspired in part by Pierre Herigone, dates from 1656.[1] But his mathematical career began in earnest following his return from a four-year tour of the Continent and study in Italy. Finally seated in 1660 as Regius Professor of Greek at his alma mater, Trinity College, he began in 1662 to supplement his living through the professorship of geometry at Gresham College. His lectures from that time have not survived, and it would be difficult to say with any confidence how much of what he later presented as Lucasian Professor stemmed from them. It was in the latter capacity, from 1663 to 1669, that he composed the bulk of his work, recorded in his three series of lectures on mathematics, optics, and geometry.[2]

Lucas's executors prescribed the holder of the chair as a "man of good repute and honest conversation, at least a Master of Arts, deeply learned, and imbued with expertise above all in the mathematical sciences." Barrow's correspondence with John Collins gives a picture of what the new Lucasian Professor knew about his subject on his accession. He spoke of "that little study I have employed upon mathematical businesses, being never designed to any other use than the bare knowledge of the general reason of things, as a scholar, and no further."[3] He was being modest. Beyond the corpus of the ancient mathematics, his "little study" by then included the writings of Mersenne, Descartes, Pascal (Dettonville), Huygens, Viviani, Torricelli, Renaldini (whom Barrow had met in Florence), Snel, Alsted, and Kersey. If he had not yet read Cavalieri's method of indivisibles firsthand, he soon did so, as he also learned Mengoli's algebra. By the time of the optical and geometrical lectures, he was abreast of current developments in both areas and could reconnoiter the frontiers of mathematics with some confidence.

179

Barrow enjoyed some standing among his contemporaries. Well known in London and a charter member of the Royal Society, he belonged to Collins's network and served as Newton's initial conduit to it. Henry Oldenburg's correspondence provided Barrow with an avenue to European mathematicians, some of whom already knew him from his travels there. That was how Huygens, for example, obtained copies of the *Optical Lectures* and *Geometrical Lectures*.[4] Several propositions of the *Geometrical Lectures* attracted attention in the years following its publication, in particular the method of tangents set out at the end of Lecture X. Though none of them stemmed originally from Barrow, they were subsequently associated with his name. However, despite the publication of several editions of the work, its reputation was slight and did not last; by the 1690s, references to it and its author all but disappear from the working literature.

Immediate posterity thought enough of Barrow to publish English translations of his two sets of lectures,[5] and Berkeley seems to have derived some of his ideas from the *Mathematical Lectures*.[6] By and large, however, Barrow seems to have acquired significant historical importance only at the turn of the twentieth century, when historians revived his reputation on two grounds: as a forerunner of the calculus and as a source of Newton's mathematics. J. M. Child in particular credited Barrow with the first enunciation of the fundamental theorem of the calculus, thus rooting the achievement of Newton and Leibniz in his work.[7] That view, contested by some at the time, has generally persisted until quite recently, albeit with some qualification. So, for example, Margaret E. Baron's oft-cited *Origins of the Infinitesimal Calculus* credits Barrow with the fundamental theorem based on an "intuitive understanding of the inverse relations of differential and integral processes," while denying him a place in the camp of those who "had the calculus" because of his commitment to geometrical modes of expression.[8] While withholding judgment on the latter issue, the *Source Book of Mathematics, 1200–1700* excerpts propositions from the *Geometrical Lectures* under the heading of the fundamental theorem.[9]

Yet beginning in the 1960s two lines of historical inquiry began to cast doubt on this consensus. First, D. T. Whiteside's magisterial edition of Newton's mathematical papers made clear the independent roots of his thinking and dispelled most legends surrounding his relationship with Barrow at Cambridge.[10] Second, research on the mathematics of such figures as Descartes, Fermat, Roberval, Torricelli, Cavalieri, and Wallis, coupled with a deeper understanding of the nature of Newton's and Leibniz's systems of the calculus, revealed the sources of Barrow's *Geometrical Lectures* and with them the conceptual framework within which he read them and chose from and among them.[11]

At the same time that that recent work provides a richer understanding of Barrow's mathematical work taken on its own terms, it also removes him from a major role either in the development of the calculus or in

Newton's early mathematical thinking. Barrow emerges as a widely read, competent mathematician capable of appreciating the newest developments but not disposed to embrace all of them. For example, the new symbolic algebra, understood as the art of analysis and grounded in the notion of the equation as both an object and a relation, was essential to the calculus. Barrow accepted neither the art nor the notion, and he avoided the technique wherever possible. The theorems in the *Geometrical Lectures* that historians juxtapose to form the fundamental theorem of the calculus belong to two quite different lines of inquiry, one of them characterized by its freedom from the "tediousness of calculation" and divorced in Barrow's mind from anything smacking of infinitesimals or limits. The central lectures contain a program of research, but it is Barrow's own, and it does not encompass the program that led Newton and Leibniz to the calculus.

What follows is an examination of the two series of lectures that contain the substance of Barrow's mathematics at its profoundest and most creative. In extent and depth of analysis, the *Geometrical Lectures* receive the major share of attention. It is there that, despite the undergraduate audience for which the lectures were supposedly meant, Barrow explored the state of the art and imposed his order on results taken from contemporaries and predecessors working in many different directions. It is from that order, and the foundations on which it rests, that the shape and direction of Barrow's mathematics emerge, both in its own integrity and in its distinction from that of Newton and Leibniz.

For that reason, the *Mathematical Lectures* receive less attention than they perhaps deserve as a comprehensive review of traditional philosophy of mathematics in the mid-seventeenth century by a man with some critical acumen. As revealing as a full study might be on a wide range of issues, they are for the most part issues that Barrow himself addressed for completeness' sake and about which he had little new or different to say. Some of the points he raised, however, reveal the paths along which he was prepared to depart from tradition, the tensions such deviations posed for him, and the limits beyond which he would not go. Those limits form the boundary between his work and that of his illustrious junior colleague and thus mark the quite different directions in which they were headed when Barrow left the Lucasian chair, abandoning mathematics and optics for theology.

Taken together, the two sets of lectures show Barrow poised between tradition and innovation, between ancients and moderns in mathematics. To see what kept him there, it is best to start with the *Mathematical Lectures*.

The *Lectiones mathematicae:* the way of the ancients

Barrow began his career as Lucasian Professor with a three-term series of lectures aimed at reviving interest in mathematics at Cambridge.

Just over halfway through the second year's presentation, he paused to lament the direction he was taking. John Kirkby's translation of 1734 captures the rhetorical flavor of the moment:

> But shall I never extricate myself from these Quirks and
> Trifles? Shall I always spend my time in examining what is of
> no Value?[12] In so plentiful an Harvest, so rich a Vintage, so
> great a Store of most important Disquisitions, why do I only
> glean the scattered Ears, search the neglected Boughs, and
> gather the fallen Grapes? When a Chace after the important
> and difficult Things in the Mathematics is offered, a Chace so
> full of Variety, so pleasant and so certain; wherefore do I dwell
> so long upon those little Questions, like one hunting after
> Flies? I shun things of Consequence, sport in serious, am
> gravely ridiculous, studiously seeking after, and nicely repeat-
> ing and inculcating every the slightest Matter. Shall I thus
> incessantly follow so many distant By-ways, so many uncouth
> Turnings, and shall I never return again into the beaten Paths
> of the King's high Road? Shall I grow old in these outer Courts
> of *general Matters?* Shall I perpetually tarry in the Entrance of
> the Sciences? Shall I always stick in the Threshold? Shall I only
> knock at the Doors of the Mathematics, and never enter within
> the Walls of the House, nor penetrate its more sacred Recesses?
> Shall I ever be upon the Parley, ever skirmish at a Distance,
> and never engage Hand to Hand, or come to a decisive Battle?
> What do I but raise Mists and Doubts, sow Strifes and Conten-
> tions, raise Storms and Tumults in that Science, which prom-
> ises, which boasts of nothing but what is clear and evident,
> certain and tried, calm and serene? And by disputing more
> freely, and bringing many Things to the Scrutiny, I seem to
> detract and derogate from the Certitude and Evidence of the
> Mathematics, which is so contrary to Jarring and Contentions?
> Thus am I wont to upbraid myself, and perhaps also others do
> the same; at least not without some seeming Cause, or Appear-
> ance of Justice. (*UML* 234–5)

For all the studied rhetoric, a genuine impatience shows through the out-burst, which may have taken its original audience as much by surprise as it does the modern reader. Perhaps we should take Barrow at his word. The lectures were not progressing as he had planned, nor could he see how to redirect them. In part, both his audience and the format in which he was addressing them lay at fault. But he too shared the blame. Something was pulling him off course. Questions he wished to pass by with a wave grabbed his attention, forcing themselves from the periphery into the heart of his lectures. He thought them resolved, yet they resisted summation.

The foundations he meant to lay would not settle into place. Quite apart from what his audience was learning, he himself was not satisfied.

Barrow set an ambitious, yet seemingly traditional program for his first year's lectures. He proposed to discuss the name of the subject, its object and the divisions of the science that followed from the nature of that object, the methods by which mathematics pursued its object, and the history, current state, and likely future of the subject. In outline, at least, that program belonged to a common genre of mathematical discourse. Directed to a general undergraduate audience that had at best some acquaintance with Euclid's *Elements* and the fundamentals of arithmetic, such lectures talked about mathematics as a body of learning and as a mode of reasoning, fixing its place and role in the catalogue of the sciences. They did not aim to teach the audience to do mathematics. Though they set forth its basic concepts, they avoided the presentation of its techniques.

One should have been able to march through such lectures at a steady pace, striding from one topic to the next, citing the *loci classici* and summarizing the commentators. But by the end of Lecture XIII Barrow had just reached the topic of measure, which was but a prelude to the subject of ratio and proportion, sufficiently complicated in itself for a year's lectures. At that rate he would never get to the heart of mathematics as it was actually being practiced. He therefore paused to offer his listeners a hint of what they were missing. Beyond the study of the foundations of the individual disciplines and an understanding of the method of the ancients for finding and proving theorems lay the "foremost goal and highest pinnacle" of mathematics: its methods of obtaining general solutions to whole classes of problems. Many were new and unknown to the ancients: Viète's and Descartes's algebraic analysis, Cavalieri's method of indivisibles, the method of maxima and minima, rules for determining tangents to curves, the generation and investigation of curves by means of motion, the use of infinite series to measure the areas of curves, and techniques for the determination of centers of gravity. These achievements marked the "progress and growth" of mathematics.

Slow pace was not the only barrier to presenting the material. Barrow could not be sure his audience was even "moderately initiated in the elements of geometry," and he had to adapt both the content and the style of his presentation accordingly. Moreover, one could not merely talk about such mathematics; one had to do it. But the "dry subtlety, the extreme rigor, and the mode of argument requiring the utmost attention" would put off most of the audience; for the remainder,

> I am afraid the most part would not perfectly understand my meaning from the vanishing words of an oration; at least (to deal familiarly with you) if I may be permitted to make an estimate of other people's capacities from my own, I confess myself so dull and unapt that I can much easier understand

> that which is plainly laid before my eyes, than that which is
> insinuated through the treacherous caverns of my ears. (*LM*
> 211-13; *UML* 241-3)[13]

The lecture hall, it would appear, was no forum for the teaching of technical mathematics.

So Barrow tried to make a virtue of necessity, arguing that there was some value in pursuing the fine points he had been discussing. Mighty oaks from little acorns grow; the slightest error can give rise to immense confusion. Yet he then turned on his own argument, noting that all of the discussion and debate to which he was devoting his lectures had no effect on the certainty of mathematics, which remained, as it were, above the fray. Critics might question whether Euclid had given proper form to the parallel postulate, but none doubted its truth. "At issue is not the truth of chief results but the order of some propositions, not the certainty of knowledge but the method and mode of knowing; it is a matter only of some exterior things, philosophical more than mathematical" (*LM* 209). They contribute to a full understanding of mathematics, but their absence does not threaten its stability. Yet, and again he turned, there are things in mathematics worth noting, paradoxes that whet the mind, ideas current in the learned world. It is good to have some accurate notion of them – but not to the point of actually learning mathematics. And so on.

All this suggests a man at cross-purposes, both with his audience and with himself. Lecturing to a general audience required that he hew closely to traditional lines of mathematics as laid out in classical texts familiar to his listeners. At the same time, he wanted to engage the students' interest in new directions of mathematical inquiry. He evidently believed he could do both at the same time by focusing on classical concepts and alluding to recent developments. The former encompassed the latter, and so the move from classical foundation to modern superstructure involved no radical disjuncture. Not all practitioners and proponents of the new mathematics shared that view. On the contrary, many believed that it called for new foundations and employed concepts at odds with those of classical tradition. To the extent that their arguments impinged on the subjects of Barrow's lectures, he had to respond to them. But their arguments presupposed some familiarity with the new problems and methods that lay behind the concepts in question. Given the nature of his lectures, Barrow could not present that new material. He had to argue within the confines of traditional themes.

But not all of Barrow's debates were with other writers, nor were his limits entirely set by his genre. At several points he seems to have been at cross-purposes with himself and unsure of the grounds of his own understanding. As he noted at the start of Lecture IX, such basic notions as magnitude, extension, space, and motion are "things which it is most difficult not to make more perplexing by explaining, not to obscure by illustrating." They brought to mind Augustine's conundrum about time: "If no one demands of me what time is, I know; if someone asks, I do not know" (*LM* 131). At issue was the nature of mathematical intelligibility itself. Mathematics had many new "chief results," most of them the fruit of new methods unknown to the ancients. Whether the concepts underlying those methods were equally unknown or were old notions clothed in new garb, trying to explicate them brought out the extent to which conviction was rooted in successful practice and "metaphysical niceties [were] reduced to common sense" (*UML* 239; *LM* 209). While defensible at the level of intuition and experience, not to say success, the new concepts eluded definition in classical terms. In some cases, Barrow was still on the chase.

Thus, the traditional form of Barrow's *Mathematical Lectures* masks an often untraditional content. A new concept of number, a new basis for comparing magnitudes with one another, and a corresponding shift of emphasis in the concepts of ratio and proportion constitute recurrent themes. Barrow tried to incorporate them into a classical pattern, linking them to traditional notions. But they did not arise from classical concerns. Rather, they reflected the need of current practitioners to accommodate numbers that were the sums of infinite series, congruences that matched only at the infinitesimal limits of magnitude, piecemeal transformations that mapped different sorts of figures into one another, and ratios and proportions among infinite aggregates. While Euclid's theory of ratio and proportion retained its primacy, his arithmetic gave way to Wallis's arithmetic of infinites, his application of areas to Cavalieri's method of indivisibles. Barrow may have felt his audience was not yet ready to hear the *Geometrical Lectures,* but he could not ignore their implications for what he had to say about mathematics in general.

Number as symbolic magnitude: the identity of arithmetic and geometry

The new shape of mathematics loomed early in Barrow's lectures. Lecture II began by setting out both traditional and recent classifications of the mathematical disciplines but ended with the rejection of the foundations on which they rested. Whatever the disagreements in detail among Plato, Aristotle, and their successors, they had all distinguished between "pure" and "mixed" mathematics, that is, between arithmetic and geometry, on the one hand, and such subjects as astronomy, optics, and harmonics, on the other. The pure subjects dealt with mathematics as applied to intelligible things, the mixed subjects with the mathematics of sensible things. Barrow took the distinction to be specious. On the one hand, it is only by virtue of their instantiation in sensible objects that mathematical objects are accessible to the imagination, and hence intelligible. On the other, these objects are mathematical only to the extent that they involve quantity and hence "are rather so many examples only of geometry, than so many distinct sciences separate from it." In essence, the classification of mixed mathematics rests on nonmathematical criteria. Insofar as all the sciences involve some aspect of quantity, mixed mathematics is coextensive with natural science, and its branches belong properly to the corresponding disciplines of the latter.

Barrow's rejection of mixed mathematics rested on the principle that

> there exists in fact no other quantity different from that which
> is called magnitude, or continuous quantity, and, further, it
> alone is rightly to be counted the object of mathematics, which
> investigates and demonstrates first of all its general properties,
> then its nearest forms and the affections congruent to them.
> (*LM* 39–40)

Initially adduced to differentiate mathematics from the other sciences, the principle "contained and circumscribed all of mathematics within the bounds of geometry." It thereby entailed an even more radical departure from tradition, which Barrow took in Lecture III. Maintaining that arithmetic had no independent standing in mathematics, but was simply an adjunct of geometry, Barrow set out a number of "probable and suitable grounds for removing arithmetic from the roll of mathematics and lowering the most noble science in the rank it has occupied for so many centuries" (*LM* 46). His aim was not to deny a science of numbers but rather "to restore it to its lawful place, as being removed out of its proper seat, and ingraf and unite it again into its *native geometry,* the stock from which it has been plucked" (ibid., *UML* 29).

Barrow might have claimed the role of *restaurator,* but his rationale appealed to utility rather than to authority. The republic of mathematics, he insisted, would benefit from recognizing that number is simply a means of expressing continuous quantity, and hence that arithmetic and geometry "demonstrate practically step-for-step properties common to one and the same subject." For then the axioms and theorems of geometry would also hold for arithmetic and conversely, thus eliminating the need for separate treatments, as in the doctrine of proportions. One could use geometry to establish theorems about numbers that would otherwise be difficult or impossible to prove within the confines of arithmetic – for example, those concerning the infinite sums involved in methods of quadrature. Conversely, geometrical proofs would be subject to refutation by arithmetical computation.

Laying the groundwork for his own concept of number, Barrow rejected the abstract numbers that were traditionally the object of arithmetic. He saw no need to posit the existence of numbers independent of things counted: $2 + 2$ cannot be 4 unless the addends are two each of the same things and those things can be combined. The combinatorial properties of numbers are rooted in those of the objects being combined, not in the numbers themselves:

> How then (you will say) comes the Determination of Numbers to certain Uses? Are they not used to denote a definite Ratio and be added to and subtracted from one another? I answer, that these things do only now and then agree with Numbers, according to the Condition of the things to which they are attributed. If the things which the Numbers are brought to denote be homogeneous, of the same Name, and when compared have a mutual Proportion to one another, and consequently the one can be increased by the Addition or diminished by the Subtraction of the other; then they impart the same Attributes, Proportions, and Increments or Decrements to the Numbers by which they are denominated. As *ex. gr.* because a Line of two Feet obtains a like *Nature* and *Denomination* with a Line of three Feet, therefore the Number 2 by which the one is signified denotes a certain sesquialter Proportion to the Number 3 by which the other is denominated; the same indeed which the Lines themselves have that they denominate: and consequently the former Number 2 added to the other Number 3 makes the Number 5 (*i.e.* five Feet) for the Sum, and that subtracted from this leaves Unity for the Excess or Difference: And by almost the same Method, whatsoever Attributes among Arithmeticians are proved to agree with Numbers, they agree not with Numbers taken abstractly and of themselves, but concretely according to the Condition of the Things they are attributed to. (*UML* 36–7; *LM* 52–3)

Citing the authority of Plato's *Philebus,* Barrow distinguished between "mathematical" and "transcendental or metaphysical" numbers. The latter, which are used to count collections of items of generally like sort, such as mountains, angels, chimeras, or even numbers themselves, are in a sense premathematical. They belong to common discourse, which follows no firm rules on which to base a mathematical science.

Mathematical numbers, by contrast, presuppose a uniform basis of measure. They enumerate collections of units precisely equal to one another or they measure magnitudes with reference to a common unit. But numbers are not abstract quantities in themselves. Their existence depends on the units underlying them, because that is where the criteria of equality and the means of measure reside. A change of unit means a change of number. Numbers should be understood merely as symbols:

> I say that a mathematical number has no existence proper to itself, and really distinct from the magnitude it denominates,

but is only a kind of note or sign of magnitude considered after a certain manner; *viz.* as we conceive it either as altogether incomplex, or as compounded of certain homogeneous parts, every one of which is taken simply, and denominated an unit; or lastly as intimating the ratio it has to other magnitudes, in like manner composed by a certain method. For in order to expound and declare our conception of a magnitude, we design it by the name or character of a certain number, which consequently is nothing else but the note or symbol of such magnitude so taken. This is the general nature, import, and notion of a mathematical number. (*LM* 56; *UML* 41)[14]

Numbers, then, symbolize magnitudes conceived of as units, collections of units, or ratios of such collections.

Barrow's definition of number was aimed at widening the traditional scope of arithmetic. He was prepared to extend the meaning of "ratio" and "composed by a certain method" to encompass fractions and surds together with integers within the purview of arithmetic:

Integers are the names of symbols of magnitudes, showing their composition from certain equal parts, every single one of which is named unity.... Fracted numbers are the symbols of any magnitude equally composed by a certain method, exhibiting its proportion to another magnitude, which is composed of the same equal parts.... Lastly, radical or surd numbers are notes showing a magnitude to be any way in a mean proportion between some assumed homogeneous magnitude equally composed according to the exigence of the number applied, whether integer or fracted, and its part supplying the place of unity; or which is the same, between any magnitude taken simply and undivided and consequently holding the place of unity, and another so multiplied as the adjoined number requires. (*UML* 43–4; *LM* 58)

In this view the domain of arithmetic was subsumed under that of geometry. Arithmetic, too, became a science of magnitude, expressing itself in symbols rather than in figures and proceeding by computation rather than by construction.[15] The precise extent of arithmetic's realm, and the nature of the magnitudes reserved to geometry, were left unclear. Barrow never raised the question.

The concept of number as symbolic magnitude was hardly original.[16] Indeed, it had become a commonplace of the algebraic literature of the period. But here it served his own peculiar ends. He would not follow those who used symbolic magnitude to restructure the whole of mathematics by subordinating arithmetic and geometry to analysis. Viète, Descartes, and those who followed their lead proposed algebra as a general science of quantity expressed symbolically. As Viète put it, algebra as the analytic art "exercises its logic not in numbers... but by means of a formal logistic [*logistice sub specie*] to be newly introduced, [which is] much more fruitful and powerful than numerical [logistic] for comparing magnitudes with one another."[17] The emphasis lay on the comparisons rather than on the magnitudes. Focusing on the structure and transformation of equations, this new view of algebra shifted attention from the quantities themselves to the combinatory relations among them and to the relations among those relations.[18] Arithmetic and geometry applied the "rules of the art" to their proper objects, numbers and magnitudes, as particular instantiations of those relations. Thus, Descartes opened his *Géométrie* by showing how one could add, subtract, multiply, divide, raise to powers, and take the roots of line lengths, in each case maintaining homogeneity between terms and factors, on the one hand, and products, on the other.[19] The purpose was not to reduce geometry to arithmetic, but rather to show how algebra could be applied to geometry. For that reason, he added, "I shall not fear to introduce these arithmetical terms into geometry in order to make myself more intelligible."[20]

By contrast, algebra had no place in Barrow's scheme of mathematics, which centered on magnitude as its proper object rather than on relations among magnitudes. His notion of number as symbolic magnitude subordinated arithmetic to geometry and ultimately erased any distinction between them. It thus formed his answer to "that great man we have sometimes mentioned, who to show that arithmetic is much more extensive than geometry, entirely transfers the whole of algebra from geometry to arithmetic, determining algebraical equations to ascend higher than geometrical ones, and denying that geometry affords as many dimensions as arithmetic exhibits degrees." He was referring to John Wallis, who in his *Univeral Mathematics* had contended that arithmetic is the more general of the two branches of mathematics, owing to the wholly abstract nature of number. Theoretically viewed, Wallis maintained, "geometry is the science of magnitude insofar as it is measurable, and arithmetic the science of number insofar as it is countable." Viewed practically and expressed in terms made famous by Peter Ramus, the former is the science of measuring well; the latter, the science of counting well. Each has full standing as a science, with its proper subject, principles, and attributes. But they are not coextensive or equally general, as their application to algebra shows:

> I prefer to explain [algebraic powers] by different arithmetical degrees, rather than by geometrical dimensions, for these reasons: 1° Because universal algebra is really arithmetical, not geometrical, and therefore is to be explained by arithmetical rather than geometrical principles. For although many geometrical things may be found or elucidated by algebraic principles,

it does not follow from that that algebra is geometrical, or indeed based on geometrical principles (as those who proceed in this manner seem to imagine). Rather, it arises from the intimate affinity of arithmetic and geometry; or, rather, because geometry is as it were subordinate to arithmetic and therefore applies the universal statements of arithmetic in particular fashion to its own matters. For, if someone affirms that a three-foot line added to a two-foot line makes a line five feet long, because the numbers two and three added together make five, that calculation is not geometrical but clearly arithmetical, even though it serves geometrical measure. That assertion of the equality of the number five to the numbers two and three together is a general assertion, applicable to other things no less than to geometrical objects. For two angels and three angels also make five angels. But the same argument holds for all operations, whether arithmetical or specially algebraic, which proceed from more general principles rather than being restricted to geometrical measures.[21]

Having eliminated abstract numbers by rooting their combinatorial properties in the magnitudes they denoted, Barrow had no need of them to support a subject he did not recognize as mathematical. Number as magnitude encompassed all algebraic powers and arithmetical degrees. Subordinated to geometry, arithmetic could accommodate the demands of the new algebraic methods.

What sounds at first like what others, such as Wallis and later Newton, called "universal arithmetic" turns out to point in quite different directions. They used that name not to denote an extended concept of arithmetic, which they preserved as the science of number conceived as a collection of units, but as a synonym for algebra, taken as the general science of quantity. In Barrow's view, to banish irrational numbers from arithmetic and assign them, as some do, to algebra is to "demutilate arithmetic of its noblest and most profitable member, for while we are measuring magnitudes and comparing them together, we oftener in our computations light upon these surds than upon rationals, which they call numbers."

Another, final look at Barrow's definition of number shows how he was poised between traditional paths and new directions. Focusing on the objects of mathematics, he insisted that numbers are signs denoting magnitudes. Yet the meaning of those signs lies in the relations they describe. By his account, except for the assumed unit, numbers identify magnitudes by reference to one another. The number $\frac{3}{5}$ names a magnitude that stands in the ratio of 3:5 to a unit magnitude. What is of interest about numbers resides in their meanings, not their denotations. That is, to say that $\frac{3}{15} + \frac{7}{3} = \frac{9}{15} + \frac{35}{15} = \frac{44}{15}$ is to talk not about the magnitudes being added, but about how the names of the addends are manipulated to produce that of the sum. Yet while disavowing the ancient distinction between the discrete "how many" and the continuous "how much," Barrow declined to follow the modern reorientation of mathematics from objects to relations. Unwilling to recognize algebra as mathematics and denying that ratios could be considered quantities and hence proper objects of mathematics, he could quote Descartes to the effect that ratio and proportion lie at the very heart of mathematics, but he would not accept the implications Descartes drew from that conclusion.[22]

More than a year passed before Barrow got to the topic of ratio and proportion. In the intervening lectures he explored a range of issues that also brought him to the line marking the gradual separation of traditional and contemporary mathematics. It will help to examine some of those issues before returning with Barrow to measure, ratio, and proportion.

Divisibility, congruence, and equality

Lectures IX–XII show how Barrow got caught up willy-nilly in what he claimed were distracting philosophical issues. Their subject is magnitude, and, constrained by the format of his lectures to approach it from the outside, he began in Lecture IX with a listing of its essential attributes: termination, extension, composition, and divisibility. They constituted a standard listing, but in some cases Barrow offered a new perspective. The boundedness of magnitude could be known not only by abstraction of dimension, but also by means of physical phenomena that occur on the boundaries, such as reflection and refraction, or moments.

Divisibility posed special difficulties. The infinite divisibility of magnitude is essential to the notion of quantity, and hence to mathematics itself. However difficult that notion may be or however clever the arguments adduced to the contrary by the atomists, it rests on the agreement of philosophers both ancient and modern, as well as on the established ratios of mathematical quantities and the proved existence of irrationals and asymptotes. More recently, convergent infinite series and Cavalieri's method of indivisibles, so inaptly named, have confirmed the infinite divisibility of magnitude, while adding to its difficulties. For example, avoiding the paradoxes lurking in Cavalieri's method requires the notion that the infinitesimal parts of infinite ensembles may themselves differ in size; "for what prevents a small thing from having as many smaller parts as another larger one has greater?" (*LM* 145). Moreover, divisibility played back on composition, as convergent infinite series showed.

Equality might seem another notion liable to be obscured by explanation. Yet as Euclid's definition of equality of ratios showed, common sense may not suffice to determine whether two things are equal, particularly when direct comparison is not possible. Some of the new methods on Barrow's mind sharpened the point. Comparing infinite aggregates and curvilinear figures required more than counting things or laying them side by side. It called for a sophisticated notion of congruence.[23]

Barrow started on familiar ground. "Congruence is usually described as the occupation, possession, or filling of the same place or space. It may be conceived as happening in three ways: by application, by succession, by mental penetration." Although congruence in all three modes is an abstract operation, the first two modes have direct physical counterparts in the measuring stick and the measuring bowl. The first applies primarily to lines and surfaces, which coincide in the dimensions not being compared. The third employs the resources of the imagination to effect the physically impossible merging of two bodies in the same place at the same time. So far, this was the standard theme of superposition, with slight variations.

Then Barrow moved to a more recent theme, as he distinguished among degrees of congruence. The first and simplest is the simultaneous, rigid congruence of whole to whole, as in superimposing lines and planes on one another. The second is the successive, but rigid congruence of part to part in order, as in laying out the sides of a polygon on a straight line. In the third, "all the indivisibles of both magnitudes successively occupy the same place and neither varies the position of its parts," as in rolling a curve along a line. (This form of congruence, Barrow notes in passing, establishes the possibility of the quadrature of the circle.) Fourth, the parts of a single magnitude change position while retaining their order and size, as in unwrapping a circle into a line or wrapping a line into a circle. Congruence of this sort underlies the rectification of curves and the determination of the areas of curved surfaces.

To this point Barrow remained within the confines of traditional mathematics and could point to classical sources for examples. But more recent results pointed to a new form of congruence:

> The fifth mode is when the congruence is carried out so that the position of some parts is varied and their order changed. This mode of congruence, which is the most imperfect of all and the most difficult to understand, applies to homogeneous figures which are wholly dissimilar to one another. For example, a triangle is congruent to a circle, or a cone to a sphere, no otherwise than by transposing the parts of each, applying a part of one to a part of the other, and a part of the remainder of the one to a part of the remainder of the other, and so on continually until the thing is exhausted and all the parts of the one end up applied to parts of the other. Those wishing a clearer exposition of the matter should consult the aforementioned places in Cavalieri.

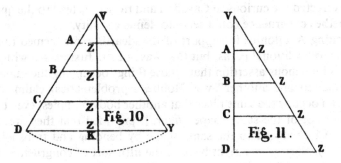

Here Barrow again reached the juncture that frustrated his intentions. This last form of congruence raised all sorts of questions pertinent to the issues he had discussed so far. In what ways is the congruence "imperfect"? What are "all the parts" of a magnitude? Given the infinite possibilities of division, how does one choose which parts to compare? What does it mean to "apply" a part of one magnitude to a part of another, especially if, as parts of dissimilar magnitudes, they are themselves dissimilar? Or does division to a certain level make all parts similar? At what point is "the thing exhausted"?

A digression in Lecture II of the *Lectiones geometricae* suggests that these questions were on Barrow's mind. Andreas Tacquet had attacked the method of indivisibles by arguing that it failed to produce the known surface areas of the right cone and sphere. If, for example, triangle *VDY* is cut by indivisible sections $AZ, BZ, CZ, \ldots$ and hence $A, B, C, \ldots$ denote the indivisible elements of the generator *VD* of the cone, then, on the one hand, the sections together constitute the area of the triangle *VDK* while, on the other, the conical surface should result from summing the circumferences generated by rotating each element about axis *VK*. But the latter is not the case. In rebuttal Barrow pointed out that the two areas require different constructions. To get the conical surface, one must erect each section $AZ, BZ, CZ, \ldots$ perpendicular to the generator *VD*. The area of the new triangle will be to the surface of the cone as the radius of a circle to its circumference (*LG*, 183)[24] (Figures 10 and 11).[25]

It is a subtle point: Despite the infinite divisibility of magnitude, "indivisibles" differ in size. Yet the point is essential to Cavalieri's method and to related methods of transformation of curves. Although Barrow could insist on the principle, he could not pursue its implications without presenting the methods, because understanding derived in part from practice. Congruence of the fifth sort worked. Provided that one knew how to set it up, it produced the areas and arc lengths of curves. Mathematicians could see that, but Barrow could not explain it to nonmathematicians.

So he directed the curious to Cavalieri and turned instead to the question of whether congruence could serve to define equality.

In citing Apollonius in support of the idea, Barrow seemed to be returning to traditional paths, but that was not so. Juxtaposed with Apollonius's innocuous assertion that "those things occupying the same place are equal to one another" was Hobbes's problematical claim that "a body can occupy the same place that another body occupies, even though they are not of the same shape [*figura*], provided that they are understood to be reduced to the same shape by bending and transposition" (*LM* 175). Barrow was right back at the fifth sort of congruence, which he now termed "possible congruence," and made his own definition of equality. As he then shifted to the question of whether equality needed to be defined at all, the notion of congruence by transformation lurked nearby.

He rejected Aristotle's assertion that equality requires no definition because it is a self-evident notion. Barrow insisted that equality becomes evident to the senses only after the objects have been compared and that, given the many possible ways of comparing things, some criterion of equality is required. Continuing debate over the nature of the "horn angle"[26] made that clear. Things are immediately evident and need no definition only if understanding them involves no more than pointing at them and saying their name. Equality is not one of those things. It is a judgment made by comparing things, and hence its definition must include a criterion. Since all questions of equality are settled by congruence, it is the most fitting criterion on which to base a definition.

But in Barrow's notion of "possible congruence" the adjective is essential; for to explain the meaning of "squaring the circle," said Barrow, one must be able to say how a circle can be equal to a square. And what better way to do that than to take "the said quadrilateral figure as if it were waxen (that is, consisted of a soft and wholly flexible material) and somehow transform it into a circular figure by bending, transposing, or compressing some of its parts"? In the absence of mathematical examples, a rough physical model would have to suffice to suggest the range of mathematical operations subsumed under the heading of "congruence." Perhaps because of the audience, Barrow was not ready for an explanation of the mathematical meaning of "bending, transposing, and compressing." Hence, although he referred to Archimedes' mapping of a circle into a triangle (indeed, in two ways), he made no mention of the transmutation of areas as later presented in the *Lectiones geometricae,* where the corresponding segments of two areas are congruent only in the limiting case, where dissimilarity of straight and curved disappears. Apollonius and his colleagues would have been uncomfortable with both the operations and the model.

Measure, ratio, proportion

Lecture XV on measure particularly reveals the new ideas Barrow was trying to fit into the traditional mold. Euclid offered two modes of measurement: an older Pythagorean approach via aliquot parts and the more recent Eudoxean method of ratios. Barrow opened a wider prospect, evident in his statement of the primary of two senses of "measure":

> To measure means to make known or determine the quantity of some magnitude with respect to another homogeneous magnitude in some way better known to us or in some way determinate, namely by declaring, exhibiting, representing in numbers or in some other comprehensible way the ratio of the one to the other; that is, by signifying what part one is of the other, or how much of a multiple, or in what manner it is unequal, by how much it exceeds or to what extent it falls short, or in any other similar way.

"For example," he continued,

> given any length hitherto unknown to us, if in some way (whether by instrumental operation or by mental reasoning based on legitimate hypotheses or previously demonstrated conclusions), if (I say) in some manner agreeable to reason we should find what is its relation in quantity [*relatio in quantitate*] to some length set out and well understood by us (say, a foot), how much it contains this [length] or is contained in it, by how much when taken once or several times it exceeds it or fall short of it; whether it is to this [length] as some number to another, or as some straight line, which I can exhibit, so another which I can also construct; or as the root of some equation which is subject to analytic exegesis and can in some way be resolved by the art.

"Then," he concluded, "we are said to measure the longitude" (*LM* 230–1, emphasis added).

The phrase "relation in quantity," conjoined with the last example of what constituted such a relation, suggested that measurement might extend beyond ratios as traditionally understood. So too did his remark somewhat later in the lecture that, although it was an essential property of magnitude that it was measurable by any other, it is often difficult to carry out the measurement and "sometimes it cannot be accurately comprehended by us in any way" (*LM* 232). Before exploring the range of relations, therefore, he considered the properties of measure. Homogeneous with the thing measured, a measure has a determinate, unique, certain, and invariant quantity. That quantity should be known independently, whether by direct intuition (as is the circumference of a circle), or by comparison with such an immediately known measure, or by a number expressing such a relationship, or by its general nature, even if the individual measure is unknown (as in the case of chords of circles).

The methods of measuring essentially derived from these properties. One could compare unknown with known quantities, "searching out their ratios by reason alone," or one could use instruments, or a combination of the two, as in measuring the circumference of the earth by gnomon and the proportionality of arcs to radii, or by some form of construction. So far, Barrow remained in the range of traditional measures. The last method took him beyond.

> The fifth way is that by which an unknown quantity is made known (*declaretur*) by some equation which expresses in whatever manner its relation to other known quantities and therefore presents it to the mind to be thus comprehended. The artful (*artificiosa*) resolution of that equation according to certain rules suited to the proposition and prescribed in the analytic doctrine wholly completes the measurement and renders the sought quantity, whether geometric or arithmetical, fully known.

As examples Barrow pointed to the equation expressing the tangent of the double arc in terms of the tangent of the arc and to the equation of the chord of the triple arc.

But just as Barrow seemed to be stepping onto new mathematical terrain, he reversed direction, turning briefly from the primary, extended sense of measure to a second, namely, exact measure by aliquot parts, and to traditional questions of commensurability and incommensurability, before drawing Lecture XV and the current term to a close. He thereby left open the question of how far he was prepared to go in recognizing magnitudes defined implicitly by equations for which explicit solutions could not be found. On the one hand, his embracing of surds as full-fledged numbers suggests that he was ready to extend the traditional domain. On the other, accepting relations that could not be expressed in direct terms would have committed him to a class of quantitative relations definable and accessible only through the theory of equations and the techniques of algebraic analysis.[27] That commitment in turn would have meant the recognition of algebra de facto as an independent, irreducible branch of mathematics, which Barrow denied.

If Barrow even saw the question, he offered no answer. Rather, with Lecture XVI, opening his third year (1666) of introductory lectures, he stepped back onto classical ground to treat proportion, "the very soul of mathematics, on which depends just about anything marvelous or abstruse that is demonstrated anywhere in mathematics." In light of the marvelous and abstruse developments he described to his students at the end of Lecture XIII, both the importance he attached to the doctrine of proportions and the content of the ensuing lectures are a bit puzzling. Although many of the newest techniques were couched in the language of proportions, the more algebraically oriented among them pressed the limits of the concepts underlying that language and, as Lecture XV shows, were pushing mathematicians toward a broadening notion of equation or of relation in general. Yet for the most part, Barrow set out a traditional commentary on the definitions of Book V of the *Elements,* distinguished only by his rejection of the notion of ratios as quantities and by his unwavering defense of the Euclidean theory against the criticism of other commentators.[28] One has to listen carefully for hints of the challenges posed by the new infinitesimal methods.

They are there, indeed right at the outset. Since proportion involves comparison of ratios,[29] which in turn involves comparison of quantities, and since comparison is possible only among things of the same kind, Barrow began with a review of the notion of homogeneity, already discussed in the previous lecture. Combinatory operations set the criterion: Quantities are homogeneous if they can be combined in a single whole or subtracted from one another to form a new remainder.

By that standard, quantities can be inhomogeneous and hence incompatible for several reasons. They may lack a common measure, as in the case of magnitude, weight, velocity, and time:

> The quantity of magnitude is continuous and simultaneous,
> absolute, tangible to the senses and conspicuous; [the quantity]
> of time is flowing, successive, imaginable only to the mind,
> consequent on motion and connoting it; the quantity of
> velocity depends on conjoint ratios of time and space; weight,
> force, and resistance involve certain actions and are known
> from certain effects. They disagree with one another so as to
> shrink from being combined. (*LM* 254)

Quantities may also be incomparable because they are dimensionally different; one cannot compare a line with a point, an instant with a time, or a tendency with a motion. Finally, the relation between two quantities may be "indefinite or incomprehensible," as for example between finite and infinite things. Here Barrow had to take account of recent work. The "cleverness of modern geometers," primarily Torricelli's, has confounded Aristotle's dictum that there is no ratio between the finite and the infinite by determining the finite areas under curves extending to infinity along one dimension, provided that there is a compensating diminution toward zero along the other (*LM* 255).

Absent from the discussion so far is any mention of Cavalieri's method of indivisibles, and how a ratio may be said to obtain between two infinite, or indefinite, aggregates or between their indefinitely small components. Barrow was coming to it, but he turned next to Definitions 3 and 4 of *Elements* V, in which Euclid captured the significance of homogeneity to the concept of ratio: (3) Ratio is a certain relation (*schesis*) of two homogeneous magnitudes with respect to size, and (4) magnitudes are said to have a ratio to one another if their multiples are capable of exceeding one another. The latter definition, Barrow insisted in opposition to Clavius, determined not what homogeneous quantities have a ratio but rather what quantities are homogeneous. For example, horn angles, or angles of contact, if they are angles (or even quantities) at all, are not homogeneous with rectilineal angles, precisely because they fail the test of Euclid's fourth definition. By the same criterion, however, all finite lines are homogeneous, whether they are curved or straight, even if the exact ratio between them is unknown; a circle's diameter taken four times exceeds its circumference. Here again, recent geometers have determined ratios between straight and curved lines previously thought impossible.

Barrow continued at some length on this theme. Something was evidently on his mind. After he turned to Viète, "the greatest teacher of the analytic art," and to his "law of homogeneity" for confirmation of the importance of homogeneity to comparison, it came out:

> It is true that among those who in solutions of problems or demonstrations of theorems use that excellent method of indivisibles there occur expressions like "all these lines are equal to such and such a rectangle" or "the sum of these parallel planes constitute such and such a solid." But they explain their thinking and say that they understand by "lines" nothing but parallelograms of a rather small and (pardon the expression) inconsiderable height, and by "planes" similarly prisms or cylinders of a hardly computable height. Or at least by the sum of lines or planes they do not denote some finite and determinate sum, but an infinite or indefinite one equal in count to the points of some straight line.

As debatable as the notion of infinite sums may be, the method of indivisibles, properly stated, agrees at least with the general rule that heterogeneous magnitudes cannot be compared. To hammer the point home, Barrow concluded Lecture XVI with a detailed refutation of Hobbes's claim that all quantities are homogeneous because one can compare their abstracted measures. "For because in motions of equal speed the distances traversed are as the times, 'By permutation,' he says, 'as a time is to a distance, so a time is to a distance.'" The error, Barrow insisted, lay in conflating the representation of measure with the measure itself. That is, though one can speak of lines measuring time in the representative sense that the ratios of the lines correspond to those of the times, one cannot measure a time by a line in the strict sense of determining what multiple the one is of the other. Ratios hold only among quantities that measure one another.

Lectures XVII–XIX hewed closely to classical lines as Barrow reviewed the terminology, definitions, and classification of ratios. If he argued with other authors, it concerned interpretation of the ancient sources themselves. Lecture XX brought him back to the present and into direct confrontation with contemporary developments. Many writers, among whom he named Gregory of St. Vincent, Hobbes, Borelli, Meibom, and Mersenne, claimed that ratios constitute a genus or species of quantity and hence are in fact as well as by way of speaking subject to combinatorial operations and relations. It was a difficult subject to treat, Barrow noted, but misunderstanding of the nature of ratios was a source of error and obscurity, which he meant to correct.

Barrow took a strictly classical position: Ratio is not a quantity in itself, nor can one attribute quantity to it except metonymically. How could one think otherwise? he asked. A ratio is a relation and therefore does not belong to the category of quantity. Quantities are concrete things, whereas ratios are abstract: "How can what is abstractly a ratio be concretely a thing related?" (*LM* 318). To make quantities of ratios is to confuse concrete and abstract names. To predicate relations of ratios, to speak of one ratio as greater than another, is to relate relations, as if one greater could be greater than another greater. Moreover, comparing ratios as quantities leads to ratios of ratios, among which one can then establish ratios, and so on in infinite regression.[30] If ratios of quantities constituted a distinct genus of quantity, then ratios of ratios would form yet another distinct genus, and so on in an unimaginable multiplication of entities. Finally, all quantities are known directly and of themselves, and no ratio is known in that way; ratios are known only through comparison of quantities.

Ratios need not be quantities in their own right to account for their apparently quantitative properties. The quantities of which they consist suffice. For example, one cannot compare ratios except by reducing them to a common consequent. Hence, the seeming comparison of ratios is in fact a comparison of their antecedents. Moreover, although ratios seem not to belong to the genus of their homogeneous terms, one cannot compare, say, a ratio of weights to a ratio of times directly, but only after expressing each of them as a ratio of lines, and that in turn means a comparison among quantities reducible to a common measure.

Here the difficulty of the subject began to take hold of Barrow. Refusing to strike out on the new path of a mathematics of relations, he began

to move in circles, contradicting his own careful arguments of previous
lectures. As he had insisted in rebutting Hobbes, lines and weights are
as heterogeneous as weights and times. If, then, ratios of weights cannot
be compared directly with ratios of times, how can one determine the
equality of ratios of lines to ratios of times? This contradiction reflects
back in turn on the previous argument, which misrepresents Eudoxus's
theory of ratio as presented by Euclid. To determine that the ratio $A:B$
is greater than the ratio $C:D$, it suffices to find a pair of integers m, n such
that $mA > nB$ and $mC \le nD$; the relation of B to D is of no import, nor
does the criterion presuppose that the two quantities are homogeneous.

The contradictions continued. Reaching back to a medieval notion that
a strict Euclidean should have abhorred, Barrow maintained that when,
as a manner of speaking, ratios are compared, the reference is actually
to their *denominations,* that is, to the fractions that name them, as $\frac{7}{3}$
names the ratio 7:3. Yet by Barrow's own account of numbers, fractions
are themselves only signs of ratios; that is, the number $\frac{7}{3}$ is a magnitude
that stands in the ratio of 7:3 to a magnitude taken as unit. How then
can such numbers be quantities? The argument here could only perplex
students who had listened carefully to the rebuttal in Lecture XVII of
Wallis's claim that ratios belonged to the genus of number. There Barrow
had said:

> I know of no genus of ratios or of numbers comprehended
> under quantity, for in my opinion ratios are not quantities, nor
> are they capable of quantity (as I shall show later in its proper
> place), but [are] mere relations based on quantity; and numbers
> I take as only names and symbols of quantitative things, as I
> have already set forth and explained several times before. (*LM*
> 281)

In the case of $\frac{7}{3}$, the name named a ratio.

As elsewhere in the *Mathematical Lectures,* Barrow here seems to have
been arguing a point other than the one at hand. His treatment of number
and of measure pointed him toward recent new directions of mathematical thinking, directions that led to new domains of quantitative relationships. To insist now on a strict, classical doctrine of relations was to move
in the opposite direction. Again, something else was on his mind. It may
have had something to do with another subject he had not raised directly,
namely, algebra.

The status of algebra

"But someone may wonder," Barrow added toward the end of
Lecture II,

when I have taken pains to review generally all (or at least
the main) parts of mathematics, why I have been silent about
algebra, as they call it, or the analytic art (*facultate*). I respond
that I have not done so without purpose. It is rather because
analysis (understood as rooted in something distinct from the
propositions and rules of arithmetic and geometry) seems to
belong no more to mathematics than to physics, or ethics, or
any other science. For it is merely a part or species of logic, or
a certain way of using reason in the solution of questions, and
in the invention and proof of conclusions, which is commonly
carried out in all other sciences. Hence it is not a part or species
of mathematics, but rather a subministering instrument; nor is
synthesis, which is a way of demonstrating theorems [which is]
opposite and inverse to analysis. (*LM* 45)

Perhaps nothing so aligns Barrow with the ancient traditions of mathematics than his refusal to recognize algebra as an independent, indeed overarching, mathematical discipline. Insofar as it represented a symbolic calculus, it was mere technique; insofar as it was an analytic tool, it belonged to logic, not mathematics.

His intransigence seems strangely at odds with other elements of his thinking. With Descartes and his successors, Barrow embraced a symbolic concept of number; indeed, he went beyond Descartes in discarding any vestige of number as a collection of abstract units and thereby ensuring that the domains of arithmetic and geometry were coextensive. Barrow's treatment of measure extended the notion of relation to encompass algebraic equations, even when the equation could not be solved. That extension in turn widened the range of symbolic representations available for expressing both arithmetical and geometrical magnitudes.

Although Barrow did not say so, it may have been precisely the growing importance of symbolic representation that underlay his rejection of algebra as a proper branch of mathematics; for symbols entail abstraction, and Barrow was wary of the pitfalls of abstraction. He insisted that the combinatorial properties of numbers were rooted in the magnitudes they symbolized, that is, $2 + 3$ equals 5 only if the magnitude (a line, a time, etc.) denoted by 2 is homogeneous with that denoted by 3. Numbers in themselves are only names and hence have no inherent properties. Hobbes's error in speaking of a ratio of distance to time arose from abstracting ratios from the objects being compared and in then manipulating those ratios in ways not consonant with the objects denoted by the terms. As Barrow so urgently insisted, ratios are not quantities in themselves and hence possess no combinatorial properties of their own. In a similar fashion, algebra as the manipulation of symbolized quantities had no rules of its own, but rather derived them from the combinatorial properties of the quantities represented by the symbols. But those properties were the proper subject of geometry and arithmetic. Algebra had none of its own and hence no proper subject.

In the end, Barrow could not accept the elevation of algebra to a universal mathematics concerned with the structure of combinatory relations defined without reference to the objects being combined. It carried abstraction beyond the limits he could tolerate. Although this stance left him free to use algebra as a heuristic and expository technique, it closed off certain lines of thought that, pursued by others, led to the fundamentally new concepts of mathematics in the late seventeenth century, in particular those associated with the calculus. As the *Lectiones geometricae* show, Barrow's rejection of algebra from mathematics meant that his investigations at the infinitesimal frontier would not go beyond technique.

The *Geometrical Lectures:* the new path of the moderns

Concerned with "the general characteristics [*symptomata*] of curved lines," the *Geometrical Lectures,* purportedly delivered in 1668 and 1669, contain precisely the problems, methods, and results that Barrow listed at the end of Lecture XIII of the *Mathematical Lectures* and hence would seem to give his remarks there a programmatic air, making them a herald of lectures still to come. Yet Barrow's lament suggests caution in taking the title *Geometrical Lectures* literally. When combined with the brief preface to the 1670 edition, his earlier remarks sound less like program and more like admonition and raise, or reinforce, the question of whether the work records lectures at all.

As published, the *Geometrical Lectures* consists of three parts of distinct origins. Lectures VI–XII form the heart of the work. In his preface, Barrow said he had not intended them to stand independently, but rather had meant them as "companions or fillers" for the series of Optical Lectures delivered over the same period (*LG* 157).[31] When his publisher decided to bring them out separately, Barrow acceded to a request for additional material by adding the first five lectures, which he had originally written for a neophyte audience and which he encouraged the expert readers of the book to ignore. They may have been meant for the *Mathematical Lectures* but then dropped from them as Barrow realized the difficulty of including technical material.[32] Finally, an unnamed friend persuaded him to add a final lecture (XIII), which has nothing to do with the others.[33]

It is clear, then, that Barrow never delivered the *Geometrical Lectures* in the sequence of the published form in which they were deposited in the university archives. But beyond that, it is difficult to imagine any of them having been delivered – if at all – in anything like their published form. For one thing, there is no evidence that Barrow's audience had changed. Yet the *Geometrical Lectures* require considerable training and skill (coupled with a certain capacity for tedium) to follow the advanced material they contain. Although the first five lectures address common general themes of midcentury mathematical discourse and follow well-marked paths from one proposition to the next, the remaining lectures often seem as diffuse as they are detailed and difficult.[34] Quite specific propositions and unusual formulations appear without apparent motivation or provenance, as if Barrow expected his audience to recognize their significance and context. Although Lectures VI–X have a recognizable structure when seen as a whole, it is not at all evident as their contents are encountered *seriatim*, and Barrow's few comments, most of them metaphorical, do little to indicate its emerging shape.

Moreover, the geometrical style of the lectures makes the 215 often quite intricate diagrams essential to Barrow's arguments, as indeed he had warned. He had referred in the *Mathematical Lectures* to the "treacherous caverns of my ears." But the most loyal ear alone – or even eye restricted to the verbal text – could not follow his reasoning. Nor, for that matter, is the text, even with the diagrams, adequate to the full meaning of certain propositions, which in published form appear to lack the commentary that accompanied their presentation to an audience. That especially seems the case in Lecture VII, where Barrow established a set of relations common to both arithmetical and geometrical progressions and where his symbolism sustains an ambiguity the spoken language could not accommodate. Appreciating the full wealth of that lecture and the ones immediately following requires more discourse than the text records. In the absence of independent evidence of a community of mathematicians sophisticated enough to follow Barrow's material without supporting explanation, it seems likely that what passed over the lectern differs significantly from what has come down in the published text. How much passed over the lectern at all remains a mystery.

Magnitude and motion

Lectures I–V continue in the discursive style of the *Mathematical Lectures*. Coupled with Barrow's remarks in the preface, their style and content suggest that they may have originally formed part of that earlier series, either in fact or in intention. Lectures III–V in particular go in the direction Barrow said he was headed in Lecture XII of the *Mathematical Lectures*, embedding the properties of curves in the larger context of the generation of magnitude. There are, he began, many ways of generating magnitude: by local motion; by the intersection of magnitudes; by distances determined in quantity and position by reference to given positions; by the multiplication, division, addition, and subtraction of magnitudes with or by one another or by aggregation of magnitudes arranged in a certain order; or by mechanical[35] construction. Of these, generation by motion is primary; all the others depend on it.

Barrow adorned the discussion with a classical veneer. Aristotle's *Physics* served as *locus classicus* for the assertion that *ignorato motu necessario naturam ignorari,* "ignorance of motion entails ignorance of nature." But the citation scarcely concealed the modern framework of Barrow's treatment. He did not say what nature had to do with mathematics, nor would Aristotle have recognized the role Barrow now assigned to motion in mathematics. He left to physicists such questions as the nature, definition, or causes of motion. On the "common-sense" premise that all magnitude is movable, mathematicians concern themselves with the mode of motion and the quantity of motive force. The mode could be rectilinear or circular; force accounts both for the speeds of magnitudes relative to one another and for the uniformity or difformity of motions in themselves.

As in Aristotle's *Physics,* so in this first lecture the topic of motion led to that of time. But again Barrow's considerations followed a modern line. In terms to be echoed by Newton, Barrow posited the existence of an absolute time, which is a quantity but which is inaccessible to measurement except by means of some sensible, constant, uniform motion such as the sun and the stars (if we could be sure of their constancy).[36] The mathematician measures absolute time in his imagination, which needs some perspicuous analog, such as distance traversed, to represent time and its parts. Moreover, since velocity can change, too, the imagination requires a representation of the conjoint dependency of velocity on distance and time, or of distance on velocity and time, or of time on distance and velocity. One such model results from moving one line parallel to itself over another, the length of the first varying as the velocity, that of the second varying as the time, and the distance traversed varying as the area.

The analogy between distance and time rested on a comparison between the continuous flow of a point and a moment, or instant:

> Except that time has wholly similar parts, it is reasonable to consider it as a quantum endowed with but one dimension; for we can imagine it as constituted by the simple addition of supervening moments or by quasi continuous flux of one moment, and thus we are wont to attribute only length to it; nor do we measure its quantity otherwise than by the length of the line traversed. Just as, I say, a line is thought to be the trace of a moved point, having by virtue of the point [the property] that it is to some extent indivisible but by virtue of the motion that it can be divided in one way, according to length, so time is conceived of as the trace of a continuously sliding instant,

> having some indivisibility by virtue of the instant, [and] by
> virtue of the successive flow [the property] that it can be
> divided to that extent. (*LG* 165)

Barrow knew the dangers of both models and of any correlation between
them. That infinitely divisible time consists of indivisible instants does
not imply that it is composed of those instants; similarly, a line consists
of points that do not compose it. Strictly speaking, an instant is not a
part of time, nor a point a part of a line, since the whole must equal
the sum of the parts. No number of successive instants adds up to an
interval of time, just as no number of juxtaposed points constitutes a
line. Hence, to correlate the two quantities by means of their indivisible
elements runs the risk of long-recognized logical fallacies given new prominence by Cavalieri's work.[37]

Yet that is how Barrow meant to proceed. In the two-dimensional model
of time, length, and motion, the points of time and of length correspond
to one another. To avoid the fallacies, he shifted the meaning of "instant"
and "point":

> To every instant of time, or to every indefinitely small particle
> of time; (I say "instant" or "indefinite particle" because, just
> as it matters nothing at all whether we understand a line to be
> composed of innumerable points or of indefinitely small linelets
> [*lineolae*], so it is all the same whether we suppose time to be
> composed of instants or of innumerable minute timelets [*tem-
> pusculis*]; at least for the sake of brevity we shall not fear to
> use instants in place of times however small, or points in place
> of the linelets representing timelets); to each moment of time,
> I say, there corresponds some degree of velocity which the
> moving body should be thought to have then; to that degree
> corresponds some length of space traversed (for here we con-
> sider the moving body as a point and thus the space only as
> length); ... (*LG* 167–8)

The aggregate of those momentary distances over the full time of motion constitutes the total distance traversed and is represented by the area
swept out by the line of velocities moving over the line of time. This way
of representing motion, Barrow hastened to add, does not imply that
motion takes place in an instant, nor that all motions are equal. In his
model, in order for different distances to be traversed at different instantaneous velocities, the "points" of the corresponding lines had to be understood as unequal.

Aristotle would have countenanced none of this. But then, Aristotle
was not a seventeenth-century mathematician seeking to rationalize a
"common-sense" model of motion that had already produced significant

results and promised further insight into the nature of curves. Perhaps to remind his audience of that, as much as to clarify his meaning, Barrow concluded the first lecture with examples of uniform and uniformly changing motion represented by surfaces, drawing Galileo's laws as corollaries from them.

With his basic framework established, Barrow turned to classes of motions and the curves they generate. He distinguished three categories: the simple motions of translation and rotation at constant speeds, compound motions, and the concourse of motions. Lecture II dealt with the simple motions. Translation included all motions in which no point remains fixed. Although in principle the moving object can follow any path, "an art can give no account of what is wholly irregular," and hence mathematicians restrict translation to definable paths: a point along a straight line, a line parallel to itself along the length of some other line, a line along its own length and parallel to itself along another line.

Generation of areas and solids by translation and rotation was a well-known subject, and Lecture II essentially reviewed what anyone trained in mathematics already knew from several sources. In Lecture III Barrow moved closer to new territory. Compound and concurrent motions involve variation in the speeds of motion. Indeed, the two classes differ only in the arrangement of the moving components. In compound motion, a curve is traced by a point traversing a line while the latter itself traverses another line; in concurrent motion, the curve results from the intersection of two moving lines. Although in principle any curve of one class can be expressed as a curve of the other, some curves are more conveniently defined in one way rather than the other. For example, the Archimedean spiral is readily visualized as the motion of a point along a rotating radius, while the classical quadratrix is most easily described by the intersection of two lines, one in translation, the other in rotation.

Although in theory any compound or concurrent motion can be resolved into myriad combinations of basic motions,[38] in practice rectilinear components are simplest and most direct. "Indeed," Barrow noted, "there is no species of magnitude (no line, no surface, no body) of which the generation cannot be conceived of as effected by straight motions." The framework Barrow preferred had its dual origins in Apollonius's *Conics* and in Galileo's analysis of falling bodies. Taking a horizontal line AY as main axis, Barrow imagined another line AZ to move parallel to itself along AY, while point M moves along AZ, starting at A. Different curves result from the different relative motions of the line and the point (Figure 17).[39]

For example, if the motions are such that $AB:AC = BM:C\mu$, then $AM\mu$ will be a straight line. But if they are such as to maintain a constant ratio of $(D - BM) \cdot BM$ to AB^2, where D is a given line segment then $AM\mu$ will be either a circle (if the ratio is 1 and angle ZAY is right) or an

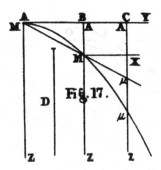

Fig. 17.

ellipse; a constant ratio of $(D + BM) \cdot BM$ to AB^2 yields a hyperbola, a constant $D \cdot BM : AB^2$, a parabola. Such ratios are most easily and clearly represented by imagining the line AZ to move uniformly along AY, while point M moves uniformly, accelerates, or decelerates according to the given relationship.

The kinematic properties of curves

Lecture IV treats the properties of the class of curves generated by the composition of a uniform horizontal motion and a constantly increasing downward motion. Included within this class are the conic sections.[40] While aiming for several striking theorems concerning tangents and areas, Barrow also touched in passing on a Cartesian theme of growing importance to advocates of the new methods of analysis, namely, the generality of those methods. The properties of concern to Barrow follow directly from the curves' common mode of generation. Apollonius's methods had required him to establish the properties singly for each of the conic sections.[41]

Any member of the class will be everywhere curved; that is, no three points will be collinear. Hence, the uniform downward motion that generates the chord of an arc is slower than the velocity reached at the lower terminus of the arc. Moreover, the chord will lie wholly within the arc between the end points and wholly outside the curve if extended in either direction. Hence, the curve will be uniformly concave with respect to AZ, and a given straight line will cut the curve in at most two points; lines parallel to AZ or AY will cut it in one point only. All chords, if produced, will cut AZ and all lines parallel to it; hence, so too will all tangents with the exception of the tangent at an extreme value. Conversely, any line that intersects AZ below A will also cut the curve. Here, in particular, Barrow pointed to the labor Apollonius had expended on proving the result for conic sections.

The composition of a uniform horizontal motion and an accelerated vertical motion lends special significance to the subtangent, which becomes

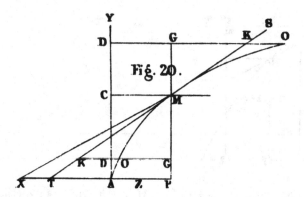

Fig. 20.

a measure of the vertical velocity at the point of tangency (Figure 20). As Proposition 11 states:

> Suppose the straight line *TMS* is tangent to the curve at point *M*; let this tangent meet line *AZ* at *T*, and draw the straight line *PMG* through *M* parallel to *AY*. I say that the [vertical] velocity that the point descending and describing the curve by its motion has at the point of contact *M* is equal to the velocity with which the straight line *TP* is described in the same time in which the straight line *AZ* is carried over *AC* or *PM* (or, in other words, I say that the velocity of the descending point at *M* is to the velocity at which the line *AZ* is moving as *TP* to *PM*. (*LG* 195)

Barrow's proof couples kinematic intuition with an implicit principle of continuity. Take any point *K* on the tangent to the left of the point of tangency, and draw *KOG* parallel to *TZ*. During the time of uniform motion of *TZ* over *DC* (= *GM*), the point on the tangent traverses *KG*, and the point on the curve traverses *OG*. Since *KG* > *OG*, the vertical velocity of the point on the tangent is greater than that of the point describing the curve at *O*; for even while accelerating between *O* and *M* the latter does not travel as far as does *K* uniformly. By contrast, for any point on the right, the point *K* would travel less far vertically in moving toward *M* than would point *O*, even though *O* would be decelerating. Since the vertical velocity of *O* is less than the vertical velocity of *K* for all points above *M* and greater for all points below, the velocities must be equal at *M*.

It is worth noting that Barrow's demonstration makes no appeal to infinitesimals or limits. In the style of Fermat's original derivation of his methods of maxima and minima and of tangents, Barrow takes an arbitrary interval *GM* and derives a relationship that holds for all values within that interval.[42] He takes as given a principle akin to Rolle's theorem

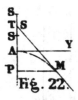

Fig. 22

that, in going from positive to negative values, a quantity must pass through 0. The validity of such a principle – and how it can be demonstrated – became a staple theme of analysis in the eighteenth and nineteenth centuries, but Barrow apparently thought it self-evident.

Although Barrow points out that the converse is true and hence that all curves in the class have only one tangent at any one point, his real interest lies in the "profit not to be spurned" that follows from this relationship between motion and tangents, namely, several propositions concerning inverse tangents. The question dated back to Descartes and Fermat. Both had speculated how one might invert the method of tangents so that, given the tangent, one could determine the curve. But despite Descartes's ingenious construction of de Beaune's problem,[43] neither man could provide a general technique. In the mid-1660s the problem remained essentially unexplored. Barrow had some things to say about it in Lecture XII, and he shared his approach and results with several contemporaries, but the breakthroughs did not come until the work of Newton, Leibniz, and the Bernoullis toward the end of the century.

Barrow certainly had no general solution to offer here in Lecture IV. Rather, he remarked, "one of this sort of inverse propositions is often far more quickly found and more easily demonstrated than another." Examples would have to serve, after one general observation, namely, that the ratio of the vertical velocities of any two points on a curve is compounded of the ratios of the subtangents and the applicates[44] to those points.

"In passing," Barrow now offered an easy and general solution to a problem "over which Galileo made so much fuss and on which he expended so much effort, and which Torricelli thought him so excellent and ingenious to have found." Not having Galileo at hand, Barrow took the formulation of the problem from Torricelli (Figure 22):

> (IV:15) Given a parabola with vertex A, find the point above
> [A] such that if a heavy body were to fall from that point to A
> and were there redirected horizontally with the impetus then
> acquired it would describe the proposed parabola (note that the
> downward motion describing the parabola is taken to begin not
> at the point above but at point A). (*LG* 198)

Barrow's solution goes beyond the problem as stated, since it depends neither on the particular rate of acceleration nor on the trajectory to be

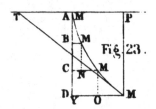

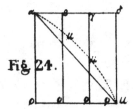

followed after redirection. One finds a point P on AS such that the ordinate PM is equal to the subtangent PT. If one then sets $AS = AP$, S will be the desired point.

The proof took on a Galilean tone. If $SA = AP$, then the impetus acquired at A in falling from S is equal to that acquired at P in falling from A (since once the body's motion is redirected horizontally at A, it starts falling from rest again). But that is the downward component of the body's motion at point M, and by Proposition 11 the body moving uniformly at that speed would traverse TP during the time it covers PM horizontally. But $TP = PM$; hence, the speeds at which they are traversed are equal, and the speed at which PM is traversed is the speed acquired in falling from S to A.

In applying that general solution to Galileo's particular case of a parabola generated by a uniform horizontal motion and a uniformly accelerated vertical motion, Barrow turned for the first time to analysis. If R is the latus rectum of the parabola, then $R \cdot AP = PM^2 =$ (by the above construction) TP^2.[45] But by a known property of the parabola, $TP^2 = 4AP^2$. Therefore, $4AP^2 = R \cdot AP$, or $R = 4AP$, whence $\frac{1}{4}R = AP = SA$. In this case, T and S coincide. But if, for example, the body's acceleration were such that the vertical distance increased according to the cube of the time, then the trajectory would take the form $R^2 \cdot AP = PM^3 = TP^3$. For this sort of parabola, $TP = 3AP$,[46] whence $27AP^3 = R^2 \cdot AP$, or $R^2 = 27AP^2 = 27SA^2$, or $SA = \sqrt{R^2/27}$.

Barrow had no doubt that many similar kinematic propositions of Galileo's followed from this particular application of the tangent property of curves generated by motion, but he was content for the moment with these examples. He turned now to another class of relations between the tangent and the curve (Figures 23 and 24).

> (IV:16) If to a straight line a plane area be applied, of which the individual parts cut off by parallel applicates to this line are proportional to lines (parallel to AZ)[47] applied to a line AY similarly[48] divided, the ratio of this area to the parallelogram of equal height erected on the same base will represent the ratio of AP and TP, cut off by vertex P and by the tangent [respectively].

That is, if $BM:CM = $ area $\alpha\beta\mu :$ area $\alpha\gamma\mu$ for all similarly corresponding B, C, β, γ, then area $\alpha\delta\mu :$ parallelogram $\alpha\delta\mu\phi = AP:TP$.

Barrow again appealed to kinematic intuition for his demonstration. Suppose $\alpha\delta$ represents the common time during which AD is traversed by a uniform motion and DM by a uniformly accelerated motion. Then $\delta\mu$ represents the maximum velocity attained at the lowest point M of the curve. By Proposition 11, that is the velocity at which TP is traversed uniformly over the same time. Hence, TP is represented by the area $\alpha\delta\mu\phi$, which is the distance traversed at that final speed during the time $\alpha\delta$. Since area $\alpha\delta\mu$ represents $DM = AP$, the proposition follows immediately.

Again, also, Barrow needs examples to clarify the meaning of the proposition. Suppose curve $AMMM$ is a quadratic parabola, that is, $BM:CM = AB^2:AC^2 = \alpha\beta^2:\alpha\gamma^2$. Then $\alpha\delta\mu$ is a triangle, since $\alpha\beta\mu: \alpha\gamma\mu = \alpha\beta^2:\alpha\gamma^2$. In that case $\alpha\delta\mu = \alpha\delta\mu\phi/2$, whence $AP = TP/2$, the well-known property of the subtangent of the parabola. If $AMMM$ is a cubic parabola, $BM:CM = AB^3:AC^3$, then $\alpha\mu\mu\mu$ will be a quadratic parabola. That is, by a property demonstrated by Pappus and others and derivable from Archimedes' *Measure of the Parabola*, $\alpha\beta\mu:\alpha\gamma\mu = \alpha\beta^3:\alpha\gamma^3$. In that case, $\alpha\delta\mu = \alpha\delta\mu\phi/3$, whence $AP = TP/3$.

It is tempting to see in Propositions 11 and 16, especially the latter, a harbinger of the calculus and of its fundamental theorem. Indeed, Newton may have been referring to the former when he said that Barrow's lectures had "put him in mind" of the analysis of curves by motion. For that reason it is important to note how specific and limited Proposition 11 is. To the retrospective eye, it may appear to contain the elements of the differential triangle defined kinematically, but in fact it does not. In Lecture IV Barrow posited no general framework for generating curves by concurrent motions. His method of generation followed Galileo's kinematic model by making the horizontal component a measure of the uniform velocity at which it is traversed in a given time. The proposition yielded a corresponding measure of vertical velocity through a distance traversed uniformly in the same time. Barrow's approach fitted a long-standing tradition of reducing nonuniform to uniform motion. He showed no hint of allowing both of the component motions to vary over time and to relate the resulting distances through an equation. That is why he nowhere drew the seemingly obvious corollary that the tangent is the resultant of the parallelogram of velocities of the ordinate and abscissa at the point of tangency. Nor *a fortiori* did he translate the proposition into a method of tangents by reducing it to rules by which, given such an equation, one calculates the ratio of the component velocities at any point. When he did present a method in Lecture X, it bore no relation in its conceptual structure to the kinematical generation of curves.

Proposition 16, for its part, does indeed draw a relation between the area under a curve and the tangent to that curve's quadratix, and to that extent it does forge a link that had escaped many others who investigated both tangents and quadrature as separate classes of problems. But it is an ad hoc relation, tied immediately to the geometrical configuration rather than to an algebraic framework such as that underlying the method of tangents presented at the end of Lecture X. In particular, that method does not yield the segment *PT* if it is applied to the curve of ordinates *BM, CM, ...*, but rather the subtangent on axis *AY*. But that segment is not labeled in Barrow's diagram, and hence even to rewrite the ratio *AP:TP* as subtangent:*AD* would be to add to the proposition the very structure that it is supposed to reflect.

Lecture V continues Barrow's analysis of the properties of curves generated by composition of uniform and accelerated motion, focusing now on tangents, normals, and extreme values to offer a general treatment of the properties worked out specifically for the conic sections in Book V of Apollonius's *Conics*. The first few propositions seem to lay the groundwork for an approach by means of infinitesimals. Because any curve so generated is uniformly concave, the tangents drawn to any two points intersect those points. Let *ME* and *NH* be segments of tangents drawn to *M* and *N* and intersecting the extended ordinates *QN* and *PM*, respectively (Figure 26). Then *ME* > arc *MN* > *NH*. "This proposition is most useful," Barrow observed, "for carrying out demonstrations of tangents; for it follows therefrom that, if arc *MN* is posited as indefinitely small, either particle *ME* or *NH* of the tangent may be safely substituted in its stead" (*LG* 203).

Yet Barrow let the remark stand undeveloped, digressing instead in Proposition 6 to offer "a general method for determining the tangents of all cycloids and of curves described in a similar manner." Evidently borrowed from an unnamed source,[49] the method fits neither in its mode of generating curves nor in its analysis of the tangent with the particular approach Barrow is pursuing at the moment. Yet it carries the treatment of tangents in the preceding lecture in a direction one would have expected to find there but does not.

Let *AY* move uniformly and parallel to itself along a uniformly concave or convex curve *APX* (Figure 27). Simultaneously let a point move uniformly from *A* along the line, generating curve *AMZ*. To find the tangent to *AMZ* at any point *M*, draw *MP* ∥ *AY*, construct *PE* tangent to *APX* at *P*, draw *MH* ∥ *PE*, and, taking any point *R* on *MH*, draw *RS* ∥ *PM* such that (arc *AP*):*PM* = *MR*:*RS*. *MS* will be the desired tangent.

The demonstration uses the inequalities of Proposition 5 to show that any point *K* on *MS* above or below point *M* will lie outside the curve. In structure, the proof is quite similar to that of Proposition 11 in the preceding lecture, and nothing in it rests on infinitesimal assumptions. For that reason, perhaps, Barrow assigned no special importance to the triangle *MRS*, which instantiates the ratio of the motions generating the

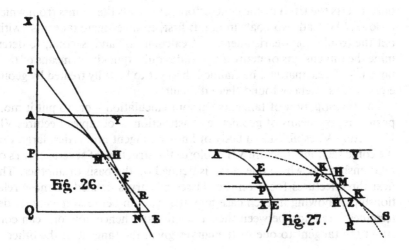

curve. Even as a finite configuration, that triangle had clearly not yet assumed any significance for him. Nonetheless, it is there, as it had not been in Proposition 11 of Lecture IV.

As abruptly as Barrow entered on this digression, he returned from it to continue his general treatment of normals and tangents to the class of curves under discussion. As Apollonius began to determine the content of Barrow's propositions, so too the style moves closer to classical geometry, even to the exclusion of material one might think pertinent. After demonstrating that parallels to a curve's tangent on the side toward the axis intersect the curve in two points and noting that Apollonius had tried laboriously to show this (*Conics* I:27 and 28), Barrow pointed out that "for the rest, fully determining the points of intersection requires knowing the specific mode or ratio of the descending and tranverse motions; then Analysis immediately yields [the points]" (*LG* 205). The question of how that analysis might yield tangents and normals does not arise.

Geometrical analysis: tangents without calculation

In the *Mathematical Lectures* Barrow had promised a survey of the most recent developments in mathematics. Lectures I–V redeemed that promise, couched in the intuitively accessible style of the generation of curves by composition of motion. He "raised his hand from the board," confessing that he had exhibited only a "specimen of a certain general doctrine comprehending the properties of curves, which surely, more fully, and more perfectly seems to offer no small benefit to geometry (which is mostly occupied with the affections and properties of curves)" (*LG* 208). In Lecture VI, he shifted to a fuller and more sophisticated mode of analysis that seemed to him particularly scientific in that "it not only asserts the truth of the conclusion but reveals the founts from which it flows." He had two goals in mind: first, to investigate tangents "without the trouble or wearisomeness of calculation" and, second, to determine the dimensions of many magnitudes quite quickly by means of their tangents. These matters, he claimed, had yet to be fully treated by geometers, who somehow found them difficult.

The investigation of tangents without calculation – or, to put it more positively, by means of geometric constructions – occupies Lectures VI–X. Lecture VI establishes a body of known tangent properties, largely of the conic sections. Lecture VII explores the structure of several classes of relationships by which one curve is defined on the basis of another. The next three lectures focus on the classes of curves defined by those relationships, showing in each case how the relation between the curves determines a relation between their tangents and hence how one can construct the tangent to one of the curves given the tangent to the other.

Analog curves. Although intricately detailed, Lecture VI remained within the generally familiar terrain of Apollonius's *Conics*. Lecture VII, by contrast, moved into still largely unfamiliar realms reconnoitered by Pappus in Book VII ("The Treasury of Analysis") but not yet common elements of current mathematical discourse. There Barrow established a series of lemmas concerning two sorts of relationships between curves, introducing in particular the notion of an *analog* curve based on an extended sense of proportionality. Proposition 6 is deceptive in its apparent simplicity, especially given the ambiguity with which Barrow seems to have been willing to invest his notation:

> Let there be four series of continuous proportionals with the same number of terms (as you see laid out), and let the first antecedents and the last consequents be proportional to one another ($A : \alpha :: M : \mu$ and $F : \phi :: S : \sigma$). Any four [terms] of the same order will also be proportional to one another (say, for example, $D : \delta :: P : \pi$).
>
> $$A \quad B \quad C \quad D \quad E \quad F$$
> $$\alpha \quad \beta \quad \gamma \quad \delta \quad \epsilon \quad \phi$$
> $$M \quad N \quad O \quad P \quad R \quad S$$
> $$\mu \quad \nu \quad o \quad \pi \quad \rho \quad \sigma$$

The proof is straightforward, or so it seems at first. From $A : B = B : C = \cdots = E : F$ and $\mu : \nu = \nu : o = \cdots = \rho : \sigma$, it follows that $A\mu : B\nu = B\nu : Co = \cdots = E\rho : F\sigma$, whence $A\mu, B\nu, \ldots, F\sigma$ forms a continuous proportion. Similarly, $\alpha M, \beta N, \ldots, \phi S$ are continuously proportional. Since $A\mu = \alpha M$ and $F\sigma = \phi S$, the two product series are clearly the same, term for term; hence, the corresponding terms $D\pi$ and δP are equal, or $D : \delta = P : \pi$.

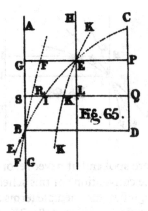

Fig. 65.

However, Barrow notes almost in passing that "this conclusion pertains equally to either proportionality (both arithmetical and geometrical)." Interpreting his notation so that ":" corresponds to subtraction, and juxtaposition to addition, yields the following argument: $A - B = B - C = \cdots = E - F$, and $\mu - \nu = \nu - o = \cdots = \rho - \sigma$, whence $A + \mu - B + \nu = B + \nu - C + o = \cdots = E + \rho - F + \sigma$. Similarly, $\alpha + M, \beta + N, \cdots, \phi + S$ forms a continuous arithmetical progression. But we are given that $A + \mu = \alpha + M$ and $F + \sigma = \phi + S$. Hence, the two progressions are the same, and $D + \pi = \delta + P$. Therefore, $D - \delta = P - \pi$.

Proposition VII:7 indicates what this result, curious in both its content and its form, has to do with curves (Figure 65):

> Let lines *AB* and *CD* be parallel, and let line *BD*, given in position, cut them. And let the lines *EBE*, *FBF* be so related that, for any line *PG* drawn parallel to *DB*, *PF* is the mean proportional in the same order between *PG* and *PE*. Then, through any designated point *E* of the line *EBE* draw *HE* parallel to *AB* and *CD*, and let *KEK* be another curve such that, for any *QL* also drawn parallel to *DB*, *QK*[50] is the mean always in the same order between *QL* and *QI* (the same [order], I say, as that in which *PF* was the mean between *PG, PE*). I say that lines *FBF* and *KEK* are *analog*, that is ordered (as are *QR, QK*) always to have the same ratio to one another, viz. the same as that which *PF* has to *PE*.

Barrow's schematic method of comparing continuous progressions makes short work of the proof. Indeed, he merely has to set down the matrix:

$$QS^*QR^*QI$$
$$QL^*QK^*QI \qquad \text{are} \;\#, \text{ whence } QR:QK = PF:FE$$
$$PG^*PF^*PE$$
$$PE^*PE^*PE$$

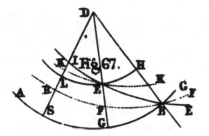

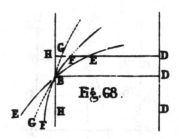

Here is where one listens for words that were spoken but never recorded on paper. Barrow's text explains none of the conventions of this schema, including its ambiguity. The "*" evidently signifies intermediate terms and thus translates the notion of "in the same order"; that is, *QR*, *QK*, *PF*, and *PE* are corresponding terms in four continuous progressions of equal length. Since $QS = PG$ and $QL = PE$, clearly $QS:QL = PG:PE$; the final terms are identically proportional. Hence, the corresponding middle terms are proportional. But they are proportional in two senses, both geometrically, as the notation suggests directly, and arithmetically; that is, if $QS - AR = QR - QI$, and so on, then $QR - QK = PF - FE$.[51]

After noting that the straight lines *AB*, *HE*, *CD* can be replaced by any parallel curves, Barrow proceeded in the next several propositions to explore the invariance of the analog relation. In Proposition 8, *AB*, *CD*, and *HL* converge on a common point, and in 9 *AGB* becomes a circle about *D*, with which *P* and *Q* coincide. Proposition 10 shows the transitivity of the relation; that is, let *AGBG*, *EBE* be any curves, define *FBF* by the relation $DF = \text{mean}(DG, DE)$,[52] let *HEL* be analog to *AGB* (*DS*:*DL* is fixed), and define *KEK* by means of $DK = \text{mean}(DL, DI)$ (Figure 67). Then *FBF* is analog to *KEK*, or $DR:DK = DF:FE$. All the propositions follow in a straightforward manner from the above schema.

From the homomorphic properties of arrays of arithmetical and geometrical progressions considered separately, Barrow then turned to a comparison of arithmetical and geometrical progressions with one another, noting that, if they start from a common first term and if the second term of the geometric progression is greater than that of the arithmetical, all succeeding terms of the geometrical series will be greater than their counterparts in the arithmetical. That result enabled him then to show in Proposition 17 that, if for a given curve *EBE* and a given line *DDD* curves *FBF* and *GBG* are constructed such that, for any *DH* ∥ *DB*, $DF = \text{geom. mean}(DH, DE)$ and $DG = \text{arith. mean}(DH, DE)$ in the same order, then *GBG* and *FBF* are mutually tangent (Figure 68). Point *B* is the common starting point of the two progressions; at any other point the term *DF* of the arithmetical is less than the corresponding term *DG* of

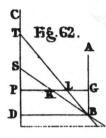

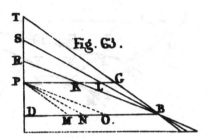

the geometrical. Hence, except at that point *GBG* lies wholly outside *FBF* with respect to *DDD*.

Projective relations. In the propositions just discussed, Barrow discerned the structural similarities between arithmetical and geometrical progressions and fashioned a general proof scheme applicable to a variety of configurations. To take full advantage of the notion of analog curves required another set of propositions, Barrow's treatment of which revealingly lacks that insight. Propositions 3 (Figure 62) and 4 (Figure 63) exemplify both the class of relations and the contrast in treatment, and it is worthwhile to consider them in tandem.

> (VII:3) Let *BA* and *DC* be parallel lines; further, let *BD* and *GP* be parallel lines, and through point *B* draw any two lines *BT* and *BS* cutting *GP* at points *L* and *K*. I say that $DS:DT = KG:LG$.

> (VII:4) Let *BDT* be a triangle, and let any two lines *BS*, *BR* drawn through *B* intersect some line *PG* parallel to base *DB* at points *L, K*. I say that $(KG \cdot TD + KL \cdot RD):KG \cdot TD = RD:SD$.

Barrow's demonstration of Proposition 3 is quite brief and uses only the elements of the initial configuration: $KG:LG = (KG:GB)(GB:LG) = (PK:PS)(PT:PL) = (DB:DS)(DT:DB) = DT:DS$.[53] By contrast, the proof of 4 begins with the construction of auxiliary lines *PM, PN, PO* (parallel, respectively, to *BT, BS, BR*) and then involves a long chain of computations of ratios, in essence piecing together the result. Nothing in the statements of the two propositions or in their demonstrations would suggest that the configurations or the relations among their elements are structurally related. In the one case Barrow saw two parallel lines and two transversals, while in the other he saw a triangle *BDT*, a parallel through it, and two oblique lines running from a vertex to the opposite side. Yet, viewed projectively, the configuration of 4 becomes that of 3 by making *BT* parallel to *DS*, and both propositions follow directly from the invariance of the cross-ratios; in 3, $(DST\infty) = (\infty KLG)$, and in 4,

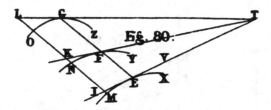

Fig. 80.

$(DRST) = (\infty KLG)$. Propositions 5 and 19 (and in retrospect VI:6 and 7, on which the latter depends) are similarly related yet treated diversely.

As D. T. Whiteside has pointed out, Barrow's ad hoc treatment of these "projective" theorems reveals his ignorance of recent French work in synthetic geometry – most notably that of Desargues and Pascal – based on Apollonius's derivation of the general conic and on Pappus's lemmas concerning the cross-ratio.[54] Euclid reigned supreme in the English classroom, and Barrow was unusual for even having included in his lectures the wider range of problems represented by the above propositions. While providing no deep insight of his own into them, he nonetheless called them to the attention of Newton, who did discern their underlying structure and the possibilities they offered.[55]

Families of tangents. Barrow introduced Lecture VIII with a question that must have been on every auditor's mind, to wit, where is all this leading? "I seem to myself (I think I will seem to you) to have done what the wise jester was ridiculing, namely, to have raised such immense gates before a rather small city. For up to now we have done nothing but move somewhat closer to the matter at hand. Let us to it." The matter at hand was the geometrical analysis of the properties of tangents to three classes of curves. The first class consists of pairs (and occasionally triples) of curves defined in various frames of reference by a fixed relation of the ordinates. The task is to construct the tangent to one of the curves given the tangent of the other(s). Proposition 5 establishes the pattern (Figure 80):

> Let *VEI* be a straight line, and *YFN* and *ZGO* be two curves so related to one another that, if line *EFG* is drawn anywhere parallel to *AB* given in position, the intercepts *EG*, *EF* will always have the same ratio to one another; and let line *TG* be tangent to one of the curves, *ZGO*, at *G* (and meet line *VE* at *T*). *TF* will be tangent to the other [curve] *YFN*.

Here, as in many of the propositions to follow, Barrow employs the "projective" relationships explored in the previous two lectures to establish alternative proofs. On the one hand, draw any line *IL* ∥ *EG*. By the relationship of the curves $IO : IN = EG : EF$, and by similar triangles, $EG : EF =$

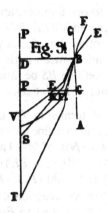

$IL:IK$. But $IO < IL$. Therefore $IN < IK$. Since this inequality holds for any line IL other than EG, curve YFN has only point F in common with line TFK, which therefore is tangent to it. On the other hand, consider the proportionality of corresponding segments delimited on any parallel LM by the curves and their tangents. That is, $IL:IK = IO:IN = (IL - IO):(IK - IN) = OL:NK$. The first four propositions of Lecture VIII provide the lemmas by which to conclude that, if GL and GO are tangent, so too are FN and FK.

The remaining propositions are variations on the basic configuration. Proposition 6 replaces the straight line VEI by a curve XEM while preserving the fixed ratio of coincident ordinates erected on it to YFN and ZGO. In that case, the tangent to YFN at F passes through the meet of the tangents to XEM, ZGO at E, G, respectively. Proposition 7 replaces the parallel ordinates GE, LM by lines emanating from a fixed center D, and Proposition 8 establishes the correlate of Proposition 6 for that configuration. Propositions 9 and 10 fix the product, first of parallel ordinates referred to a given line and then of ordinates referred to a given center. Proposition 11 holds the difference of parallel ordinates constant (i.e., parallel translation), and Proposition 12 treats the difference of convergent ordinates, obtaining as a corollary the tangent to the conchoid of Nicomedes in the case where one of the curves is a straight line.[56] And so on. Where the tangent sought does not result from rectilinear constructions, Barrow obtains it by means of circles or hyperbolas drawn tangent to the curve, hence reducing the problem to the known construction of tangents to those curves. It is here that the bulk of the theorems of Lecture VI come into play.

Barrow then moves in Lecture IX to curves related to one another through the series of arithmetical and geometric means investigated at the beginning of Lecture VII. The first proposition again establishes the basic theme for the variations to follow (Figure 94):

> Let the straight lines *AB*, *VD* be parallel to one another, and let
> *DB* given in position cut them. Also let the lines *EBE*, *FBF*
> pass through *B* and be so related to each other that, for any
> *PG* drawn parallel to *DB*, *PF* is the arithmetic mean in the
> same designated order between *PG*, *PE*. Let line *BS* be tangent
> to *EBE*. It is required to draw the tangent to *FBF* (at *B*).

The construction employs two quantities, N and M, termed in VII:12 the "exponents" of the proportionals *PF* and *PE* and meant to define the notion of "in the same designated order." That is, if *PF* is the Nth of $M - 1$ arithmetical means between *PG* and *PE*, then $PF = PG - N[(PG - PE)/M]$. Since $PE = PG - M[(PG - PE)/M]$, it follows (VII:11) that $(PG - PF)$: $(PG - PE) = N:M$. That ratio is the basis for the construction of the desired tangent. Set $DS:DT = N:M$, and *BT* will be tangent to *FBF* at *B*. For, taking any *PG* parallel to *DB* and intersecting *EBE*, *BS*, *FBF*, and *BT* at *E*, *K*, *F*, and *L*, respectively, then $FG:EG = N:M = DS:DT = LG:KG$. But $KG < EG$, since *BS* is tangent to *EBE*. Hence, $LG < FG$, and *BT* is tangent to *FBF*.

Pursuing the structural similarities between arithmetical and geometrical means leads directly to the first variation in Proposition 2, namely, determining the tangent to the curve defined by the Nth of M geometric means between *PG* and *PE*. Rather than devising a derivation analogous to the one just given, however, Barrow has only to invoke VII:17 to show that the same construction suffices. That is, to the configuration above, add another curve *F'BF'* such that *PF'* is the geometric mean in the same designated order as *PF* is the arithmetic mean. By VII:17 the two mean curves are mutually tangent at *B*. Hence, the tangent to the arithmetical curve is also the tangent to the geometrical.[57]

Combining these results with those of Lecture VIII produces then in Proposition 3 a general construction of the tangent to *FBF*, whether arithmetically or geometrically defined, at any point *F*. Through *F* draw *PG* ‖ *DB*, cutting *EBE* at *E*. Through *E* draw *HEH* analog to *FBF* and use IX:1 to construct the tangent to *HEH*. By VII:7, the line *FY* drawn from the meet of that tangent *PY* and the axis *DY* will be tangent to *FBF*. Since in the case that *EBE* is a straight line, the curves corresponding to the geometric means constitute the family of simple and higher-order parabolas (IX:4) with common vertex at *S*, Barrow now has a general construction of their tangents. "What has been observed of these in passing (deduced from calculation and confirmed by some induction, but I do not know whether ever shown geometrically) flows from an immensely more fertile source, spreading out to innumerable curves of other sorts" (236). One particular property, which follows directly, will shortly prove valuable in linking areas and tangents, to wit, if ordinate *PE* is as the nth power of ordinate *PF*,[58] then the subtangent to *FBF* is n times the subtangent to *EBE* (IX:5).

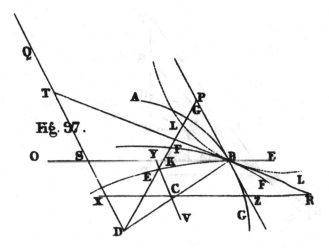

Fig. 97.

Following the pattern of Lecture VIII, Barrow now transformed the basic configuration, converting the straight line *ABG* into the arc of a circle about a point *D*, from which the ordinates defining the curves *EBE, FBF* emanate. Proposition 6 is the correlate of 1 for the arithmetical case of this configuration, showing that on a perpendicular to *DB* the segments *DS* and *DT* cut off by the tangents to the base curve and the mean curve are in the ratio *N*:*M*. Proposition 7 argues the same conclusion as 2 for the geometric mean, 8 shows how to generalize for any point on the mean curve via its analog and thus to determine the tangent to a wide variety of spirals, and 9 links the ratio of the subtangents to the degree of the mean (Figure 97).

Propositions 10–15 center on yet another variation of the basic configuration. Here the ordinates *PEF* are again perpendicular to a rectilinear axis *DT*, but the line *ABC* now meets that axis at *T*, rather than lying parallel to it. The strange form of VII:4 now becomes clear, as Barrow sets the ratio *RD*:*SD* of the subtangents to $[N \cdot TD + (M-N) \cdot RD]:M \cdot TD$.[59] As might be expected from the preceding propositions, the same construction works for geometric means, and from there one can move to finding the tangent to *FBF* at any point (Figure 100). If *EBE* is a straight line and *R* and *T* lie on the same side of *D*, then *FBF* belongs to the family of hyperbolic curves; if *D* lies between *R* and *T*, *FBF* is one of the elliptic curves, and the tangent results from setting the ratio of the subtangents equal to $[N \cdot TD - (M-N) \cdot RD]:M \cdot TD$. In either case, one has the general solution to finding the tangent to a family of curves. That is also the case with the final configuration in Proposition 16, which yields the tangent to the cissoids.

Lecture X, finally, opens with the class of curves of which one of a pair is defined in terms of the arc length of the other, the most famous

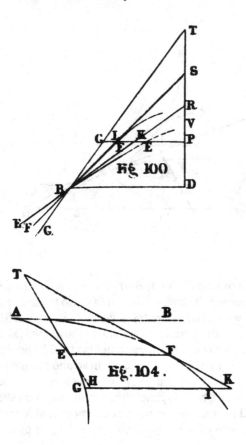

Fig. 100

Fig. 104.

example being the cycloid. Again, Proposition 1 sets up the sequence of transformations (Figure 104):

> Let *AEG* be any curve and *AFI* another curve related to it such that, for any line *EF* drawn parallel to *AB* given in position (which cuts curve *AEG* at *E* and curve *AFI* at *F*), *EF* is always equal to the arc *AE* of the curve *AEG* measured from *A*. Also, let line *ET* be tangent to the curve *AEG* at *E*, let *ET* be equal to arc *AE*, and connect line *TF*. [*TF*] will be tangent to curve *AFI*.

The proof follows by way of VII:22, and VIII:6 allows in Proposition 2 an immediate generalization from *AE = EF* to *AE* : *EF* = const. "We showed this earlier in another way," Barrow noted, "but this demonstration seems somehow simpler, clearer, and more fitting to the method we are introducing" (241). Similar results for the class of epicycloids follow by shifting from parallel to radial ordinates (X:3–4). If *EF* is measured

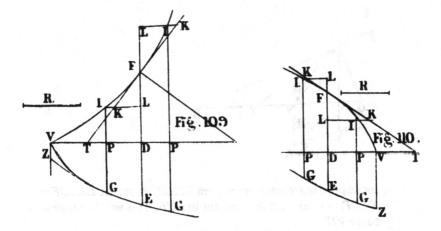

Fig. 109

Fig. 110.

not from the curve *AEG* but from an axis perpendicular to *AB*, then *AFK* is the *rectificatrix* of *AEG* and can be considered either simply or proportionally (X:5–6) and for either parallel or radial ordinates (X:7–8). Finally, the circular spiral and the generalized quadratrix (in the ancient form) also belong to this class of curves (X:9–10).

Tangents and areas: the measure of curves

At this point, Barrow announced the completion of his first task of determining tangents without calculation. He concluded the lecture with four general theorems forming the transition to the second task of investigating the magnitudes of curves by means of their tangents. Since it is his treatment of this subject that has formed the basis of his reputation for contributing to the invention of the calculus, it is worth looking closely at these theorems to see their intimate relation with the finite, geometrical reasoning that precedes them, rather than with the notions of infinitesimals or limits that underlie Lectures XI and XII.

The propositions are grouped in pairs, the first two dealing with curves defined by ordinates drawn perpendicular to a fixed rectilinear axis, the second setting forth correlate results for curves defined by ordinates measured from a fixed center (Figures 109 and 110). Since they proceed in similar fashion, it suffices to examine the first pair. Proposition X:11 reads:

> Let *ZGE* be any line, to the axis *VD* of which are first applied perpendiculars (*VZ, PG, DE*) which continuously increase in some way from the first at *VZ*. Further, let line *VIF* be such that for any straight line *EDF* drawn perpendicular to *VD* (which will cut the curves at points *E, F,* and *VD* itself at *D*) the rectangle from *DF* and some fixed *R* is always equal to the

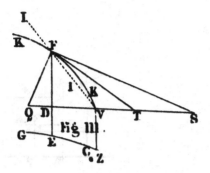

Fig 111.

correspondingly intercepted area *VDEZ*. Also, set $DE:DF = R:DT$, and connect the straight line *TF*. This will be tangent to curve *VIF*.

As in the previous propositions, the proof views the tangent geometrically as a straight line intersecting the curve in only one point and lying wholly outside it elsewhere. Take, then, some ordinate $PI < DF$, and draw $IL \parallel PD$. By the nature of the curve *VIF*, $PI \cdot R =$ area *VPGZ*. By construction $DT:DF = R:DE = KL:FL$. Therefore, $FL \cdot R = DF \cdot R - PI \cdot R =$ area $PDEG = KL \cdot DE$. But by the nature of curve *ZGE*, area $PDEG < PD \cdot DE = IL \cdot DE$. Hence, $KL \cdot DE < IL \cdot DE$, or $KL < IL$. Mutatis mutandis, for any ordinate $PI > DF$, $KL > IL$. As a corollary it follows that the area *VDEZ* is equal to the rectangle $DE \cdot DT$, that is, to the product of the ordinate of the curve *ZGE* and of the subtangent to its quadratrix at the corresponding point.

Combining this basic relationship with the previous results leads then to a theorem (X:12), the converse of which serves as the basis for Lecture XI (Figure 111). Taking curve *ZGE* as before, let curve *VKF* have the property that $DF^2 = 2 \cdot$ area *VDEZ*, and let *FS* be its tangent. On the axis set $DQ = DE$. Then *FQ* will be the normal to *VKF*. The construction follows from pairing *VKF* with the quadratrix of *ZGE* to obtain the configuration of IX:5. That is, let curve *VIF* be such that $DF \cdot R =$ area *VDEZ*, as in X:11, and draw its tangent *FT*. Then $DT \cdot DE =$ area *VDEZ*, or $2DT \cdot DE = 2 \cdot$ area $VDEZ = DF^2$. By IX:5, however, $SD = 2TD$.[60] Hence, $2DT \cdot DE = SD \cdot DE = SD \cdot DQ = DF^2$. But the latter proportion means that angle *QSF* is right.

The method of tangents. Barrow's geometrical treatment of tangents in Lectures VIII–X proceeds by a form of reduction analysis linking the unknown tangent of one curve to the hypothetically known tangent of the other by means of the relation between the curves, thus establishing the tangent properties of families of curves.[61] Although that suffices for theoretical purposes, the analysis provides an effective practical technique

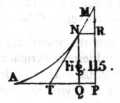

only if one is in fact able to determine the tangent to the base curve of the family. At the close of Lecture X, Barrow offered a glimpse of how to do that, turning from geometry to calculation, and turning as well from finite to infinitesimal modes of analysis (Figure 115):

> Thus we have carried out what we indicated as the first part of our proposition. To complete it by way of a small appendix, we shall add on the method used by us for finding tangents by calculation [ex calculo]. Although I don't know whether it will prove of much use, what with so many widely known and commonplace methods of the sort, yet I do so on the advice of a friend [Newton?], and all the more freely because it seems more encompassing and general than the others I have treated. I proceed in this manner: Let *AP, PM* be straight lines given in position (of which *PM* cuts the proposed curve at *M*), and suppose *MT* to be tangent to the curve at *M* and to cut the line *AP* at *T*. Now, to find out the quantity of this line *PT*, I posit the arc *MN* as indefinitely small. Then I draw lines *NQ* parallel to *MP* and *NR* [parallel] to *AP*. I call *MP* = *m*, *PT* = *t*, *MR* = *a*, *NR* = *e*; the remaining lines determined by the special nature of the curve and useful to the proposition I designate by names. But *MR* and *NR* (and by means of them, *MP* and *PT*) I compare to one another by an *equation* expressed in terms of calculation [ex calculo]. In doing so I observe these rules:
>
> 1. In the computation I reject all terms in which a power of *a* or *e* occurs or in which these are multiplied by one another (because these terms will count for nothing).
> 2. After the equation has been set up, I reject all terms consisting of letters designating known or determinate quantities, or in which *a* or *e* does not occur (because these terms, when brought to one side of the equation, will equal [*adaequabunt*] nothing).
> 3. I substitute *m* (or *MP*) for *a*, and *t* (or *PT*) for *e*. From this finally the quantity of *PT* itself is determined.
>
> And if an indefinitely small part of any curve should enter the calculation, substitute in its place a suitably chosen small part

of the tangent, or any line equivalent [*aequipollens*] to it (by
virtue of the indefinite smallness of the curve). (*LG* 246–7)

There was nothing new in any of this. Behind the procedures stands
the method of tangents originally worked out by Pierre de Fermat around
1630 and circulating in various versions by the 1650s. Barrow's rules strip
that method to its bare bones, thus removing from view the two basic
relationships on which it rests and which themselves rest on the assumption of infinitesimal differences.[62] If from the point of tangency *M* (which
lies on both the curve and the tangent) one lays off an "indefinitely small"
arc *MN*, then point *N* can be viewed counterfactually as a second point
common to both the curve and the tangent.

Considered as lying on the curve, its abscissa AQ ($= AP - e$) and ordinate NQ ($= PM - a$) satisfy the curve's defining conditions as generally
expressed by an equation. When substituted into the equation, $AP - e$
and $PM - a$ will generate an expression with three main parts: first, terms
containing only *AP*, *PM* and constants; second, terms containing either
e or *a*; and, third, terms containing powers or products of *e* and *a*. Because those quantities are indefinitely small, their powers and products
are negligible: "These terms will count for nothing." Hence, the third
part of the expression disappears (Rule 1).

Moreover, the first part consisting of all terms not containing *e* or *a*
will simply express the relationship that places point *M* on the curve.
Hence, those terms add up to 0 and may be eliminated (Rule 2). What
remains is an equation that establishes a value for the ratio of *e* to *a* in
terms of the elements of the curve. If, then, point *N* is considered to be
lying on the tangent, it also establishes a value for that ratio in terms of
the subtangent and the ordinate, namely, $PT:MP = t/m = NR:MR =
e/a$. Hence, one can set the value derived by Rule 2 equal to t/m, thus
determining the subtangent *t*. Substituting *t* for *e* and *m* for *a* (Rule 3)
achieves the same end. Finally, to adapt the method to a curve defined
with reference to another curve, one may substitute for the indefinitely
small segments of the latter the corresponding segments of their subtending tangents.

Barrow offered his auditors no derivation of the method itself and hence
no indication of the theoretical foundations, if any, on which it rested. Instead, he offered a series of five examples of its application. By the standards of the day, none was simple and straightforward, although the first
three ("Gutschoven's curve," $py = x\sqrt{x^2 + y^2}$; a "cubic circle," $x^3 + y^3 =
r^3$; and *la galande*, $x^3 + y^3 = pxy$) were algebraic.[63] The fourth example,
of the quadratrix of the circle, and the fifth, of the curve of trigonometric tangents, brought out the resources of the method for nonalgebraic
curves. The last perhaps shows the method at its most resourceful. The
curve *AMO* is defined by reference to the circular quadrant *DEB* by setting the abscissa $AP = $ arc *BE* and ordinate $PM = BG$, which is the tangent

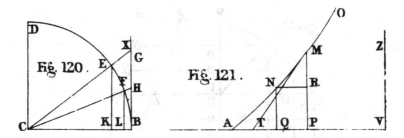

of arc *BE* (Figures 120 and 121). Setting arc *BF* = *AQ* and *BH* = *NQ*, Barrow then relates the interval *e* to the interval *LK* by taking arc *EF* as if it were a straight line perpendicular to *CE* and setting *CE* : *EF* = *EF* : *LK* = *QP* : *LK*.

But the examples also revealed the method's limitations. None of the curves was referred to a preferred axial system and hence to a standard pair of ordinate and abscissa. Applying algebra to geometry rather than employing algebraic geometry, Barrow chose his points of reference as best suited each individual problem. Indeed, as the last example shows, he operated between two frameworks, linking them by the relation *MP* : *CB* = *BG* : *CB* = *EK* : *CK*. Although this may have facilitated his solutions, it also made the method seem less methodical. Deprived of derivation or justification of the rules themselves, the uninitiated could scarcely find enlightenment in their application to special cases, and the initiated could learn nothing new about the structure of the relation of tangents to their curves.

The transmutation of curves. Presented as an afterthought to Lecture X, the method of tangents seems to be only a digression for Barrow, a technical appendix to an essentially different subject treated in a different manner. Yet it leads to a major conceptual shift. Through it, the "indefinite," introduced in principle in Lecture V but left hanging in what was, after all, a separate work, makes its operational entrance into Barrow's geometrical analysis. When in the eleventh lecture he turned back to his main theme, taking up the measurement of curves by means of tangents, the indefinite remained among his analytical tools.

In Lecture XI Barrow takes up what had recently come to be called the "transmutation" of curves. It constituted yet another body of techniques of reduction analysis. Just as the configurations of Lectures VIII–X link families of curves to basic constructions of their tangents, so here a family of related results reduce problems of quadrature to canonical form, transforming the unknown area under a given curve to a known area by means of an auxiliary curve. Neither the techniques nor their name were original. Both are central to Gregory's *Geometriae pars universalis*,

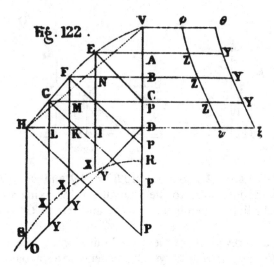

Fig. 122.

inserviens quantitatum curvarum transmutationi et mensurae, and Fermat had covered essentially the same material in a treatise titled *De aequationum localium transmutatione et emendatione ad multimodam curvilineorum inter se vel cum rectilineis comparationem...,* which he appears to have composed around 1658 in response to Wallis's *Arithmetica infinitorum* (1657) but which dated in substance from the 1640s.[64]

In the lecture, Barrow presented four means of transformation: the curve of subnormals, the curve of subtangents, change of axes, and the curve of tangents. The first two and the last consist of element-by-element mappings of the indefinite sections of the given curve to those of the known area, whereby the equality of the limit sums of the wholes follows from the equality of the corresponding elements. By contrast, the technique of change of axes, derived from considering two different means of generating the same solid of revolution and then extended by means of general relations between curves and their subtangents, transforms the limit sums as wholes without reference to the relation of corresponding elements. Here the notion of equality via possible congruence assumed its full shape, stretching to the limit the traditional concepts to which Barrow had sought to link it in the *Mathematical Lectures.*[65]

Transmutation via the curve of subnormals rests on the converse to Proposition X:12, which the notion of indefinite section transforms into another style altogether (Figure 122):

(XI:1) Let *VH* be any curve with axis *VD* and applicate *HD* perpendicular to *VD*, and let line $\phi Z\psi$ be such that, if from an arbitrary point of the curve, say *E*, the line *EP* is drawn perpendicular to the curve, and the line *EAZ* perpendicular to the axis, then line AZ is equal to the intercept AP. Then the area $VD\psi\phi$[66] will be equal to half the square of line DH. (*LG* 251)

Divide VD equally but "indefinitely" by points $A, B, C, \ldots$, and draw $EAZ, FBZ, GCZ, \ldots$ perpendicular to VD and $EIY, FKY, GLY, \ldots$ parallel to VD and intersecting line DO drawn at 45° to HD. Then triangle HLG will be similar to triangle PDH, since "owing to the indefinite section the curvicle[67] GH can be taken as a straight line." Therefore, $HL:LG = PD:DH$, or $HL \cdot DH = LG \cdot PD$, or by definition and construction $HL \cdot HO = DC \cdot D\psi$. Similar analysis of the indefinite triangles GMF, FNE, and so on, leads to the equalities $LK \cdot LY = BC \cdot CZ$, $KI \cdot KY = AB \cdot BZ$, $ID \cdot IY = AV \cdot AZ$. But the sum of the products $HL \cdot HO + LK \cdot LY + \cdots + ID \cdot IY$ "differs minimally" (*minime differt*) from triangle HDO, while $DC \cdot D\psi + BC \cdot CZ + \cdots + AV \cdot AZ$ "likewise differs minimally" from area $VD\psi\phi$. Hence, the triangle is equal to the curved area. But the triangle is equal to $\frac{1}{2}HD^2$.

Dividing the curve's axis into an "indefinite" number of segments, each of which must then be "indefinitely small" serves two purposes. First, on the premise that the corresponding segments of the curve itself may be viewed in effect as straight lines, it leads to the use of the differential triangle as the keystone of an element-by-element transformation of the segments constituting area $VD\psi\phi$ into those making up triangle HDO, although Barrow does not call special attention to the triangle and its role. Second, when applied to the elements of area constructed on the segment, the "indefinite" brings Barrow back around to the point of his commentary in Lecture II on Tacquet's objection to Cavalieri's method of indivisibles. For the indefinite, as opposed to the indivisible, makes clear how the accurate comparison of aggregates of elements rests on establishing the relation of the base segments.

Barrow's concluding remark reinforces this point. "A longer demonstrative discourse could be set forth here," he noted, "but to what end?" That he could put the question rhetorically suggests that he took, and felt his audience would take, the notion of "differs minimally" as unproblematic. In each case, what "differ minimally" are the area under the curve of ordinates and the step figure consisting of rectangles formed by an indefinite number of ordinates and the indefinitely narrow segments of the abscissa between them. As the number of subdivisions increases and hence their size decreases, the difference between the curvilinear area and the step figure decreases, until the latter coincides with the former viewed in Cavalierian terms as "all the ordinates."[68] The "longer demonstrative discourse" surely refers to a classical double *reductio ad absurdum* in the Euclidean or Archimedean mode to show that any assumed finite difference would lead to a contradiction.

Barrow left that classical demonstration for a later appendix. The sums of the two sets of indefinitely narrow rectangles were less on his mind

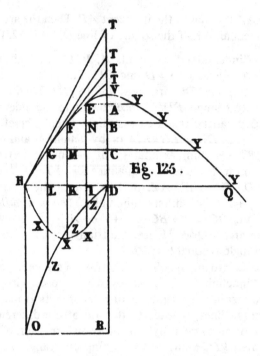

Fig. 125.

than the relationship between the elements of the sets. The equality of
the sums rested not on the process of summation but on the one–one
correspondence between the elements being summed. When seeking to
relate the ordinates of one curve to those of another it is essential to
base the correspondence between them on the relationship between the
axes on which they stand. In Proposition 1, the curve *VEH* serves that
mediating purpose between the ordinates to the curve $\phi Z\psi$ and those to
the line *DYY* by establishing the relation between the equal subdivisions
of *VD* and the unequal subdivisions of *DH*. Propositions 2–9 build on
that configuration, mapping the sum of the products $AZ \cdot AE$, $BZ \cdot BF$,
and so on, into the pyramid that is one-third of the cube on *DH*, that
is, $DH^3/3$, and the sum of the products $AZ \cdot AE^2$, $BZ \cdot BF^2$, into $DH^4/4$.
If $VD\psi\phi$ is any area, then $VAZ\phi \cdot AZ + VBZ\phi \cdot BZ + \cdots = \frac{1}{2}VD\psi\phi^2$, and
$AZ \cdot \sqrt{VAZ\phi} + BZ \cdot \sqrt{VBZ\phi} + \cdots = (\frac{2}{3})\sqrt{VD\psi\phi^3}$. And so on.

Proposition 10, credited to Gregory, shifts from the curve of subnormals to the curve of subtangents (Figure 125). Given any concave curve
VH over an interval *VD*, construct curve *DZO* such that for any point
E on *VH* the segment *IZ* on *EZ* drawn parallel to *VD* is equal to the
subtangent *AT* determined by the tangent drawn to *E*. Then the area
DZOH = area *VEHD*. The demonstration follows the pattern for Proposition 1. Cut *DH* into indefinitely many equal parts by points $I, K, L, ...$,
and draw $EIZ, FKZ, GLZ, ... \parallel VD$. "Now because of this indefinite division the little arc *GH* can be thought of as straight and to that extent

as coincident with the tangent *HT*." Therefore, $LG:LH = TD:DH$, or $LG \cdot DH = TD \cdot LH = CD \cdot DH = HL \cdot HO$. Similarly, $BC \cdot CG = LK \cdot KZ$, $AB \cdot BF = KI \cdot IZ$, and so on. But the sum of the left-hand sides "differs minimally" from area *VDH*, as does that of the right-hand sides from area *DHO*, whence the proposition follows.

It is in drawing the corollaries to Proposition 10 that Barrow invokes the third means of transformation. He has already laid the groundwork in Proposition 8. There he constructs a second curve of subnormals *RXS* referred to ordinate *DH* as axis, that is, for *E* on *VH* and $EX \parallel VD$, $IX = AP$ [$= AZ$] (Figure 122). The two curves are related to one another by way of the solids of revolution generated by revolving them about the common axis *VD*. For, by construction, $HL:LG = DP:DH = D\psi:DH = D\psi^2:(D\psi \cdot DH) = D\psi^2:(HS \cdot DH)$, whence $HL \cdot HS \cdot DH = LG \cdot D\psi^2 = CD \cdot D\psi^2$. Similarly, $LK \cdot LX \cdot DL = CB \cdot CZ^2$, and so on. Each of the left-hand products, multiplied by 2π,[69] constitutes a thin cylindrical shell about axis *VD*, and hence their sum forms the solid of revolution of the whole area *DRSH*. Similarly, each of the right-hand products, multiplied by π, forms a thin disk centered on *VD*, and hence their sum constitutes the solid of revolution of area $VD\psi\phi$. But the sum of the left-hand sides times 2π is twice the sum of the right-hand sides times π.

Proposition 11, which forms the correlate to Proposition 8 for the curve of subtangents, shows that Barrow is concerned here not with the measure of solids, but with the relationship they instantiate among the aggregates of the indefinite elements of the two areas in the limiting, or Cavalierian, case. That is, ignoring the factor of π, Proposition 8 asserts that "all the products $LX \cdot DL$" is twice "all the squares CB^2." The focus becomes clear through the introduction of a deceptively similar transformation that holds for the aggregates but not for the constituent elements.

To show in Proposition 11 that the solid formed by revolving area *DHO* (Figure 125) about axis *VD* is twice that of *VDH* about the same axis, Barrow begins as he did in Proposition 8 with a comparison of corresponding elements via the indefinite triangles connecting them. Here, where *DZO* is the curve of subtangents, $HL:LG = DH:DT = DH:HO = DH^2:(DH \cdot HO)$, whence $HL \cdot DH \cdot HO = DH^2 \cdot LG = DH^2 \cdot CD$, and so on for the other elements. But instead of multiplying the two sides by 2π and summing the elements, he invokes as "commonly known" (*vulgo notatum*) that "the sum $CD \cdot DHq + BC \cdot CGQ + AB \cdot BFq + VA \cdot AEq$ is double the sum $DI \cdot IE + DK \cdot KF + DL \cdot LG$, &c." Both the proposition and the form in which it is cast are alien to Barrow's discourse up to this point. Although the terms of the first sum contains indefinite factors, those of the second do not; nor is it clear how the two sums are related on a term-by-term basis.

The context of solids of revolution suggests the provenance and purpose of the proposition, while it also supplies the missing indefinite factors. Consider the solid formed by *VDH* about axis *VD* as constituted in

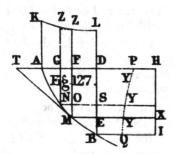

Fig. 127.

two different ways, either as a stack of disks or as a nest of cylinders. The terms (times π) of the first sum correspond to the disks; those (times 2π) of the second, when each multiplied by the corresponding segment of DH (e.g., $DK \cdot KF \cdot KI$), to the cylinders. Hence, the first sum is twice the second. But the relation obtains only between the sums, not among their terms. That is, in general $BC \cdot CG^2 \neq DK \cdot KF \cdot KI$. Thus, the use of the transformation in Proposition 11 breaks the term-by-term comparison with which it starts. The *sum* of the terms $HL \cdot DH \cdot HO + \cdots$ equals the *sum* of the terms $DI \cdot IE \cdot DI + \cdots$, but the corresponding terms themselves are not equal. The proposition itself follows by simply multiplying the two sums by 2π.

Henceforth in Lecture XI Barrow dropped the vehicle of solids of revolution and spoke instead of sums of terms, interweaving global and term-by-term equalities. Moreover, he also expressed the sums without reference to their indefinite elements. Proposition 12 asserts what was expected in 11, namely, that "the sum $DI \cdot IZ + DK \cdot KZ + DL \cdot LZ$ &c. is equal to the sum of the squares of the applicates to VD, i.e. to $AEq + BFq + CGq$ &c.," which is true term-by-term, while 13 sets the sum of the products $DI^2 \cdot IZ + \cdots$ equal to three times the sum of the products $DI^2 \cdot IE$ and thus to three times the sum of AE^3, which is true only globally. Barrow asserts the latter equality as if it too were *vulgo notatum* and goes on to generalize it as $\Sigma(DI^{n-1} \cdot IZ + \cdots) = n \cdot \Sigma(AE^n + \cdots)$. It is not as easy to reconstruct the basis for the generalization, since it cannot be expressed through solids of revolution (see Figure 127). It was not original with Barrow. Fermat had made it the keystone of his treatise on quadrature, though he too left no direct evidence of its derivation.[70] Proposition 19 sets out the final class of transformations; Barrow suggests no particular provenance:

> Furthermore, let AMB be any curve of which AD is the axis and BD a perpendicular to it. Then let KZL be another line such that, if M is taken anywhere on curve AB and through it are drawn line MT tangent to curve AB and line MFZ parallel to DB (which will cut line KL at Z and line AD at F), and if R

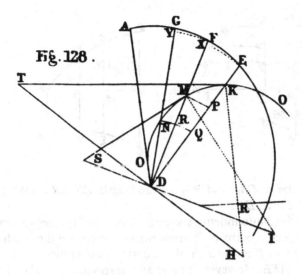

Fig. 128.

is some given line, then $TF:FM = R:FZ$. Area $ADLK$ will be equal to the rectangle [formed] by R and DB.

Rewriting the last ratio in the form $FZ = R \cdot FM/FT$ reveals the nature of the curve KZL. It is the curve of tangents. In relating the area under it to the product of the final ordinate DB and the constant R, Barrow focuses squarely on the small triangle determined by an "indefinitely small particle MN of the curve AB." By reasoning already introduced in earlier propositions, the ratio of the sides of that triangle is the same as that of the subtangent to the ordinate drawn to point M, that is $NO:MO = FT:FM$, which in turn is equal to $R:FZ$. Hence, $NO \cdot FZ = FG \cdot FZ = R \cdot MO = EX \cdot ES$. Then, while "all rectangles $FG \cdot FZ$ differ minimally from area $ADLK$ . . . all rectangles $ES \cdot EX$ compose the rectangle $DHIB$."

Barrow drew no special attention to this last family of transformations but simply extended it in the same ways as he had the earlier ones. So Proposition 20 maps the curve of tangents KZL into the curve PYQ referred to the ordinate BD as axis and establishes that "the sum of the squares FZ (computed with respect to line AD)" is equal to the product of the area under the new curve times the constant R, while Proposition 21 generalizes the result for higher powers of FZ.

As might be expected from the structure of the earlier lectures, Barrow concluded with a short investigation into the correlate forms of his propositions for curves defined by ordinates radiating from a common, fixed center. For example (Proposition 22), let DOK be a given curve, D a fixed point on it, and DK a fixed chord (Figure 128). Define curve $AFGE$ such that for any ordinate DMF, where M lies on DOK, MS is tangent to DOK,

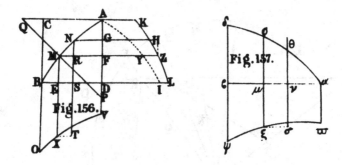

DS perpendicular to DM, and R is a given length, $DS:2R = DM^2:DF^2$. Then area $ADE = R \cdot DK$.

With DK divided into indefinite segments, much of the derivation proceeds along expected paths as Barrow worked to equate the product of each segment times R to the area of the corresponding indefinitely small sector of curve AFE. However, it takes an unexpected turn when for the first time he explicitly neglects an indefinitely small difference before any summation. Consider the first interval, where KT is tangent to DOK at K. "By indefinite section," triangle $KPM \approx$ triangle KDT, whence $MP:PK = DT:DK$. By similar sectors, $DP:PM = DE:EX$, where PM and XE are arcs of circles centered at D. "For the reason given," Barrow then noted, evidently referring to indefinite section, $DK:PM$ may be substituted for $DP:PM$. The step is tantamount to ignoring the indefinitely small difference PK between DP and DK. Barrow offered no further explanation or comment,[71] even though in the very next step he combined the two proportions, retaining PK in the first, to get $(MP \cdot DK):(PK \cdot PM) = DK:KP = (DT \cdot DE):(DK \cdot EX)$. From that point, the derivation resumes its now expected course.

With the groundwork of quadrature laid in Lecture XI, Barrow could make straightforward work of rectification in Lecture XII. It remained only to transform the curve by making its arc length the axis of a new curve and preserving the area constructed on the corresponding indefinite segment. A standard configuration served as preface for the propositions to follow (Figures 156 and 157):

> We proceed with the business at hand. To keep it as brief
> as we can and to save words, observe that everywhere in the
> following line AB is any curve; AD is its axis; all lines
> BD, CA, MF, NG are perpendicular applicates to [that axis];
> ME, NS, CB are parallel to it; point M is taken arbitrarily; arc
> MN is indefinitely small; straight line $\alpha\beta$ is equal to curve AB,
> $\alpha\mu$ to curve AM, $\mu\nu$ to arc MN; and the applicates to $\alpha\beta$ are
> perpendicular to it.

Proposition 1 then sets the style:

> Let MP be perpendicular to curve AB, and let lines KZL, $\alpha\phi\delta$ be such that FZ is equal to MP and $\mu\phi$ to MF. Area $\alpha\beta\delta$ will be equal to $ADLK$.

The problem of rectification now brings out the implications of the notion, mentioned several times in passing, that indefinitely small segments of a curve may be treated as if they were straight lines coincident with their tangents. Rectification also completes the gradual process by which the indefinitely small triangle MNR has become the focus of Barrow's analyses. In this instance it is similar to the normal triangle PMF, whence $MN:NR = PM:MF$, or $MN \cdot MF = NR \cdot PM$, or $\mu\nu \cdot \mu\phi = FG \cdot FZ$. The proposition follows by summation of corresponding equal rectangles over bases AD and $\alpha\beta$.

The bulk of the propositions of Lecture XII follow this basic pattern, reducing rectification to quadrature. In essence they transform an area defined by parameters of the given curve into a new area referred to the arc length of the curve as axis. Thus Proposition 1 maps the area defined by the curve's normals applied to its axis into the area defined by its ordinates applied to its arc length; Proposition 6, the normals applied to the final ordinate into the subnormals applied to the arc length; Proposition 10, the tangents applied to the final ordinate into the ordinates applied to the arc length; and Proposition 14, the tangents applied to the axis into the subtangents applied to the arc length.

Rectifying a curve means finding a transformation such that the area referred to the arc length is a rectangle and the area defined by the curve's parameters is known. Although there is no systematic way of doing that for a given curve, Proposition 20 offers a means of generating classes of rectifiable curves from curves of known area. In algebraic translation, it asserts that if $z = z(x)$ and $v = v(x)$ are curves of known area and so related that $z^2 = v^2 + c^2$ for some constant c, and if $c \cdot w = c \cdot w(x)$ is the quadratrix of v (i.e., if c times ordinate w at any point x is equal to the area under v to that point), then c times the arc length of w over a given interval is equal to the area under z over that interval.

Lecture XII contains no hint of its sources, nor by the same token does Barrow claim any originality, noting in Proposition 3 that it was common knowledge (*evulgatum*). But so too was the basic idea of reducing rectification to quadrature. It formed the subject of the one treatise Fermat published in his lifetime, and he in turn was responding to the work of Heuraet and Wren, among others.[72] Yet, though not new, Barrow's treatment is striking for its generality and for its emphasis on the transformation of curves. This quality points to Gregory's *Universal Part of Geometry* as the most immediate source.

Barrow and the calculus

Eyes conditioned by Newton and Leibniz recognize in the methods of quadrature and rectification presented in Lectures XI and XII the essence of the calculus viewed as a class of relations among the elements of a curve and a body of techniques for transforming those relations. Hindsight magnified by anachronistic symbolism readily translates Barrow's geometrically couched propositions into familiar forms (see Figure 122). If $y = y(x)$ denotes curve *VEH*, and $u = u(x)$ curve $\phi Z\psi$, then $u = y \cdot dy/dx$. Hence, $\int u \, dx = \int y \, dy = \frac{1}{2}y^2$, and Proposition 1 becomes an almost trivial transformation. So too do the next two propositions, which proceed inductively toward the generally demonstrable result that $\int uy^n \, dx = \int y^{n+1} \, dy = y^{n+2}/(n+2)$. The relation $t \, dy = y \, dx$ for the curve *DZO* of subtangents $t = t(x)$ makes similarly short work of Proposition 10 and its corollaries.

But the conditioning is essential, enabling recognition only through hindsight, and such facile translation risks seeing in Lecture XI the outline of concepts that are not in fact present.[73] It also may relocate emphasis and rearrange the structure of Barrow's material according to later canons. Barrow evidently did not consider its propositions as constituting a distinct new form of mathematical analysis. He called no special attention to them here, nor, when Newton's method of fluxions began to make its public appearance in the early 1670s, did Barrow ever lay claim to authorship or even inspiration.

Both the order and style of the propositions show that Barrow thought of them as serving special purposes rather than elaborating a central, organizing concept. Viewed in retrospect, Proposition 19 ought to form either the basis or the culmination of Lecture XI. Moreover, it should refer back to X:11 as its converse. Viewed in situ, it does neither. Except for taking advantage of a gradually developing economy of expression, it neither grounds the preceding propositions nor follows from them. Barrow assigns it no special conceptual or operational significance, but simply derives several corollaries concerning the summation of powers of the ordinate before shifting to a series of propositions on the transmutation of curves defined by ordinates emanating from a fixed center. What in substance becomes part of the fundamental theorem of the calculus is clearly not fundamental for Barrow. Given the geometrical style of his mathematics, there is no reason it should be.

Consider the actual form of Proposition 19 with the form it takes when translated into differential notation. In Barrow's terms, the ordinate *FZ* of the curve of tangents at any point on the axis *AD* is determined as a fourth proportional by multiplying the fixed length *R* by the ratio *FM/FT* of the ordinate to the subtangent of the given curve at that point, which ratio expresses in finite terms the ratio *MO/NO* of the sides of the triangle determined by the indefinitely small particle of the given curve at the point. Following the pattern of the previous propositions, Barrow then transforms the aggregate of the ordinates FZ into the aggregate of ordinates EX of the rectangle formed by R, the final ordinate of the given curve. Even though by this stage in the lecture Barrow is speaking of sums of representative elements and one could rephrase his final statement to read "the sum of the rectangles $FG \cdot FZ$ is equal to the sum of the rectangles $EX \cdot ES$,"[74] the statement would not convey the essence of the claim that $\int (dy/dx)\, dx = y$. What is lacking is the crucial notion that the operation of summation is the inverse of the operation that determined the curve of tangents in the first place.

That notion might have emerged if Barrow had expressed his conclusion in the seemingly equivalent terms "the sum of the rectangles $NO \cdot (MO/NO) \cdot R$ is equal to the sum of the rectangles $MO \cdot R$." But doing that requires two conceptual steps. First, it means focusing on indefinitely small quantities as the fundamental terms of analysis, and, second, it means seeing in the two forms "$NO \cdot (MO/NO) \cdot R$" and "$MO \cdot R$" not just a trivial algebraic difference but rather the intervention of the tangent as a derived quantity. Barrow was not ready for the first step. What is missing here is precisely a reference to X:11 with its use of the tangent. But it is absent because X:11 is not about indefinitely small triangles expressing the local character of tangents but rather about the global properties of tangents. Nor was he ready for the second step. To take it, he would have to see a ratio as a quantity, and he had devoted a whole mathematical lecture to rejecting that notion.[75]

Newton and Leibniz took those steps, and beneath the differences between fluxions and differentials lies the common element of algebraic analysis. It was by positing fluxions or differentials as algebraic operations – Leibniz spoke of dy as "a certain modification [*quaedam modificatio*] of y"[76] – that they both made sense of indefinitely small quantities as the terms of analysis and brought out the power inherent in seeing ratios of those quantities as finite expressions, that is, as quantities in themselves.

Thinking algebraically counted. That is why historians of mathematics must be careful about referring to Barrow's methods as calculus in another – in geometric – guise. Conceptually the guise is everything. As Leibniz himself put it in a letter to Huygens:

> Il me semble que M. Wallis parle assez froidement de M. Newton et comme s'il estoit aisé de tirer ses methodes de leçons de M. Barrow. Quand les choses sont faites il est aisé de dire: et nos hoc poteramus. Les choses composées ne sçauraient estre si bien démelées par l'esprit humain sans aides de caractères.[77]

More than merely aiding the mind, the use of symbols reflected an approach to mathematics itself. In Leibniz's version at least, the calculus was from the outset synonymous with infinitesimal analysis and coupled not with geometrical analysis but with ordinary analysis, or algebra. By the turn of the eighteenth century (witness Reyneau's *Analyse démontrée*),[78] analysis as a branch of mathematics consisted of algebra and the calculus, differing by whether the domains of quantity involved were finite or infinitesimal (i.e., Archimedean or non-Archimedean) but sharing the common attributes of an operational symbolism expressing computational relations among abstract magnitudes, a body of rules governing the manipulation of those relations or their transformation to equivalent forms, and a focus on investigating the properties of curves expressed as relations, or equations, linking their constituent elements.[79]

Therefore, expressing the calculus in algebraic terms was an essential aspect of creating it. For that reason, Fermat's treatment of the quadrature of curves lay closer to the calculus than either Barrow's or Gregory's. As the title states, it concerned the "transmutation or transformation of the equations of loci." Yet Fermat missed the meaning of the very relation that made his transformations work, namely, the inverse relation between the area of a curve and its subtangent. He missed it because he did not recognize in each of those two quantities themselves a general relationship derived from that defining the curve. At a crucial point, even his algebraic approach fell victim to the geometrical context to which he was applying it.[80]

In that light, Barrow's treatment of the method of tangents takes on interpretive significance. As noted above, he appended it to Lecture X, almost as an afterthought or digression. Lecture XI shows how much of a digression it was, not only in his text but also in his thinking about the meaning of the propositions. At no point does he suggest that the terms and techniques of the method of tangents might be applied in some manner to the transmutations of curvilinear areas by means of tangents. In the former, the variables *a* and *e*, manipulated in ways peculiar to the indefinitely small quantities they represent, provide a symbolic vehicle for translating the geometrical construction of a tangent into an algebraic operation on an equation, the result of which is another equation. Notably absent from Lecture XI is any similar effort – much less a closely related effort – to capture the transmutation of areas in terms of operations on the equations of curves.

Also lacking is any discussion of, or even concern with, the range of applicability of the transformations presented. There are no examples of how to apply the propositions to concrete problems, especially those in which the relation between the ordinate and abscissa is not algebraic. Only the most astute reader would recall a cryptic statement in the description of the method of tangents, noting that in treating some curves it helps to treat an indefinitely small arc as if it were a straight line. Here too Leibniz provides the telling contrast. Responding to those who thought

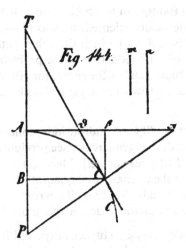

that the calculus, as initially presented in 1684, was just another method of maxima and minima similar to Barrow's, Leibniz explained in 1686 how the calculus opens transcendental relations to algebraic expression and manipulation and thus widens the range of curves that can be treated analytically.[81]

Leibniz offers further contrasting insight into how an algebraic foundation ordered the material of Lecture XI and, ironically, how only then the full power of that material became evident, although not to those who did not fully appreciate what he had achieved. Where Barrow had treated the subtangent, subnormal, and tangent as separate classes of problems, Leibniz conjoined them in a single configuration, referring to them as "functions" of a curve and generalizing the problem of inverse tangents to include them: Given the ratio of any two functions, find the curve.[82] That concern for generality was a hallmark of algebraic analysis in the seventeenth century. Its touchstone was the theory of equations, which by couching problems from different domains in the same symbolic language revealed common underlying structures and the relations of various structures to one another. Applying that touchstone to the general problem of inverse tangents similarly presupposed that the functions are couched in a form that makes their relation to the curve analytically accessible, that is, that they are expressed in the quasi-algebraic terms of the calculus (Figure 144).

Moreover, just as the theory of equations provided no general algorithm for solving equations but rather formed a framework linking classes of equations to their canonical forms, so too the calculus offered no general method of inverse tangents but rather organized and extended the range of the special solutions such as those in Barrow's *Lectures*. That is, Barrow's propositions contained the substance, but not the concepts, of

the calculus. Clear to someone who thought in algebraic terms, that relationship took someone more geometrically oriented, such as Huygens,
by surprise. Seeking from Leibniz the curve of which the subtangent is
$(y^2\sqrt{a^2-x^2})/ax$, he wondered at Leibniz's reduction of the problem to
the quadrature of $ax/\sqrt{a^2-x^2}$, the subnormal of the curve; for Huygens
knew that solution from having read Barrow's *Lectures;* he was looking
for something more from Leibniz:

> . . . je crus en le luy proposant qu'il le resoudroit independam-
> ment, et qu'ainsi sa methode des tangents renversée produirait
> la quadrature de la ligne *AH* [the subnormal]. Mais cela a
> esté autrement, et il a falu qu'il cherchast cette quadrature.
> Je ne scay pas par quel moien, mais c'est ce qu'il devroit
> m'apprendre, pour me rendre sa methode de quelque usage.[83]

Given the expressed goals of algebraic analysis, Huygens expected from
a method of inverse tangents a systematic, algebraically calculated solution, not a transformation to an equivalent form that happens to be
known. Leibniz seems to have known better. The essence of algebra is
the transformation of relations from complicated to canonical form. By
expressing a transformation in general symbolic form, algebra reveals its
structure and relates it to other transformations. The algebra of differentials extended that power of expression to Barrow's and Gregory's geometrical transmutations, revealing both the basic relations on which they
rested individually and the relations that obtained among them. Translated into Leibniz's new language, they acquired a new profundity.

   Nothing in Barrow's *Lectures* or other works suggests that he ever envisioned such a translation or, for that matter, that he shared the convictions
about the analytical power of algebra that underlay it. For that reason,
Lectures XI and XII remain historically an example of mid-seventeenth-
century extensions of geometrical analysis by way of infinitesimal elements rather than the prototype of the calculus. Competent and well informed, but not particularly original, Barrow provided in his *Lectures*
an inventory of the materials available to Newton and Leibniz. What it
reveals about the origins of the calculus says more about the nature of
their creativity than about his.

Barrow's influence

   In the end, the question of Barrow's influence on the development of mathematics comes back around to his successor in the Lucasian
chair. What did Barrow teach Newton; what did Newton take from him
by way of either substance or inspiration? Following D. T. Whiteside's
careful investigations, we may ignore the legends. Barrow did not send
Newton away to learn Euclid properly. Barrow did not resign from his

chair to make room for the more creative Newton. Barrow was not New-ton's tutor. Newton did use Barrow's editions of the classical Greek ge-ometers, but there is no evidence that Barrow ever examined Newton on the subject. Certainly, when Newton returned from his eighteen-month forced vacation, he had nothing to learn from the lectures Barrow was then delivering.[84] By the late 1660s, Newton's own method of series and fluxions outstripped anything the *Geometrical Lectures* had to offer.

More importantly, by then Newton and Barrow were pursuing quite different mathematical paths. The *Geometrical Lectures* were squarely geometrical in style and substance. As delivered, after all, they served as ancillae to his optics. Barrow clearly preferred the analysis of geometri-cal configurations to what he called (at the beginning of Lecture 6) "the trouble or wearisomeness of calculation." Although the *Principia* and other works show clearly Newton's command of that form of analysis, indeed his quite original development of its possibilities, the method of series and fluxions lay squarely in the algebraic tradition that by then had become generally identified with analysis itself. Newton's later attempt at restoring Pappus's "Treasury of Analysis" reflects the true momentum of his mathematical thought and puts the lie to the legends that would make him his age's greatest algebraist *malgré lui*. What started out as a geometrical treatise soon became an algebraic exposition of the method of quadrature.[85]

At the time when Barrow was delivering the *Geometrical Lectures,* he was collaborating with Newton, not teaching him. In Isaac Newton, Bar-row found a critical reader and creative young colleague whom he could encourage and support and to whom he could direct mathematical in-quiries as his own interests turned more to theology. Thus, when Collins suggested that Barrow might compose a commentary to accompany a translation of Kinckhuysen's *Algebra,* Barrow turned the task over to Newton, who began the task only to abandon it shortly thereafter as the market for mathematical books shrank.[86] It was natural that, when Bar-row resigned the Lucasian chair, he should recommend his young friend. And in choosing to lecture first on optics, Newton may have left a trace of his own sense of intellectual debt to Barrow.

Notes

This essay has benefited at various stages from the thoughtful reading and criticisms of Mordechai Feingold, I. B. Cohen, Antoni Malet, and H. J. M. Bos, to whom I extend my thanks.

1. *Euclidis Elementorum Libri XV, breviter demonstrati, Operâ Is. Barrow, Cantabrigien-sis, Coll. Trin. Soc.,* 1655 (O.S.); a second edition appeared in 1657 and included a sim-ilar treatment of Euclid's *Data.* Barrow continued this effort after taking up the Lucasian chair. Though published only in 1675, portions of his *Archimedis Opera; Apollonii Pergaei Conicorum Libri quatuor; Theodosii Sphaerica Methodo nova illustrata et suc-cinctè demonstrata, Per Is. Barrow, exprofessorem Lucasianum Cantab. et Societatis*

Regiae Soc. may date from the early 1660s. Newton certainly read the edition of Euclid, even laying out his own version of Book II, and he may have used the others while still in manuscript; see D. T. Whiteside, *The Mathematical Papers of Isaac Newton,* 8 vols. (Cambridge, 1966–83), 4:218, hereafter *MPN.*

2. Lectiones XVIII, *Cantabrigiae* in Scholis publicis habitae; in quibus OPTICORUM PHAENOMENΩN GENUINAE RATIONES investigantur, ac exponuntur. Annexae sunt Lectiones aliquot *Geometricae. LONDINI,* Typis *Gulielmi Godbid,* & prostant venales apud *Johannem Dunmore,* & *Octavianum Pulleyn* Juniorem, 1669. LECTIONES Geometricae: In quibus (praesertim) GENERALIA *Curvarum Linearum* SYMPTO-MATA *DECLARANTUR. LONDINI,* Typis *Gulielmi Godbid,* & prostant venales apud *Johannem Dunmore* 1670 (republished 1683, 1684). Lectiones Mathematicae XXIII; in quibus Principia Matheseως generalia exponuntur: Habitae CANTABRIGIAE A.D. 1664, 1665, 1666. Londini, Typis *J. Playford,* pro *Georgio Wells* in Coementario D. *Pauli,* 1683. The three sets of lectures were republished in a single volume edited by William Whewell, *The Mathematical Works of Isaac Barrow, D.D.* (Cambridge, 1860). References to this edition are enclosed in parentheses and consist of the designation *LO, LG,* or *LM,* respectively, followed by the page number.

3. Barrow to Collins [late 1663], in S. P. Rigaud (ed.), *Correspondence of Scientific Men of the Seventeenth Century* (Oxford, 1841), 2:33.

4. When the *LO* appeared in November 1669, Oldenburg informed Huygens that a copy was being sent to Henri Justel, from whom Huygens could borrow it (*Oeuvres complètes de Christiaan Huygens,* 6:520; hereafter *HOC*). Huygens did so and responded to the work toward the end of January (*HOC,* 2:2–3). The *LG* reached Huygens along the same route the following September (*HOC,* 7:38). Illness prevented Huygens from reading it at the time (ibid., 41, 43), but in his later correspondence with Leibniz, Fatio de Duillier, and Hubertus Huighens, he discussed Theorem X:1 in some detail (*HOC,* Vol. 10, passim). Huygens's marginal notes on articles in the *Acta eruditorum* also display some familiarity with the *LG.*

5. *The Usefulness of Mathematical Learning Explained and Demonstrated: Being Mathematical Lectures Read in the Publick Schools at the University of Cambridge by Isaac Barrow,* trans. John Kirkby (London, 1734), hereafter *UML,* and *Geometrical Lectures,* trans. Edward Stone (London, 1735).

6. See Helena M. Pycior, "Mathematics and Philosophy: Wallis, Hobbes, Barrow, and Berkeley," *Journal of the History of Ideas* 48 (1987), 265–86. According to Pycior, Berkeley read Barrow's *Geometrical Lectures* with some care and relied "heavily on Barrow's theory of numbers." While he followed Barrow and Hobbes in "developing arithmetic as the science of signs, [he] did not share their distrust of algebra"; rather, following "Wallis's enthusiasm for symbolic reasoning and algebra," he redirected their approach to make algebra itself the science of signs.

7. J. M. Child, *The Geometrical Lectures of Isaac Barrow* (Chicago, 1916). See his précis of the main argument in "The 'Lectiones Geometricae' of Isaac Barrow," *The Monist* 26 (1916), 251–67. Philippe E. B. Jourdain and Florian Cajori published reviews critical of Child's translation in *Isis* 3 (1920–1), 283, and *American Mathematical Monthly* 26 (1919), 15–20. Cajori's critique, "Who Was the First Inventor of the Calculus?" drew a belated rejoinder from Child in "Barrow, Newton, and Leibniz, in Their Relation to the Discovery of the Calculus," *Scientific Progress* 98 (1930), 295–307. In "Neue Einblicke in die Entdeckungsgeschichte der höheren Analysis" (*Abhandlungen der preussischen Akademie der Wissenschaften,* Jahrgang 1925, Physikalisch-mathematische Klasse, Berlin 1926), based on a detailed study of the manuscripts, Dietrich Mahnke defended Leibniz's independence of any Barrovian influence.

8. Margaret E. Baron, *Origins of the Infinitesimal Calculus* (Oxford, 1969). By casting Barrow's geometrical results in the modern notation of the calculus while placing Barrow himself squarely in a geometrical tradition of Italian origin, Baron effectively begs the

question of the precise nature of such "intuitively understood" processes and of their relation, both historical and conceptual, to the explicitly defined operations denoted by the symbolism.

9. *Source Book of Mathematics, 1200–1700,* ed. D. J. Struik (Cambridge, 1969). Sel. IV, 14, pp. 253–63. The work, of course, is a revised and expanded edition of D. E. Smith's original source book, which included the same selections from Barrow under the same title (and taken from the same source, i.e., Child). Interestingly, Struik tempered the implications of the selection with a caution against the dangers of translating Barrow's mathematics from its original geometry to analytic form. "They [Barrow and his contemporaries] saw geometric theorems in the sense of Euclid, where we see operations and calculating processes. At the same time, just because these mathematicians applied their geometric notions in an attempt to transcend the static character of classical mathematics, their geometric thought has a richness that may easily escape observation in the modern transcription" (p. 263).

10. *MPN,* passim, but especially Vols. 1 and 3. On the question of Barrow's influence on Newton's concept of a fluxion, Whiteside's findings confirmed J. E. Hofmann's conjecture ("Studien zur Vorgeschichte des Prioritätsstreits zwischen Leibniz und Newton um die Entdeckung der höheren Analysis," *Abhandlungen der Preussischen Akademie der Wissenschaften,* Jahrgang 1943 [Mathematisch-naturwissenschaftliche Klasse, Berlin, 1943]) that the *Mathematical Lectures* had provided some initial inspiration in mid-1665. Whiteside pointed as well to the influence of the *Geometrical Lectures* as Newton was composing the *Method of Series and Fluxions* in 1670; see ibid. 3:70.

11. Josef E. Hofmann paved the way for much of this work through a series of studies aimed at clarifying the background of Leibniz's mathematics. See in particular his *Die Entwicklungsgeschichte der Leibnizschen Mathematik während des Aufenthalts in Paris (1672–1676)* (Munich, 1949).

12. Lat. *quisquiliae,* literally "trash, rubbish."

13. The passages quoted appear in slightly different order in Barrow's lecture.

14. In light of Barrow's defense of the Euclidean theory of ratio and proportion, his definition of number raises the question of formulating that theory without reference to the prior laws of arithmetic. That is, unless Barrow meant to use "metaphysical numbers" to establish the integers, he could hardly define equality of ratio in terms of equimultiples.

15. Later in Lecture IX Barrow brought out explicitly an implication of making surds numbers, namely, that some numbers correspond to infinite series: "Nay it is plainly taught and demonstrated by arithmeticians, that an infinite series of fractions, decreasing in a certain proportion, is equal to a certain number, or to unity, or to a part of a unit; *ex. gr.* that such a series of fractions decreasing in a subsesquialter proportion is equal to two, in a subduple proportion to unity, in a subtriple to one-half; from whence it is not inconsistent for something finite to contain in it an infinity of parts: especially since nothing agrees with number, which does not with more right agree with magnitude, which number represents and denominates" (*UML* 157; *LM* 145).

16. On the development of the concept in the sixteenth and early seventeenth centuries, see Jakob Klein, *Greek Mathematical Thought and the Origins of Algebra* (Cambridge, Mass., 1968) (originally, "Die griechische Logistik und die Entstehung der Algebra," *Quellen und Studien zur Geschichte der Mathematik, Astronomie und Physik,* Abteilung B: *Studien,* 3, no. 1 (1934), 18–105; 3, no. 2 (1936), 122–235.

17. François Viète, *In artem analyticen isagoge,* ed. F. van Schooten (Leiden, 1646), Chap. 1, p. 1. "Forma autem Zetesin ineundi ex arte propriâ est, non jam in numeris suam Logicam exercente, quae fuit oscitantia veterum Analystarum: sed per Logisticen sub specie noviter inducendam, feliciorem multò & potiorem numerosâ ad comparandum inter se magnitudines." See Chap. 4, p. 4: "Logistice numerosa est quae per numeros, Speciosa quae per species seu rerum formas exhibetur, ut pote per Alphabetica elementa."

18. See Michael S. Mahoney, "The Beginnings of Algebraic Thought in the Seventeenth Century," in *Descartes: Philosophy, Mathematics and Physics* (ed. S. Gaukroger, Sussex and Totowa, 1980), Chap. 5.

19. In Viète's system, it was the function of *exegetic* to translate the abstract general relations of quantity to concrete terms. Arithmetical exegesis requires that one show how to add, subtract, multiply, and divide numbers; geometrical exegesis calls for corresponding constructions.

20. René Descartes, *La géométrie* (Leiden, 1637), 298.

21. John Wallis, *Mathesis universalis, sive Arithmeticum Opus Integrum, Arithmeticam tum Numerosam, tum Speciosam, complectens; tum etiam Rationum & Progressionum traditionem; Logarithmorum item doctrinam* (Oxford: Lichfield, 1657), Chap. 2.

22. See the start of Lecture XVII (*LM* [1860], 268-9), where Barrow quoted the Latin version of the original French: "Voyant qu'encore que leurs [les sciences, qu'on nomme communément mathématiques] objets soient différentes, elles ne laissent pas de s'accorder toutes, en ce qu'elles n'y considèrent autre chose que les divers rapports ou proportions qui s'y trouvent, je pensai qu'il valait mieux que j'examinasse seulement ces proportions en général" (*Discours de la méthode*, ed. E. Gilson [Paris, 1954], 67-8). Cf. Isaac Newton: "By *Number* we understand not so much a multitude of Unities, as the abstracted Ratio of any Quantity to another Quantity of the same kind, which we take for Unity" (*Universal Arithmetick*, 2d ed. [London, 1728], p. 2).

23. Whewell's outline uses the term "coincidence" to translate Barrow's *congruentia*, but that may miss precisely the point he is making.

24. Tacquet's argument appears in his *De solidis cylindricis et annularibus libellus*.

25. The figures in this chapter are reproduced from Whewell's 1860 edition of Barrow's *Works*, which in turn appears to have reproduced them from the original copperplate drawings of 1670; both give the name of the engraver as Cross. Thus, what the reader will see is what Barrow's readers saw. Because I have taken the figures from the original text, I have preserved the numbering. The sole exception is the final figure, numbered 144, which is taken from Gerhardt's edition of Leibniz's works.

26. The "horn angle" is the apparent angle formed by a circle and its tangent; its counterpart is the angle formed by the circle and its radius, which Euclid showed in *Elements* 3:16 to be greater than any rectilineal angle less than a right angle. As the prime example of a non-Archimedean magnitude, the horn angle has attracted the attention of mathematicians to the present day.

27. In omitting any substantive discussion of algebra in general and the theory of equations in particular, Barrow also left unclear the nature of the relationship between equations and proportions and his position on the then quite open question of whether every equation could be reduced to a proportion.

28. For a detailed discussion of the classical issues treated by Barrow, see Chikara Sasaki, "The Acceptance of the Theory of Proportion in the Sixteenth and Seventeenth Centuries: Barrow's Reaction to the Analytic Mathematics," *Historia Scientiarum* 29 (1985), 83-116.

29. For some reason Barrow opened his discussion using the medieval terminology of *proportio* and *proportionalitas* and then abruptly switched to the classical forms, *ratio* and *proportio*.

30. In rejecting the notion of ratios of ratios Barrow made no mention of the doctrine of *proportiones proportionum*, which was introduced into English and French universities in the later fourteenth century and had since become a standard subject of the scholastic curriculum. Yet he knew of it, since he alluded to it later in the lecture when debunking the logomachy of "multiple" and "multiplicate" ratios (*LM* 328-9). That he otherwise ignored it is strange, since at the heart of the doctrine lay the notion of exponential arithmetic so fundamental to the concept of logarithms, another of the new developments Barrow was eager to present to his students. See M. S. Mahoney,

"Mathematics," in D. C. Lindberg (ed.), *Science in the Middle Ages* (Chicago, 1978), Chap. 5, esp. pp. 162–8.

31. The original Latin for the quoted phrase is *comites vel mantissas.* He also characterized them as *quisquiliae* (literally "trash, rubbish"), but this may have signified no more than ritual modesty.

32. The agenda laid out at the start of (Mathematical) Lecture XVII concludes with a treatment of "the motion by which magnitude is generated" (*LM* 269).

33. Antoni Malet has suggested to me that the person in question was Collins.

34. Shapiro notes the same quality in Barrow's optical lectures: "They are often repetitive and tedious, and the lectures on spherical mirrors, especially, could have been substantially condensed by not treating so many cases separately. Yet their contents, especially the principal contents of image location, are so sophisticated that they were surely above the head of any student." See Shapiro, Chapter 2, this volume.

35. Barrow uses the term "organic" in its original meaning of "pertaining to instruments" or "instrumental."

36. "Ergo tempus absolute quantum est; ut quantitatis admittens (modo suo) praecipuas affectiones aequalitatem, inaequalitatem, proportionem; . . . Quantitatis igitur particeps esse tempus communis sensus agnoscit, pro modo permanentiae rerum in suo esse" (*LG* 161). Whiteside observes that, in the *Methods of Series and Fluxions* (1671), "Newton was led, under Barrow's guidance, to postulate a basic, uniformly 'fluent' variable of 'time' as a measure of the 'fluxions' (instantaneous 'speeds' of flow) of a set of dependent variables which continuously alter their magnitude" (*MPN* 3:17). For a detailed comparison, cf. ibid., 70–2, notes 80–3, where Whiteside finds evidence of influence from both the *Mathematical Lectures* and the *Geometrical Lectures*.

37. On the traditional fallacies, see the *sophismata* literature of the fourteenth-century Oxford logicians. For a discussion of those raised in criticism of Cavalieri's *Geometria nova indivisibilibus promota* (Bologna, 1653), see Kirsti Andersen, "Cavalieri's Method of Indivisibles," *Archive for History of Exact Sciences* 31 (1985), 291–367.

38. Barrow here cited Descartes's *Principles of Philosophy*, 2:31–2.

39. Note that this framework differs from the one used on Lecture I to analyze distance, time, and motion as analogous quantities. Here both the axis and the ordinate are conceived of as distances traversed at various velocities over the same time.

40. Or at least portions of them. Barrow's method of generation restricts him to a quadrant of the circle and ellipse and to a half of the parabola and hyperbola. As he later observes in Proposition 18 (*LG* 202), "From the above it follows that if the circle, ellipse, and other recurrent curves of that sort are thought of as engendered in this manner, the point describing them will have infinite velocity at the turning point." Curves like the parabola and hyperbola will have finite speeds of generation at any given point.

41. Of course, Barrow is relying on physical intuition to make his argument; Apollonius had to make the argument explicit and mathematical.

42. On Fermat's method, see Mahoney, *Fermat,* Chap. 4, esp. pp. 163–9.

43. For a survey of treatments of the problem from Descartes to the Bernoullis, see Joseph E. Hofmann, "Ueber Auftauchen und Behandlung von Differentialgleichungen im 17. Jahrhundert," *Humanismus und Technik* 15, no. 3 (1972), 1–40. For a detailed account of Descartes's solution, see Jules Vuillemin, *Mathématique et métaphysique chez Descartes* (Paris, 1960), Chap. 1.

44. Barrow's mathematical language is interesting here. The term "applicate" for the ordinate reflects the continuing influence of the classical Apollonian terminology that went along with viewing a curve as generated by the "application" of a line to an axis. Barrow shifts to the term "ordinate" in later lectures. He uses no technical term for the subtangent, referring to it as the "[line] intercepted by the tangent to the point."

45. Here and throughout the essay I have translated Barrow's algebraic notation into modern form, most notably by employing superscripts where he used literal abbreviations

of the powers; thus, PM^2 for *PMq*. Except in Lecture XIII, Barrow used algebra as little more than shorthand for essentially geometrical arguments. Nothing hangs on the particular symbolism employed.

46. Here Barrow takes the result as given and familiar, although he later establishes it in Lecture IX; see below, p. 220. In general, for curves of the form $y^n = px$, the subtangent is *nx*. Fermat established the property in his study of such "higher parabolas." It follows easily from his method of tangents, which Barrow adopted from him and set out at the end of Lecture 10; see below.

47. The diagram in *LG* omits *Z*, which, following the pattern of the preceding diagrams, lies indeterminately on *TMP* extended.

48. *LG* has *simpliciter*, but clearly *similiter* is meant. See Proposition 17, which follows as a corollary (*LG* 201): "Si ad rectam aliquam lineam (hoc est ad ejus singula quaeque puncta) applicentur rectae lineae parallelae, ad rectam AD *consimiliter* divisam applicatarum differentiis proportionales" (emphasis added).

49. Gregory's *Geometriae pars universalis* seems the likely source of Propositions 5 and 6.

50. Whewell's text has *QX*, which makes no sense.

51. Barrow explored the structural analogy between geometrical and arithmetical proportion in Lecture XVII of the *Mathematical Lectures* (*LM* 274–82) but there concluded that the notion of ratio be reserved for the geometrical case.

52. The notation is mine, not Barrow's; it abbreviates his phrase *eodem ordine media proportionalis inter DG, DE.*

53. I use parentheses here to avoid Barrow's confusing, yet suggestive symbolism for the composition of two ratios, e.g., $KG : GB + GB : LG$. The implied analogy between composition of ratios and the addition of numbers places Barrow in a tradition dating back to the late Middle Ages. See Mahoney, "Mathematics," pp. 166–8. It appears to get Barrow into trouble later in Lecture IX; see n. 59.

54. Pappus, *Mathematical Collection,* Vol. 7, propositions, 129, 136, 137, 140, and 142.

55. D. T. Whiteside, "Patterns of Mathematical Thought in the Later 17th Century," *Archive for History of Exact Sciences* 1 (1961-2), 179–388, at 271-2 (concerning Barrow) and 275-81 (for Newton's work in this area).

56. Proposition 17 similarly yields the tangent to the cissoid.

57. Given the tendency of some historians of mathematics to see in Barrow some sort of cryptoanalyst, it is perhaps worth comparing these theorems with their analytic counterparts to see how very differently the lines of thought run. In IX:1 take $y = PE$ and $z = PF = PG - n[(PG - PE)/m] = [(m - n)/m]PG + (n/m)y$ as functions of *x* measured along the axis *DT* with respect to some origin. Then $z' = (n/m)y'$, and the ratio of the subtangents DS/DT evaluated for the common ordinate $DB = PG$ is $(y/y')/(z/z') = z'/y' = n/m$. In the case of the geometric mean, $z = PF = PG/(PG/PE)^{n/m} = PG(PE/PG)^{n/m} = y^{n/m} \cdot PG^{1-(n/m)}$. Then $z' = (n/m)y^{(n/m)-1} \cdot y'PG^{1-(n/m)}$, which evaluated at $y = DB = PG$ yields $(n/m)y'$, and the result follows as before. Note that these derivations presuppose a complex of concepts of which Barrow's lectures so far offer no hint. Those concepts are unnecessary in reaching the results Barrow seeks, for the purposes for which he seeks them.

58. In the case of the simple geometric mean, $PF^2 = PE \cdot PG$, where *PG* is constant.

59. Although the proposition is correct, Barrow's proof is invalid. He appears to have succumbed to the ambiguity of his notation. Working toward the inequalities needed to show that, since $EG > KG$ (*BR* tangent to *EBE*), $EF < EL$, he wrote $FG \cdot TD + EF \cdot RD > LG \cdot TD + KL \cdot RD$ and from it concluded $FG : EF + TD : RD > LG : KL + TD : RD$. Evidently he meant the last expression to be read as composition of ratios; that is, according to the usage in VII:3, $(FG : EF)(TD : RD) > (LG : KL)(TD : RD)$. But how he thought that composition followed from the previous expression is not at all clear.

60. Again, the text shows evidence of having served as notes for a fuller oral presentation, for the line of argument needs explanation. Given are the curves *ZGE* and *VKF*, and a

point F on *VKF*. *VIF* is a distinct, auxiliary curve; its parameter R is chosen so that it passes through the given point F, i.e., $R = DF/2$. In general, for ordinates $D'K, D'I$ on a common abscissa, $D'K^2 = 2 \cdot R \cdot D'I = DF \cdot D'I$. If $D'K > DF$, $D'I > D'K$; if $D'K < DF$, $D'I < D'K$. Hence, the two curves intersect only at points of which the ordinate is DF.

61. On reduction analysis, see M. S. Mahoney, "Another Look at Greek Geometrical Analysis," *Archive for History of Exact Sciences* 5 (1968), 318–48, and Mahoney, *Fermat*, passim.

62. On the origins and conceptual structure of Fermat's method of tangents, see M. S. Mahoney, *The Mathematical Career of Pierre de Fermat, 1601–1665* (Princeton, N.J., 1973), Chap. 4. The method came to the attention of Parisian mathematicians as the result of a dispute between Fermat and Descartes over its relationship to Fermat's method of maxima and minima and its effectiveness in comparison with Descartes's method of determining the normal as set forth in *La géométrie* (Leiden, 1638). A copy of an exposition of the method written by Fermat during that debate accompanied Mersenne on his trip to Italy in 1644–5. Torricelli obtained that copy and turned it over to Ricci to make another. Barrow may in turn have learned of the method during his visit to Italy (Mahoney, *Fermat*, p. 145, n. 3). Another version of the method, probably the work of Mersenne, appeared in the 1644 *Supplément* of Pierre Hérigone's *Cursus mathematicus* (Paris, 1634), but it is so highly abbreviated as to convey little to someone who does not already understand it.

63. On the first of these, see *MPN* 3:286, n. 602. The third curve was also known as *folium Cartesii* and first appeared as a challenge problem in Fermat's and Descartes's correspondence over the method of tangents; see Mahoney, *Fermat*, Chap. 4, 181.

64. Ibid., Chap. 5, pp. 243–64. How much of Fermat's work in this area had reached England either directly or indirectly is not clear. Collins reported to Gregory in March 1672 that Pell had seen Fermat's papers, including those on rectification and quadrature, and had plans of publishing them. Nothing came of the plans. See J. E. Hofmann, "Studien zur Vorgeschichte des Prioritätsstreits zwischen Leibniz und Newton um die Entdeckung der höheren Analysis," *Abhandlungen der Preussischen Akademie der Wissenschaften,* Jahrgang 1943, Mathematisch-naturwissenschaftliche Klasse (Berlin, 1943), p. 74, n. 311; p. 82, n. 367. See also *MPN* 3:7.

65. See the subsection on divisibility, congruence, and equality.

66. 1860 has $AD\psi\phi$, which is clearly a misprint.

67. *Curvula;* the Latin term is as artificial as the English.

68. In the sense established in Lecture I concerning "linelets" and "timelets," namely, that the "lines" referred to in the phrase "all the lines" are in fact indefinitely narrow rectangles. Whatever the phrase meant originally, both Fermat and Leibniz, among others, understood it as a sum of infinitesimals of the same dimension; see Mahoney, *Fermat,* Chap. 5, 254–64.

69. Barrow expresses π as π/δ, where π is the circumference, δ the diameter of a circle.

70. On Fermat's use of the relation and for a reconstruction of its derivation, see Mahoney, *Fermat,* Chap. 5. That reconstruction overlooked the possible role of solids of revolution in establishing the lowest-order case.

71. With no help from Barrow, one may note that the area of the indefinitely small sector *KDM* can be expressed in two ways, either as a circular sector $\frac{1}{2}DP \cdot PM$ or as a triangle $\frac{1}{2}DK \cdot PM$; the latter takes both *MK* and *PM* as straight lines. Since the areas are ultimately equal, so too are *DP* and *DK*.

72. *De linearum curvarum cum lineis rectis comparatione dissertatio geometrica,* published as Appendix I to Antoine de Lalouvère's *Veterum geometria promota in septem de cycloide libris* (Toulouse, 1660). On Fermat's method of rectification and its relation to the work of others, see Mahoney, *Fermat,* Chap. 5.

73. Propositions 4 and 5, which fit the Leibnizian mold less comfortably than their predecessors, also indicate the difficulties of too modern a reading. In 4, taking $VD\psi\phi$ as

any given area, Barrow asserts that, "if the individual areas $VAZ\phi$, $VBZ\phi$, $VCZ\phi$, etc. are thought of as multiplied by their ordinates AZ, BZ, CZ, etc. respectively, the sum that results will be equal to the semi-square of the area $VD\psi\phi$ itself." To translate this symbolically requires multiplying $\int u\,dx$ over the interval $(0, u)$ by its last ordinate u and then taking the integral of the product over the entire interval VD, itself variable. Writing the expression to show what is going on requires a double integral and its transformation internally by the rules of the calculus to a single integral. Both the symbolism and the transformation are conceptually more refined than any proposed by Leibniz or Newton. The proposition fits Barrow's geometrical framework, since he envisioned a series of small blocks raised on the segments of the curved area to the height of the final ordinate, but it does not translate easily into the algebraic framework of the calculus.

74. Or, e.g., as he has it in Proposition 20: *summa quadratorum ex FZ (ad rectam AD computata)*. Yet even here the goal is economy of expression, rather than any sort of operative symbolism.

75. Indeed, in this context one passage in that lecture (XX) takes on special meaning: "How, without a ratio, does one successfully carry out a comparison of quantities comprehensible in themselves, and [how] will a known relation spring from unknown quantities? Moreover, a ratio conjoined with another ratio forms only an inconstant and arbitrary quantity, for, insofar as one assumes an arbitrarily variable common consequent, what are termed the quantities of the conjoined ratios will vary. Therefore these ratios will have an inconstant and indeterminate quantity, if they have one; that is, they have none. For whatever is is determinately; what is in any way is not" (*LM* 323). That quantities which in themselves are indeterminate can form a determinate ratio was a fundamental assumption of the early calculus; see M. S. Mahoney, "Infinitesimals and Transcendent Relations: The mathematics of Motion in the Late Seventeenth Century," in D. C. Lindberg and R. S. Westman (eds.), *Reappraisals of the Scientific Revolution* (Cambridge, in press).

76. "De geometria recondita et analysi indivisibilium atque infinitorum," *Acta Eruditorum* (1686), 292–300; rpt. in Gerhardt, ed., *Leibnizens mathematische Schriften*, 5:226–33, at 230.

77. Leibniz to Huygens, 14 September 1694, in *Oeuvres complètes de Huygens*, 10:675.

78. *Analyse demontrée, ou la méthode de resoudre des problèmes de mathématiques et d'apprendre facilement ces sciences, expliquée et demontrée dans le premier volume et appliquée, dans le second, à découvrir les propriétés des figures de la géométrie simple et composée, à résoudre les problèmes des sciences physico-mathématiques, en employant le calcul ordinaire de l'algèbre, le calcul différentiel et le calcul intégral. Ces derniers calculs y sont aussi expliqués et demontrés* (Paris, 1708).

79. On the nature of algebraic thought in the seventeenth century, see M. S. Mahoney, "Anfänge der algebraischen Denkweise im 17.Jahrhundert," *Rete* 1 (1972), 15–31.

80. See Mahoney, *Fermat*, Chap. 5, pp. 278–9.

81. "De geometria recondita..." (see above, n. 75).

82. Soit donnée la raison, comme *m* à *n*, entre deux fonctions quelconques de la ligne ACC, trouver la ligne. J'appelle *fonctions* toutes les portions des lignes droites, qu'on fait en menant des droites indéfinies, qui répondent au point fixe et aux points de la courbe, comme sont AB ou Aβ abscisse, BC ou BΘ sous-tangente, BP ou Bπ sous-perpendiculaire, AT ou AΘ *resecta* ou retranchée par la tangente, AP ou Aπ retranchée par la perpendiculaire, TΘ ou Pπ sous-retranchées, *sub-resectae a tangente vel perpendiculari*, TP ou $\Theta\pi$ *corresectae*, et une infinité d'autres d'une construction plus composée, qu'on se peut figurer. "Considérations sur la différence qu'il y a entre l'analyse ordinaire et le nouveau calcul des transcendantes," *Journal des sçavans* (1694), rpt. in Gerhardt, ed., *Leibnizens mathematische Schriften*, 5:306–8, at 307–8.

83. Huygens to Fatio de Duillier, 18 December 1691, *Oeuvres complétes de Christianne Huygens*, 10:211.

84. D. T. Whiteside, "Isaac Newton: Birth of a Mathematician," *Notes and Records of the Royal Society,* 19, no. 1 (1964), 61, n. 25. Newton said he attended the lectures in 1665, "but would not allow that they were helpful to him, a plausible statement when we look at the non-technical nature of those lectures." The same note adduces the evidence against Newton's ever having been a student of Barrow's.

85. See *MPN* Vol. 7, Part 2, passim but esp. 402, n. 2.

86. See *MPN* 3:7.

4

Isaac Barrow's academic milieu: Interregnum and Restoration Cambridge

JOHN GASCOIGNE

Apart from the four years that Barrow spent traveling abroad, his entire adult life was linked with Cambridge. Even as a traveler in distant lands his close association with his alma mater was apparent: While other travelers toasted their mistresses, Barrow drank to Trinity College.[1] To understand the diverse concerns that shaped Barrow's work we must therefore know something of the academic milieu in which he lived and moved. This is particularly true since Barrow was associated with Cambridge in one of the most turbulent periods of the university's history. In the years between Barrow's matriculation at Peterhouse in 1643 and his death in 1677 Cambridge underwent a number of upheavals that were to affect its political and religious character as well as its standing in the eyes of the larger community.

In the first place, Cambridge was under parliamentary rule from virtually the beginning of the Civil War. Cromwell occupied it in 1642; in the following year many of its "monuments of superstition or idolatry" were destroyed,[2] and in 1644–5 there was a wholescale purge of its members following the imposition of the Covenant, a declaration of support for an antiepiscopal church order. This purge resulted in the ejection of 217 dons (about half the fellowship of the university)[3] and their replacement by those who were more sympathetic to the parliamentary regime. This purge was later followed by a more minor culling of dons following the Engagement of 1650, an oath of loyalty to the "Commonwealth of England, as the same is now established, without a King or a House of Lords."[4] At least forty-seven fellows (and possibly as many as fifty-six) were ejected. Most of these ejections dated from the pre–Civil War period, though nineteen of those ejected had been parliamentary appointees, the vacancies thus created being largely filled with Independents.[5] Among the ejected were the masters of five colleges and, ironically, the university's former chancellor, the earl of Manchester, who had supervised the purge of 1644–5. The return of Charles II meant yet another purge, with forty-seven members of the university being ejected between 1660 and 1662.[6]

Such changes had consequences for both the political and religious character of the university. Pre–Civil War Cambridge had been characterized by a considerable degree of religious pluralism. Although it was reputed to be a breeding ground for Puritans (a description that was particularly applied to Emmanuel College), it also had pockets of high church sentiment, notably at Peterhouse. There both Matthew Wren and John Cosin (the masters from 1625 to 1635 and 1635 to 1644) were strong supporters of Archbishop Laud and his campaign to establish within Anglicanism an emphasis on ritual and religious decorum that would bear witness to the "beauty of holiness" – an ambition that, to judge from the widespread decoration and use of ritual practices such as bowing to the east in college chapels, had considerable support within Cambridge.[7] After the Parliamentary Visitation of 1644–5, which enforced subscription to the Covenant, such centers of Laudian sentiment were eradicated and the Presbyterian party became dominant within the university. Symptomatically, of the twelve new masters who were appointed at the direction of the Parliament, nine were members of the Westminster Assembly and seven were graduates of Emmanuel.[8] Yet the triumph of Presbyterianism either in Cambridge or in the nation as a whole was far less complete than its advocates might have expected as the godly divines began their deliberations at the Westminster Assembly in 1643. In religion as in physics every action is inclined to produce an equal and opposite reaction, and the apparent victory of Presbyterianism strengthened the influence of the Independents;[9] it also stimulated the growth of a new movement led by the Cambridge Platonists, who sought to overcome the bitter wranglings about church government and doctrine by an emphasis on the mystical experience of the individual believer. The diversity of Cambridge's religious life was further increased by the fact that the parliamentary regime, having failed to produce an acceptable religious settlement for the nation as a whole, largely left the universities to themselves after the imposition of the Engagement Oath of 1650 (which was primarily political rather than religious in character), despite strong rumors in 1653 and 1659 that there was to be wholesale reform or even abolition of Oxford and Cambridge.

The Restoration religious settlement, then, was imposed on a university many of whose members had already lived through a number of religious upheavals and who had grown increasingly tolerant of the various forms of church government that changing political circumstances brought in their wake. The failure of the Presbyterians to inculcate a belief in the essential nature of a presbyterian church order is evident from the relatively small number of university members (a mere forty-seven) who were ejected for refusing to conform to the restored episcopal order. But those restored Anglicans who dominated the upper echelons of the university after 1660 took a rather less relaxed view of the nature of the newly restored church order than many members of the university who had lived through the religious upheavals of the Interregnum. To men like Peter Gunning (master of Clare and Lady Margaret Professor of Divinity, 1660; master of St. John's, 1661-9; and Regius Professor of Divinity, 1661-74), Joseph Beaumont (master of Jesus, 1662; master of Peterhouse, 1663-99; and Regius Professor of Divinity, 1674-1699), and Anthony Sparrow (president of Queens', 1662-7) - all of whom had earlier been ejected from the university for their loyalty to the Royalist cause - the restored episcopal church was not simply another form of church government to be justified on grounds of utility but rather a divinely ordained order that it was the university's mission to defend and propagate. Hence, when the latitudinarian-inclined Gilbert Burnet visited Cambridge in 1662, he described these three men as carrying "things so high that I saw latitude and moderation were odious to the greater part even there." [10]

The period following the Restoration was marked, then, by clashes between men such as Gunning, Beaumont, and Sparrow who were zealous to restore the Church of England after its time of troubles and other members of the university who were willing to live under both the parliamentary and Royalist regimes since they lacked the high churchmen's conviction that the episcopal church order was divinely ordained. In particular, the views of Henry More and Ralph Cudworth - the two surviving Cambridge Platonists - were subject to close theological scrutiny, and for a time Henry More contemplated leaving Cambridge. [11] Though the zeal of the restored high churchmen abated somewhat in the years following the Restoration, they were largely successful in implanting within Cambridge their religious ideals - their success owing much to the fact that, unlike the Presbyterians of the 1640s, they could count on the support of the state in the form of the restored monarchical regime. Thanks to the restored high churchmen's activities within the university, a new generation of clergy was produced that was committed to the ideals of an episcopal church characterized by passive obedience to the Lord's anointed (in the unlikely guise of Charles II). Indeed, after 1688, it was Cambridge, not Oxford, the reputed "home of lost causes," that produced the greater number of Nonjurors, the Cambridge total being forty-two, that of Oxford a mere fourteen. [12]

Along with the political and religious upheavals that were associated with the Interregnum and the Restoration, the Cambridge that Barrow knew was also changing in its curriculum and general intellectual orientation. When Barrow first came up to the university, much of the traditional scholastic curriculum remained intact and was, despite considerable adaptations and revisions, still in continuity with that established in the universities in the High Middle Ages. A growing number of dons within Cambridge were taking an interest in scientific developments that would call into question the still largely Aristotelian-based premises on which the traditional scholastic curriculum was based, but this, as yet, had had little impact on the undergraduate curriculum. By the time of Barrow's death in 1677, however, the scholastic natural philosophy had been largely replaced by Cartesianism, thus destroying the neatly integrated character of the traditional curriculum, the longevity and tenacity of which had owed much to its ability to use the same method for all areas of knowledge. Moreover, within a few decades of Barrow's death, Cambridge's curriculum in natural philosophy was to change again, with Cartesianism giving way to Newtonian physics and an emphasis on mathematical inquiry that has remained one of the most notable characteristics of Barrow's alma mater.[13]

Not only was seventeenth-century Cambridge changing politically, religiously, and intellectually, but its significance within the wider community was also changing as the size of its student body contracted. Pre-Civil War Oxford and Cambridge educated a larger proportion of the national population than were to go to university until after the First World War.[14] During the Interregnum the number of students did fall below pre–Civil War levels, but not as drastically as one might have expected in such a troubled period: The average annual number of matriculations in the 1630s was 373; in the 1640s it was 262; and in the 1650s 254. After the Restoration there appeared to be a trend back to pre-1642 levels, with an annual average of 304 in the 1660s, but thereafter there was a steady decline – 290 in the 1670s, 226 in the 1680s, and 191 in the 1690s – the university reaching its nadir in the 1760s with an annual figure of 112.[15]

The reasons for such changes are obscure. Perhaps they reflect a tendency on the part of the post-Restoration governing classes to restrict opportunities for higher education after the upheavals of the mid-seventeenth century. The duke of Newcastle, for example, advised Charles II to reduce the universities' intake lest they produce more of the Puritan lecturers who "preached your Majesty out of your kingdom."[16] The drop in the number of students was probably also related to a decline in the upper classes' interest in the religious issues that had so dominated the mental outlook of the early seventeenth century. Whereas in the early seventeenth century an education at the clerically dominated universities was a means of coming to terms with the burning theological issues of the time, in the late seventeenth century and, even more markedly, in the eighteenth century the laity showed less and less interest in religious problems and considerably more antagonism toward the clergy. As fewer of the sons of the well-born came to the universities for a general education, their social role became increasingly (though not exclusively) restricted to that of seminaries for intending clergy.[17] Such developments, however, were only beginning to become apparent by the time of Barrow's death in 1677; thus, his career corresponds largely to a period when the number of students (relative to population) and the influence of the university were near their pre-twentieth-century peak.

Barrow's time at Cambridge, then, was remarkable for the number and extent of the changes, whether political, religious, intellectual, or social, that were to alter considerably the character of the university. A detailed examination of Barrow's Cambridge career will help to illustrate these basic transformations, which, in turn, mirrored the tensions and innovations that were reshaping the outlook of England during the turbulent seventeenth century.

Barrow's original decision to enroll at Peterhouse in December 1643 was significant – even though he subsequently matriculated at Trinity – since it is an indication of the strong Royalist and episcopal sympathies of his father, a London merchant who later went into exile with the future Charles II. Peterhouse had been singled out for particular attention by Parliament as a center of Laudian innovations in religion. On 22 January 1640/1 the House of Commons had declared that its master, Cosin, was "guilty of bringing superstitious innovations into the Church tending to idolatry,"[18] and at Laud's trial in 1644 one of the charges brought against him was that he had countenanced superstitious practices at Cambridge and, in particular, had permitted Peterhouse to set up "a glorious new Altar...to which the Master, Fellowes, Schollers, bowed."[19] Not surprisingly the college was purged with particular vigor by the parliamentary visitors in 1644–5 – indeed, all save one of its fellows were ejected.[20]

Among the ejected was Barrow's uncle and namesake, Isaac Barrow (1614–80), who was to have acted as his tutor. After Isaac Barrow (the elder) lost his Peterhouse fellowship, he left for the Royalist stronghold of Oxford with his close friend, the influential high churchman Peter Gunning, who had been ejected from Clare. At Oxford both men were made college chaplains by Robert Pinck (1573–1647), the warden of New College and a close associate of the recently executed Laud. There Barrow remained until Oxford fell to the parliamentary forces in 1646 and thereafter lived quietly until the Restoration, when his loyalty was rewarded with his reinstatement to his Peterhouse fellowship in 1660 and appointment to the lowliest of English sees, that of Sodor and Man, in 1663, the sermon at his consecration being preached by his nephew. As bishop of Sodor and Man (where he remained until he was translated to the only slightly more prestigious see of St. Asaph in 1670) Barrow gained the reputation of being one of the most conscientious and forceful of English bishops. He was also known for his strong high church views on both theology and church order and, within the limited sphere of the Isle of Man, carried out some of the reforms that Laud had vainly attempted to implement in the nation as a whole. In particular, he regained for the church control of church revenues that had come into lay hands.[21]

Though the scanty biographical records of the younger Barrow's life say little about the role of his uncle, one may suspect that he played an important role at a number of critical points in Barrow's career, and he probably did much to maintain and strengthen his nephew's Royalist and episcopal sympathies even while the younger Barrow was immersed in Interregnum Cambridge, where such views were under constant attack. Significantly, the main contemporary life of Barrow by Abraham Hill notes that it was his uncle who, in the early 1650s, persuaded him to take the critical step of specializing in divinity rather than continuing with medicine.[22] After he became bishop of St. Asaph in 1670, Bishop Barrow provided his nephew with a small sinecure in Wales and may have played a role in prompting influential episcopal patrons to look kindly on the younger Barrow.

When young Isaac Barrow's plans to enter Peterhouse were aborted by his uncle's ejection, Barrow (the elder) probably played a role in recommending his nephew to Henry Hammond, the most important figure in preserving and reformulating the Anglican tradition during the Interregnum.[23] As part of his plans to revitalize Anglicanism Hammond provided financial assistance to support promising young Anglicans in their studies.[24] Among his beneficiaries was Barrow, who, thanks to Hammond, was able to complete his studies at Trinity when he enrolled there in February 1645. In gratitude to Hammond, Barrow composed his epitaph in 1660[25] – a piece that was effusive in its praise of Hammond's virtues even by seventeenth-century standards.

Superficially, Trinity might have seemed a strange choice of college for a young Royalist. In the same academic year in which Barrow enrolled at Trinity, the parliamentary visitors ejected the master (Thomas Comber) along with forty-nine fellows.[26] Thus, more than two-thirds of the fellowship[27] was replaced by new fellows, most of whom were chosen for their Presbyterian sympathies. The new master, Thomas Hill, a graduate of Emmanuel and a foundation member of the Westminster Assembly, was among the university's most stalwart champions of Calvinism. Two years before, in 1643, he had urged the Commons to institute a "purging and pruning" of Laudian innovations within the university. Evidently Hill himself had been a victim of Laudian intolerance since, in the dedication to his *Six Sermons,* he thanked the earl of Manchester for "protecting me in the exercise of *my Ministry* from *Prelaticall Tyranny* in the worst of times."[28] He also praised Manchester for "your new modelling Cambridge" – a reference to the purge of 1644–5 – than which "there was never a greater change with fewer mistakes."[29] When he became master, Hill found that there were still "divers [fellows] of *those* [principles] *opposite* to the *intended* Reformation,"[30] though most of these were removed for failing to take the Engagement Oath of 1650 to judge by the remarks of the influential high churchman William Sancroft, who commented to his brother in November 1650 that "our friends of Trinity are out, and others in their places."[31] Nonetheless, in March 1653 Sancroft's correspondent, Paman, could still report that someone had had the temerity to use the Anglican Book of Common Prayer in the Trinity Chapel, to the immense annoyance of Hill.[32]

Despite such intermittent opposition Hill appears to have been largely successful in making Trinity a virtual seminary of Presbyterians. Oliver Heywood, who came up to Trinity two years after Barrow, spent much of his time there studying the works of Calvinist theologians such as Perkins and Preston;[33] true to his teaching Heywood became a Presbyterian teacher after the Restoration. When Hill died in 1653 his Calvinist zeal was maintained by his successor, John Arrowsmith, a fellow founding member of the Westminster Assembly. Although Arrowsmith opposed any deviation from Calvinist orthodoxy and regarded "every variation from unity" as "a step to nullity,"[34] he was rather more conciliatory in his methods than Hill, whose imperious manner of dealing with fellows who did not share his political and religious views had caused considerable dissension within the college.[35] Arrowsmith's greater caution owed much to his experiences as master of St. John's from 1644 to 1653, where the political and religious conflict among the fellows was so great that in October 1647 the House of Commons formed a committee to "examine the Information given in concerning Malignants chosen Fellows in St. John's College, or any other College in the University of Cambridge; and the other Informations concerning the praying for Bishops, and using the Book of Common Prayer."[36] The success of Hill and Arrowsmith in implanting their religious views within the college is evident from the fact that after the Restoration it was Trinity that produced the greatest number of nonconformists within the university, its total being sixteen, while no other college produced more than five.[37]

Why, then, did a young Royalist like Barrow choose Trinity when even in 1645 its strongly Presbyterian character must have been evident? According to John Aubrey, Barrow enrolled at Trinity because of an offer of support from an (unspecified) member of the Walpole family, one of his former school friends[38] – an offer that was perhaps prompted by the Walpoles' sympathy for Barrow's adherence to the Royalist cause.[39] The Walpoles not only influenced Barrow's choice of college, but also probably played a part in placing the young Barrow under James Duport. The Walpole family's close and continuing links with Duport are evident in a letter addressed by him in November 1654 to "my very worthy and much honoured Friend Edward Walpole."[40] Appropriately, Edward Walpole later entrusted his son, Robert, to the care of Barrow, thus maintaining the family's links with Trinity. The Walpoles were but one example of

Duport's close connections with a number of prominent landed families who had sent their offspring to Trinity because of Duport's considerable reputation as a tutor. Such connections, together with his high reputation as a scholar, help to explain why the college continued to tolerate Duport despite his overt Royalist and Anglican sympathies.[41] Indeed, Duport was elected vice-master in 1655 despite having been deprived of the Regius Professorship of Greek in the previous year for refusing to subscribe to the Engagement.

Along with his uncle, Duport was the young Barrow's most influential intellectual mentor, and his presence in Trinity helps explain why someone with Barrow's political and religious sympathies nonetheless developed considerable affection for the college. Duport's vain attempt to obtain the Regius Professorship of Greek for Barrow in 1654 – a post Barrow was denied because of his Arminianism[42] – is an indication of his high regard for Barrow's abilities. Barrow's affection for his teacher is, in turn, evident in a number of his academic addresses and, in particular, in his inaugural address as Regius Professor of Greek (to which post Barrow was finally appointed in 1660), in which he comments with heavy, but affectionate facetiousness on Duport's diminutive size in contrast with the large scope of his learning.[43] It was the protection of Duport – together with Barrow's own abilities – that probably accounts for the fact that Barrow was spared the normal requirement of subscribing to the Covenant and, subsequently, the Engagement (though five other fellows of Trinity were ejected for refusing to make the latter undertaking).[44] It may also have been Duport who persuaded the master to make light of Barrow's Royalist outburst in his college oration of 1651. Certainly, Hill's reported remark to Barrow – "Thou art a good lad; 'tis pity thou art a cavalier"[45] – is out of keeping with the picture of a Puritan martinet that emerges from other sources. In 1646, for example, he imprisoned a fellow by the name of William Wotton for declaring that Parliament was more rebellious than the Irish and had him tried by Parliament, which deprived Wotton of his fellowship. However, Hill may have been more tolerant of those with Royalist sympathies than this incident might suggest, since in 1649 the parliamentary visitors took him to task for permitting two fellows to remain at Trinity who "are proved Delinquents for sending Plate for the King."[46]

Though Barrow's Arminianism and his support for the episcopal order would have set him apart from many of his Trinity contemporaries who were imbued with orthodox Calvinism, Interregnum Cambridge was far more theologically diverse than the predominantly Presbyterian character of the parliamentary appointees might suggest. Barrow's notes on the sermons he attended as an undergraduate[47] show that he was exposed to a range of theological views even in the period before the execution of the king in 1649, when the influence of the Presbyterians was greatest. As might be expected Barrow attended the orthodox Calvinist sermons of Thomas Hill and John Arrowsmith, but he also took notes from sermons by Richard Vine (the master of Pembroke from 1643 until 1650, when he was ejected for refusing the Engagement), who was to question the doctrine of predestination, and even listened to the discourses of Robert Boreman, an ejected Royalist. Barrow also heard the Cambridge Platonist Ralph Cudworth and regularly attended the Sunday afternoon sermons of Benjamin Whichcote, who greatly influenced the Cambridge Platonists.

Tillotson, a near contemporary of Barrow at Cambridge, later wrote of Whichcote that he contributed "more to the forming of the students of that university in a sober sense of religion than any man in that age;"[48] more caustically, Salter (Whichcote's eighteenth-century biographer) remarked that Whichcote's discourses served to counteract the "fanatic enthusiasm and senseless canting" then in vogue.[49] Indeed, Whichcote's success in encouraging what Salter called "a nobler, freer, and more generous set of opinions"[50] among the younger members of the university was a source of considerable concern to orthodox Calvinists and in particular to the university's preeminent Calvinist theologians Hill, Arrowsmith, and Tuckney.[51] As a result Anthony Tuckney, Whichcote's former tutor at Emmanuel and (like Hill and Arrowsmith) an inaugural member of the Westminster Assembly, began a correspondence with Whichcote in which he vainly endeavored to persuade him to return to the Calvinist fold.

The fact that Whichcote remained undisturbed as provost of King's (despite not having taken the Covenant) is an indication of the theological diversity that was still permitted within Interregnum Cambridge. A former pupil of Tillotson's from the 1650s remarked that this period had been "a time of freedom," and although "the most prevailing men were generally contra-remonstrants" (orthodox Calvinists) – of whom the most influential was Thomas Hill – "there were [also] divers young preachers came up in those times, who were of a freer temper and genius; such as were Mr. SAMUEL JACOMB, Mr. BRIGHT of *Emanuel College,* Mr. PATRICK, now Bishop of *Ely,* and others." Tillotson himself attended sermons of various theological hues and "seem'd to be an eclectic man, and not to bind himself to opinions."[52] The notebook of Charles North (who was admitted to Sidney Sussex in 1651) also indicates that students were exposed to theological opinions that diverged from the dominant Calvinism; thus, North took notes from preachers like Cudworth and Richard Love (master of Corpus Christi, 1632–61) who, by their emphasis on good works, appear to have called into question the Calvinist emphasis on *sola fide.*[53]

The fact that Interregnum Cambridge was far from being monolithically Calvinist is underscored by the extent and vitality of the debate within the university between the orthodox Calvinists and their Arminian opponents, who rejected predestination and argued that salvation was potentially open to all. This was a controversy that had been gathering force since 1626, when Charles I's favorite, the duke of Buckingham, had been appointed chancellor and, with royal support, had proceeded to attempt to silence predestinarian preaching. As Arminianism became more entrenched within Cambridge during the 1630s, so, too, did a more militant form of Calvinism[54] – a theological polarization that continued to color university life throughout the Interregnum. These debates became particularly heated in the 1650s after the eclipse of the Presbyterians' power. Simon Patrick (whom Barrow certainly knew well after the Restoration and possibly before) described this period as one when "the general Outcry was, that the whole University was over-run with Arminianisme, and was full of men of a Prelatical Spirit, that has apostatized to the Onions and Garlick of *Egypt,* because they were generally ordained by Bishops."[55] Patrick himself was particularly subject to accusations of Arminianism because he based his tutorial instruction on the work of the Laudian Henry Hammond.[56] His early opposition to predestination had been strengthened by the teaching of his mentor, the Cambridge Platonist John Smith, who "made me take the liberty to read such authors (which were before forbidden me) as settled me in the belief that God would really have all men to be saved."[57] Simon Patrick's library catalogue from this period gives some indication of what these forbidden books would have been since it includes works by prominent Arminian theologians such as Episcopius, Grotius, and Hammond.[58]

The spread of Arminian ideas produced in turn a counterattack by the Presbyterians. In his preaching Hill was said to have made a deep impression on the undergraduates of Trinity by laying "his hand upon his breast and say[ing] with emphasis, 'Every Christian hath something here that will frame an argument against Arminianism.'"[59] At his death in 1653 Hill, according to his friend and executor Tuckney, "had made fair progress in a learned confutation of the great daring champion of the Arminian errors, whom the abusive wits of the university, with an impudent boldness, would say none there durst adventure upon."[60] Hill's target was John Goodwin, a former fellow of Queens', who, though of republican sympathies, had been ejected from his living in 1645. According to that zealous defender of Calvinist orthodoxy Thomas Edwards, Goodwin was "a monstrous Sectary, a compound of Socinianism, Arminianism, antinomianism, independency, popery, yea and of scepticism."[61] Tuckney's opposition to the preaching of Whichcote owed much to the fact that he regarded his former pupil as having been infected by Arminianism and, in particular, by the work of John Goodwin. In their famous correspondence of 1651 Tuckney wrote of Whichcote's sermons that "those whose footsteps I observed were the Socinians and Arminians, the latter whereof,

I conceive, you have been everywhere reading." To which Whichcote replied that "truly I have read more Calvin and Perkins and Beza [two prominent Calvinist theologians] than all the books, authors or names you mention."[62] Indeed, Whichcote denied having read the Arminians' manifesto – the *Apologia remonstratium* – and much of his work, like that of More's, appears to have arrived at conclusions similar to those of the Arminians by an independent route.[63] Though Whichcote complained of the intolerance of the Presbyterians, "who indeede professe some zeal for that happie point of 'Justification by Faith,' yet are sensiblie degenerated into the devilish nature of malice, spright, furie, envie, revenge,"[64] the correspondence was generally conducted in a friendly spirit. Neither side, however, gave ground. Tuckney's continuing suspicion of Whichcote's popular preaching was apparent in his university sermon of 1652, in which he insisted on an orthodox Calvinist understanding of justification as against the teaching of those like Whichcote who, he claimed, maintained that one could be saved through "a *Philosophical faith,* or the use of *right Reason,* or a *virtuous morality,* too much now admired and cried up. As of old, *The Temple of the Lord, The Temple of the Lord.* So now, The Candle of the Lord, The Candle of the Lord"[65] – an obvious reference to Whichcote's phrase to describe the role of reason.

These theological debates within Interregnum Cambridge were to leave their mark on Barrow's mature theology. True to his early Arminian beliefs (which had cost him the chair of Greek in 1654) Barrow was later to devote considerable attention to an attack on the doctrine that came to be regarded as the hallmark of orthodox Calvinism: predestination. Thus, Barrow devoted three complete sermons to defending the Arminian view that salvation was potentially available to all believers in the course of which he explicitly rejected Calvin's view of faith.[66] After the Restoration, Arminianism began to lose its distinctiveness and came to be equated with a simple negation of Calvinism.[67] Barrow certainly did not shrink from attacking Calvinism, but he also retained some of the more positive aspects of Arminianism, not only in his emphasis on the doctrine of universal redemption, but also in his stress on the role of the Holy Spirit, particularly in the process of justification.[68] Barrow also made frequent use of the distinction between fundamental and nonfundamental doctrines that formed part of the Arminian attempt to reduce theological conflict.[69]

Arminianism had attempted to place a greater emphasis than Calvinism on the ethical dimension of faith, and this, along with Barrow's wariness of theological "speculation," caused him to take a rather ambivalent attitude toward the central doctrine of the Reformation: justification by faith alone. In contrast to Luther, Barrow attempted to avoid too sharp a dichotomy between faith and works in his concept of justification. He argued, for example, that justifying faith denoted not only "acts of the mind" but also "acts of the will" in the form of good works.[70] Indeed, he seemed to associate thoroughgoing Solafideism with Antinomianism, as when he commented that St. James's epistle was intended to prevent "Solafidian, Eunomian and Antinomian positions, greatly prejudicial to good practice."[71]

In his patently anti-Calvinist sermon, "The Doctrine of Universal Redemption Asserted and Explained," Barrow set out to demonstrate that salvation was available even to those who had not been exposed to scriptural revelation since they could come to a knowledge of God through natural law. Thus, he argued that "we may substitute with St. Paul, for the law of revelation engraved upon tables, the law of nature written in men's hearts; for prophetical instruction the dictates of reason."[72] Accordingly Barrow maintained that "we may at least discern and shew very conspicuous footsteps of divine grace . . . even among pagans."[73] This line of argument had been employed by those like Hooker who had set out to refute the Puritans' claim that the Church of England should be bound solely by the explicit text of scripture rather than by custom: "Doth not the Apostle term the law of nature," wrote Hooker, "even as the Evangelist doth the law of Scripture, God's own righteous ordinance?"[74] As the example of Hooker indicates, an emphasis on the theological significance of natural law was not confined to those who were influenced by the Arminians, but, thanks particularly to Grotius, the Arminian tradition became closely identified with such a position.

Barrow's Arminianism is one indication of his links with Laudian Anglicanism; another is his emphasis on the role of the episcopate. Like his early mentor, Henry Hammond, Barrow took the view that the position of bishop was not simply a useful administrative device (the commonly held view, even in Anglican circles, before the days of Archbishop Laud), but rather was an essential feature of the church. Thus, Barrow described bishops as those "who derive their authority by a continual succession from the Apostles" and who are appointed in a manner "agreeable to the institution of God, and the constant practice of his Church."[75]

Barrow also shared with the high church Anglican tradition an emphasis on the role of church tradition and, in particular, on the importance of the early church fathers, quotations from whom are lavishly sprinkled throughout his works. To one such as Barrow the Civil War served to confirm the view that scriptural doctrine should be mediated through the teachings of the church, rather than being open to a purely individual interpretation, which, in Royalist eyes, had resulted in anarchy. On the question of justification, for example, Barrow attempted to overcome the conflicting interpretations of Scripture by appealing to "the general consent of Christians, and . . . the sense of the ancient Catholic Church."[76] As the title of the sermon "Of Obedience to Our Spiritual Guides and

Governors" would suggest, it is the Puritans that he has in mind in arguing, "Can we mistake or miscarry by complying with the great body of God's Church through all ages, and particularly with those great lights of the primitive Church, who, by the excellency of their knowledge, and the integrity of their virtue, have so illustrated our holy Religion?"[77]

Another ground for the appeal to tradition arose out of Barrow's conflict with the Antitrinitarians, notably the Socinians. In his introduction to Barrow's *Treatise on the Creed* the publisher recommends the book as a means of combating "the gross Error of Socinus,"[78] an issue to which Barrow returned in another of his works, where he wrote that in "affirming the Holy Spirit to be a person, [we] do thereby intend to exclude the opinion of Socinus and his followers."[79] In arguing against Socinus's views, Barrow naturally turned to scriptural exegesis, but in an attempt to circumvent further disputes about scriptural interpretation he appealed to the *consensus ecclesiae:*

> To all the premised points no small accession of weight doth come from the authority of so many holy Fathers and Councils; and from the consent of the Church running down through so many ages; to oppose which, without very weighty and manifest reasons, doth as much recede from prudence, as it is far from modesty.[80]

Though the Protestant Reformers had at first rejected the apparatus of scholastic theology, Protestant universities, including Interregnum Cambridge, had once more assimilated this tradition into their intellectual armory as part of the continuing battle with Roman Catholic theologians whose work was couched in scholastic language.[81] As Archbishop Bramhall of Armagh (M.A. Sidney Sussex, 1616) put it, "It may be the Schoolmen have stated many superfluous questions and some of dangerous consequences but yet the weightier ecclesiastical controversies need the help of their distinctions."[82] Barrow's own training in this tradition of controversial divinity can be seen in passages such as his criticism of Bellarmine's theory of justification, in which, Barrow argues, Bellarmine "blunders extremely, and is put to his trumps of sophistry" since he confuses the "qualities" of sanctification and justification when "the inherent unrighteousness consequent upon Adam's sin is not included in God's condemn[ation]."[83]

In a speech of 1654 Barrow expressed his contempt for "puerile speculations: such as *de entibus rationis, de materia prima,* and the like scholastic chimaeras,"[84] but elsewhere he indicated that he regarded his scholastic education in a more positive light, particularly for its training in mental discipline. Thus, in his speech as moderator in the schools in 1651, he commented:

> Those unpleasant Scholastics do not deserve to be laid aside altogether.... None of them may have erected much in the way of experimental science but rather elaborate structures on flimsy foundations... [but] they will instruct you... how to debate concisely and correctly, how to prosecute an inquiry judiciously, how to experiment warily.[85]

The continuing influence of Barrow's early scholastic education[86] is apparent even in the structure of his sermons. The biblical text becomes a proposition: Its meaning is first minutely examined, and something like a definition of terms is set up; on this is based a discussion of the various aspects of Christianity to which the text is relevant; and Barrow concludes by reaffirming the text and pointing out the ways in which it may be applied. Barrow's thorough grounding in the techniques of scholastic debate is evident, too, in the readiness with which he employs terms drawn from classical logic, even in his theological writings. In his *Exposition of the Creed* he calls the scholastics overly subtle and implies that they have distorted Christianity by their methods; yet in the same work, he uses terms drawn from scholastic logic to clarify his argument with phrases such as "to prove God is veracious because he saith so, or that revelation in general must be trusted from particular revelations, are *petitiones principii,* most inconclusive and ineffectual discourses."[87] Clearly, then, Barrow's early scholastic training left a deep impression on the framework of his thought.

Though this training left its mark on his work, Barrow's praise of the scholastics was, as we have seen, rather grudging. However, his respect for Aristotle, on whose principles scholasticism had been largely, though not exclusively, based, remained undimmed. To Barrow Aristotle continued to be "the unchallenged Prince of all who have ever been or ever will be philosophers."[88] Faithful to the teaching of Duport, who had urged his students, "In your answering reject not lightly the authority of Aristotle, if his own words will permitt a favourable and a sure interpretation,"[89] Barrow departed from Aristotle's conclusions only with the greatest reluctance. When, in his *Mathematical Lectures,* he dissents from Aristotle's views on mathematical first principles, he adds that "I speak with all due Regard to the Honour of that great Man, for whom, notwithstanding, I preserve the greatest Veneration and Esteem, by reason of his most profound Wisdom and extensive Learning, and would have all Men to do the same."[90] In this same work Barrow also castigates Ramus, who, by arguing that Aristotle had given geometry a disproportionate importance,[91] had impugned two of Barrow's intellectual deities: Aristotle and Euclid. Ramus's views had caused considerable debate within the university at the end of the previous century,[92] and the reverberations of this controversy continued during the Interregnum, since Duport had warned

his students to "follow not Ramus in Logic... but [rather] Aristotle."[93]
This background helps to explain Barrow's impassioned denunciation of
Ramus as one who had "defame[d] those Builders, and first Propogators
of the Sciences... without whom perhaps this mighty Artist at Trifles had
had nothing but Fables to apply his wonderful Logic in the Resolution
of."[94]

Barrow's grounding in the scholastic tradition and his introduction to
Aristotle derived from his undergraduate studies, which would have been
closely supervised by Duport. As a candidate for the M.A. degree, however, he would have been permitted considerably more freedom to direct
his own reading; indeed, the formal requirements for the degree appear
not to have been enforced after 1608. Barrow's wide reading and intellectual daring are apparent in his choice of topic for his M.A. oration
in 1652 – "That the Cartesian Hypothesis Concerning Matter and Motion Scarcely Satisfies the Phenomena of Nature" – an examination of the
still novel and controversial philosophy of Descartes. This speech indicates the growing interest in Cartesianism within Cambridge and, more
generally, England in the 1650s. This owed much to the enthusiasm of
Henry More, who at first regarded Descartes's work as offering a bulwark
for Christian belief against the rising tide of philosophical materialism,
though, as Gabbey has shown, More's apologetical intent shaped his attitude toward Cartesianism, with the result that he at first downplayed
those mechanistic aspects of Descartes's system that later prompted him
to reject it.[95]

More's early interest in Descartes was shared by his fellow Cambridge
Platonist, John Smith of Queens' (d. 1652), and (in a more qualified manner) by Ralph Cudworth, who may have prompted More to begin his
famous correspondence with Descartes in 1648.[96] Around the same period
Henry Power, an undergraduate at Christ's, began to take an interest in
Descartes's work,[97] no doubt thanks to More, who was a fellow of the
same college. In the following year John Hall, a fellow of St. John's,
whose work had shown signs of Cartesian influence as early as 1646, published a pamphlet addressed to Parliament in which he described the university as being preoccupied with "a jejune Peripatetic philosophy; suited
only (as Monsieur Descartes sayes) to witts that are seated below medi-
ocrity."[98] By the early 1650s Cambridge's reputation as a center of interest
in Cartesianism was such that Joseph Glanvill (who took a B.A. at Oxford in 1655) lamented that his friends did not first send him to Cambridge,
because... that new philosophy and art of philosophizing were there more
than here in Oxon."[99] Among those exposed to Descartes's work in this
period was Barrow's close friend, John Ray;[100] the prominent high churchman William Sancroft also took an interest in Descartes's work some time
before he was expelled from his fellowship at Emmanuel in 1651.[101] One
of the most enthusiastic Cambridge disciples of Descartes was Gilbert

Clerke, who appears to have been acquainted with Barrow before resigning his Sidney Sussex fellowship in 1655;[102] five years later Clerke produced a defense of Descartes entitled *De plenitudine mundi brevis et philosophia dissertatio, in qua defenditur cartesiana philosophia contra sententias Francisci Baconis de Verulamio, Thomas Hobbii Malmesburiensis, et Sethi Wardi*...

Barrow, then, showed considerable intellectual independence in choosing to offer a critique of Descartes's work at a time when senior members of the university like More still viewed Descartes's work as a support for revealed religion. Indeed, in 1655 More was still claiming of Cartesianism that "there is no philosophy except perhaps the Platonic, which so firmly shuts the door against atheism."[103] Perhaps Barrow's caution about the implications of Cartesianism owed something to Duport, but, unlike the determinedly Aristotelian Duport, it is clear that Barrow had read the French philosopher with care. He described him as a "most ingenious and important philosopher" and, significantly, praised his "extraordinary skill as a mathematician." Though Barrow conceded that Descartes's philosophy offered a highly plausible explanation of natural phenomena he nonetheless had reservations about the whole attempt to explain nature in solely mechanical terms. He contrasted Descartes's work with that of the "Chemists, those who are called Natural Magicians, or those who profess to know about sympathies and antipathies." He turned, too, to Plato and Aristotle for an alternative to Cartesian mechanism. Later, he contrasted their belief that "everything is crammed with soul or that a vital spirit is diffused through the Universe to preserve and sustain all things" with Descartes's theory of matter and motion, which he took to mean that God "created one homogeneous Matter, and extended it, blockish and inanimate, through the countless acres of immense space, and, moreover, by the sole means of Motion directs these solemn games and the whole mundane comedy." Barrow also contrasted Cartesianism unfavorably with the atomistic theories that had recently been revived by Gassendi and Kenelm Digby and, with even more relish, cited Bacon, "*noster Verulamus,*" as "the supreme Champion on this battlefield...A man of indisputable judgment and repute, who damned his philosophy before it was born. For several times in his *Organon* he most carefully warned us against all such general hypotheses." In Barrow's view Descartes would have "incurred Bacon's censure" for having begun with "metaphysical truths" rather than empirical data. This argument is later given further force by Barrow's discussion of the characteristics of plants, which, he contends, cannot be explained solely in mechanical terms[104] – a line of argument that probably owed much to the influence of his friend and colleague John Ray. Whereas Descartes's philosophy presented a whole world view based on a consistent metaphysics, Bacon's empiricism had the attraction, to one such as Barrow, of having more limited goals and therefore being a lesser threat to the long-standing amalgam of Aristotelian philosophy and Christian theology to which Barrow was still committed.

Barrow's response to Descartes raises the more general question of his relationship with those involved in the best-known intellectual movement within Interregnum Cambridge: the Cambridge Platonists. Assessments of Barrow's relationship with this group vary considerably. According to Tulloch, Barrow was "almost the only great name in Cambridge at this time that remained uninfluenced, or nearly so, by the new movement [of Cambridge Platonism]."[105] Hurlbutt and Baker, in contrast, see Barrow as a member of the Cambridge Platonist school.[106] No one would question, however, that the Cambridge Platonists formed an important part of Barrow's intellectual milieu and ought to be considered in any survey of the influences that shaped Barrow's thought.

At a personal level Barrow probably knew John Smith, whose work he praised in his inaugural lecture as Regius Professor of Greek in 1660. He also regularly attended the sermons of Benjamin Whichcote, who exercised a considerable influence over the theological development of the Cambridge Platonists. Whichcote also intervened in favor of Barrow when he was a candidate for the Regius Professorship of Greek in 1654 – a position that he was nonetheless refused on the grounds of his Arminianism, the post being given to Ralph Widdrington at the direction of Cromwell[107] (presumably thanks to the influence of his brother, Sir Thomas, a member of the Treasury Commission, 1654–9). However, Hill records that Barrow "always acknowledged the favour which Dr. Whichcote shewed him on that, as on all occasions."[108] By 1654, too, Barrow was known and admired by John Worthington (master of Jesus, 1650–60),[109] who had close links with the Cambridge Platonists.

Although it is difficult to discern any obvious links between Barrow's work and that of Smith, Whichcote, or Worthington, there are some indications that the work of Henry More – the most prolific of the Cambridge Platonists – did leave its mark on Barrow's thought. As Dobbs suggests the two men may have been drawn together by their common interest in alchemy and chemistry[110] – some of the activities of the group of predominantly biomedical experimentalists that was centered around John Nidd, a fellow of Trinity, and that included Ray and, to a lesser extent, Barrow. More's influence within this group is clearly evident from the considerable intellectual debt that Ray owed the Cambridge Platonist; indeed, More's *Antidote Against Atheism* was the starting point for Ray's magnum opus, *The Wisdom of God Manifested in the Creation*.[111] Barrow and More appear to have been on friendly terms after the Restoration despite the fact that after 1660 More came under attack from some high churchmen (including a number of Barrow's friends). Thus, Worthington associates the names of More and Barrow in a letter written soon after the Restoration;[112]

both figures are also linked in a letter of Oldenburg to Newton of 1673 in which he asks Newton to deliver a book by Boyle to both Barrow and More "if they be now at Cambridge."[113]

Barrow's close study of Descartes, evident in his M.A. oration, may well have owed much to More's influence. In the same speech Barrow places considerable emphasis on the role of nonmaterialist explanations of natural phenomena, an approach that may also have been influenced by More. Thus, Barrow refers to the work of the chemists and alchemists, the natural magicians, and Plato and Aristotle's belief that there was a "vital spirit" at work in nature, an approach that has an obvious similarity with More's view that natural phenomena could be explained only by "some inward principal of life and motion" or what he elsewhere calls "a seminal form [which] is a created spirit organizing duly prepared matter into life and vegetation."[114] Barrow appears to endorse a similar view both in his M.A. oration and in one of his post-Restoration sermons, in which he asks,

> Who can imagine those admirable works of nature, the seminal propagation and nutrition of plants, and however more especially the generation, motion, sense, fancy, appetite, passion of animals to be accomplished by a mere passive agitation of matter, without some active principle distinct from matter, which disposeth and determineth it to the production of such effects?

Interestingly, however, Barrow goes on to ascribe such a view to Aristotle, indicating the basically Aristotelian frame of his thought even though on this issue he appears to be in sympathy with the Cambridge Platonists. "If God could...produce and insert such an active principle," he asks, "(such a [Form] as the Philosopher [Aristotle] calleth it) why might he not as well produce a passive one, such as the matter is?"[115]

Cudworth, whose attitude toward Cartesianism was always more qualified than that of More, also ascribed considerable importance to the active powers of nature. In his *True Intellectual System of the Universe* he, like Barrow, contrasted this organic view of nature with Cartesian mechanism, which, he argued, involved "rejecting all plastic-nature, [since] it derives the whole system of the corporeal universe from the necessary motion of matter."[116] Though Cudworth's work was not published until 1678, Barrow may have been acquainted with his reservations about Descartes's work while still a candidate for the M.A. – particularly since we know that Barrow attended Cudworth's sermons while an undergraduate. However, Barrow's invocation of Bacon and his discussion of the inadequacy of Cartesianism in explaining the behavior of animals are less characteristic of the Cambridge Platonists, who tended to regard empirical investigation as less important than metaphysics. Thus, More contrasted the "blundering Naturalist" with the "deep-searching soul,"[117] and Cudworth regarded some of Bacon's work as "tending to irreligion."[118]

The Cambridge Platonists in general, and Henry More in particular, sought to combat the drift toward philosophical materialism by attempting to show that the spiritual realm was endowed with some of the same characteristics as matter and therefore had a similar reality. Descartes had seen the characteristic feature of matter as extension. More, therefore, set out to demonstrate that extension was also an attribute of spirit. The extension of spirit, to More, is space, since space can exist without matter, and absolute space became for More an aspect of the divine sensorium – an indication of God's involvement in the natural order.[119] More first expounded his views on absolute space in 1655 in his *Antidote Against Atheism,* a work that Barrow had in his possession along with another four books by the same author.[120] This, together with the other links between Barrow and More, makes it highly probable that it was to More that Barrow was indebted when he expounded a similar theory of space in his inaugural mathematical lectures as Lucasian Professor in 1664.

In these lectures Barrow outlines a number of arguments to demonstrate that space exists independently of matter. First, "since Matter may be finite, and God in infinite in Essence, he must subsist beyond the Bounds of Matter, otherwise, he would be enclosed within its Limits or some Way bounded, and therefore could not be infinite. Therefore something is beyond, *i.e.* some sort of *Space*." Barrow next envisages the possibility of God creating new worlds, which would need space to contain them. This thought experiment then leads him to conclude that God must "have before been present in the *Space,* where they [the new worlds] are now reposited." Barrow then argues in a manner that makes plain the apologetical purposes that he shared with Henry More that "the Conceptions which we have, or ought to have, concerning the Divine Infinity, Power, and Immutability do involve some distinct *Reality* of *Space*."[121] Not surprisingly Barrow is at pains to refute both Descartes's view that matter is infinitely extended and the view of space held by Hobbes, who, consistent with his philosophical materialism, would not allow space to have a reality of its own but to be only the phantasm of reality.[122]

Barrow and More, then, had much in common in their treatment of space, though Barrow does introduce some important modifications. In the first place Barrow argues for the existence of absolute space not only on metaphysical and theological grounds but also on geometrical grounds. He argues, for example, that since two spheres can meet at only one point there must be space between them.[123] He concludes, then, that the concept of absolute space is "most agreeable, and abundantly sufficient for Geometricians."[124] Barrow also gives greater weight to experimental evidence than More. Whereas More largely dismissed Boyle's experiments with the vacuum pump,[125] Barrow regarded them as a useful confirmation of his concept of space since although a vacuum contains no matter it has space and therefore extension.

Barrow also showed his intellectual independence from More by extending the arguments for absolute space to demonstrate the existence of absolute time. As a self-proclaimed Platonist, More was less inclined to deal with the concept of time than the more Aristotelian-orientated Barrow, since in Plato's philosophy the process of change and the flow of time are given a far more subordinate role than they are in that of Aristotle.[126] But whereas Aristotle had viewed time as a consequence of motion, Barrow asserts that motion can act as a measure of time and indeed that, viewed absolutely, time is independent of motion. Thus, he argues in his *Geometrical Lectures* that time "as to its absolute and intrinsic Nature" is independent of motion since "whether things move on, or stand still . . . Time flows perpetually with an equal Tenor." Time, he continues, is a "Quantum in itself, tho' in Order to find the Quantity of it, we are obliged to call in Motion to our Assistance."[127] Barrow is here making the important and original point that science is concerned with the measurement of time rather than with its essential nature.[128] He also makes it apparent that he is extending his theory of space into his treatment of time, so that it, too, in its absolute form, can be regarded as an attribute of God. Like the Cambridge Platonists Barrow was attempting to reassert the dependence of matter and motion on God against the tendency of the mechanical philosophy to make them independent. Hence, he argued that just as there was "Space before the World was created and that there now is an Extramundane, infinite Space (where God is present)," so too "Time existed before the World began, and does exist together with the World in the Extramundane Space."[129] As Burtt and Strong[130] have suggested, Barrow's arguments for absolute time may have subsequently played a role in Newton's own formulation of the metaphysical premises of the *Principia*. Barrow's more experimentally based and geometrically inclined exposition of the concept of absolute space may also have appealed more to Newton than that of More's – particularly since Barrow was more cautious than More in avoiding any possible identification of absolute space with God in keeping with Newton's famous statement in the general scholium of the *Principia* that God "is not duration, or space, but He endures and is present."

Another feature of Henry More's work that is reflected in Barrow's thought is the Cambridge Platonist's interest in collecting information about witchcraft and occult events in general in order to provide empirical evidence for the existence of the spiritual realm. Barrow did not often touch on this issue, but in one of his sermons – significantly entitled "The Being of God Proved from Supernatural Effects" – Barrow, like More and his disciple, Joseph Glanvill, argued that, although there were certainly false reports of occult phenomena, nonetheless "no counterfeit coin would appear were there no true one current."[131] He advanced a similar argument in his *Exposition of the Creed,* in which he lists among the proofs for the existence of God "those opinions and testimonies of mankind concerning apparitions...[which] infer (at least confer much to) the belief of the Divine existence."[132]

Barrow, at times, also showed some sympathy with the Cambridge Platonists' emphasis on the religious experience of the individual rather than the formularies of the church. To the Cambridge Platonists the chief instrument by which the individual arrived at the knowledge of the divine was reason – by which they meant not simply an ability to carry out deduction or induction but rather the basic intuitive faculty, what Whichcote called "The Candle of the Lord." In the manner of the Cambridge Platonists Barrow could describe reason as that "divine power implanted in us,"[133] and one could be reading one of the Cambridge Platonists when Barrow asserted that "every man is endued with that celestial faculty of reason, inspired by the Almighty...and hath an immortal spirit residing in him; or rather is himself an angelical spirit dwelling in a visible tabernacle."[134] However, such lyrical touches are not typical of Barrow's theological works more generally. Barrow was too attached to the traditions and structures of the Established Church to be altogether comfortable with the Cambridge Platonists' stress on the mystical experience of the individual believer. In his sermon, "An Adequate Knowledge of God Attainable by Man," he sums up his qualified enthusiasm for theological Platonism by remarking of Plato's comparison between God and man: "But I will not proceed in this speculation lest I seem too Platonical, against my will and desert; only hear, if you please, how that great contemplator discourses; (whose conceptions perhaps some too much admire...)."[135]

This somewhat equivocal atitude toward Platonism is reflected in Barrow's underlying theory of knowledge, particularly on the issue of innate ideas. The Cambridge Platonists had regarded it as supremely important that knowledge should be regarded as resulting from the active participation of the mind rather than the mind being seen as a *tabula rasa* on which sensations were passively imprinted. Intellect had to precede sensation just as the spiritual had to precede the corporal. This view led them to develop a theory of innate ideas to counteract the tendency of the mechanical philosophy to make the mind an inert receiver of external stimuli. The soul, wrote Cudworth,

> is not a mere tabula rasa, a naked and passive thing, which has
> no innate furniture or activity of its own, not anything at all in
> it, but what was impressed from without; for if it were so, then
> there could not possibly by any such thing as moral good and
> evil, just and unjust; forasmuch as these differences do not
> arise merely from outward objects, or from the impressions

which they make upon us by sense, there being no such thing in them.[136]

Barrow, too, inclines toward emphasizing what he calls the "active powers of our soul."[137] But his theory of innate ideas is concerned more with giving a solid base to the first principles of mathematics than with the more religious and moral motivation of the Cambridge Platonists.[138] For the Aristotelian-minded Barrow these first principles are in fact the metaphysical basis of science since the truth of geometrical axioms is demonstrated

> by other Axioms more simple, if any such there are, which are to be drawn from some higher and more universal Science, as *Metaphysics;* I say *Metaphysics* which is, or ought to be, the Treasure of the most general and simple Notions, and is therefore by *Aristotle* named the *Mistress of all Sciences.*[139]

Once this foundation has been established by notions "implanted in us by God together with the Faculty of Reasoning,"[140] however, Barrow gives greater scope than the Cambridge Platonists to a more empirical epistemology. In his *Geometrical Lectures* he gives a summary of his theory of knowledge:

> We perceive nothing, unless so far as we may be instigated by some change affecting the Senses, or that our Souls are mov'd and excited by the internal Operations of the Mind. We esteem the Quantities and different Degrees of things according to the Extension or Intension of Motions striking upon us either interiorly or exteriorly.[141]

Here a thoroughgoing sensationalist theory of knowledge is avoided since Barrow talks of the "internal operations of the Mind," but in its implication that knowledge depends largely on the impact on the mind of external stimuli, it departs significantly from the view of Cudworth that in order to gain knowledge we must start with the innate ideas within ourselves and then through deduction come to understand the nature of particular objects about us – or, as he himself put it, "knowledge doth not begin in Individuals but ends in Them."[142]

Barrow's Aristotelian-based empiricism is apparent, too, in his rejection of the view of those, like Descartes, who maintained that knowledge obtained through the senses was uncertain and even suspect. In his *Mathematical Lectures* Barrow asserted that the sense of sight "discerns many Objects *certainly*...to doubt which would rather seem the Part of an impertinent Trifler than a sage Philosopher." Aristotle, he continues, "will have it to be the Property of *Sense, That it is always true, and is inherent to all living Creatures.*"[143] Barrow argues, too, that we ought to accept any "Proposition [which is] confirmed with frequent Experiments as universally true, and not suspect that Nature is inconstant, and the great Author of the Universe unlike himself."[144] Barrow's approach to scientific certainty was in accord with that of some of his more notable Royal Society acquaintances such as John Wilkins and Robert Boyle. As van Leeuwen[145] has argued, these men applied to natural philosophy the same common-sense criteria as those which had been developed by earlier liberal Anglican thinkers such as Chillingworth in the context of theological debate with Roman Catholic controversialists who asserted that true religion could be based only on an infallible church. Barrow did, however, add a new dimension to the discussion of this issue by suggesting that a new level of certainty could be found by combining experimental data with mathematics, thus moving away from the "hypothetical physics" of someone such as Descartes to an "Archimedian physics" or what was later called mathematical physics – an approach that, Kargon suggests, may have influenced the young Newton.[146]

In his theory of knowledge as in the rest of his thought Barrow reflects a greater involvement with the developments of the "new philosophy" than do the Cambridge Platonists. This perhaps results in his theological work missing the religious vitality of that of the Cambridge Platonists, though Barrow's type of divinity, as more closely a part of the liberal Anglican tradition, was to exercise a more important long-term influence. As elsewhere in his work, Barrow's status as a transitional figure emerges in his epistemology since he combines a sympathy for the view of knowledge as intuitive with an attempt to come to terms with the empiricism associated with the new philosophy. And this was also to be reflected in his life, since he was both an associate of the Cambridge Platonists and a close friend of John Locke, whose empiricist theory of knowledge was partly shaped by reaction against the Cambridge Platonists' emphasis on innate ideas.[147]

Barrow's theory of knowledge is a further indication that, although the Cambridge Platonists had an important influence on Barrow, he was not really a member of their school. Parts of his work were colored by their concerns, but its basic framework remained Aristotelian rather than Platonic. The imprint of the Cambridge Platonists is most pronounced in Barrow's mathematical writings – and, in particular, in the sections on absolute space and time – perhaps because the abstract nature of mathematics made him more sympathetic to a form of Platonism. Significantly, when, in his inaugural lecture as Lucasian Professor of Mathematics, he listed the merits of mathematics, he included among them the traditional Pythagorian–Platonic view that through mathematics "the Mind is abstracted and elevated from sensible Matter, distinctly views pure Forms, conceives the Beauty of Ideas, and investigates the Harmony of Proportions."[148] In his theological works, however, the Platonic element – though not completely absent – is less pronounced, possibly because he may have been less inclined to philosophical innovation in his theological works than in his mathematical.

The very nature of the Cambridge Platonists' theology with its tendency to downplay the role of ecclesiastical institutions meant that they accepted the various changes in church and state in the mid-seventeenth century without any obvious crisis of conscience. However, Barrow's continued commitment to the Anglican cause made him more vulnerable to such changes. Up to 1654 Barrow had been shown remarkable tolerance despite his overt Royalist sympathies, but his unsuccessful bid for the Greek professorship, which he was refused on the grounds of his Arminianism, made it evident that Barrow's political and religious principles were likely to overshadow his future career within the university. Furthermore, a new parliamentary commission that had been established in September 1654 to investigate the universities[149] had compelled the Royalist Duport to vacate the chair of Greek. It is not surprising, then, that Barrow took four years' leave of absence in 1655.

When Barrow returned to Trinity in 1659 the political and religious climate had changed considerably. The death of Cromwell in 1658 and the succession of his ineffectual son, Richard, as protector made many feel that the return of the monarchy and even of the episcopal church might well come to pass. It is this that probably accounts for the growing number of official complaints from 1658 to 1660 about the revival of "prelatical" worship throughout the country.[150] Within Cambridge, too, there were signs of something like an Anglican revival: Richard Kidder reported that as early as 1657 there were "great disputes" within Emmanuel, the traditional bastion of Cambridge Puritanism, about the issue of episcopal ordinations,[151] while at Trinity it was decided in 1658 to enforce once more the statute that required fellows to take orders seven years after their election, a provision that had been allowed to lapse in 1650.[152] Although the Trinity regulation did not explicitly require episcopal ordination, it appears to have been commonly interpreted in this way. Thus, Ray, who was unenthusiastic about being ordained – in a letter of 1658 he commented that "I must of necessity enter into orders"[153] – nonetheless was ordained by Bishop Sanderson. In contrast, Barrow would have positively welcomed the reintroduction of this college regulation and was ordained by Bishop Brownrigg in 1659.

A further indication that Trinity was preparing itself for another change in the religious climate was the appointment on 17 August 1659 of the latitudinarian John Wilkins as master – a break with the traditions of Calvinist orthodoxy that had been established within the college by Hill and Arrowsmith and an example of the decline of the "old precisenes" on which Pepys remarked during a visit to Cambridge just before the

Restoration.[154] Though the position of master of Trinity was in the gift of the protector (having been formerly part of the Crown's patronage) Wilkins's appointment was made after the fellows of Trinity petitioned in his favor. Later, the fellows acknowledged that they "feared lest some person either of mean sort or factious principles should be thrust upon them."[155]

There are indications that even after Wilkins's appointment tensions continued to exist between the Calvinist elements within the college and the growing number of fellows who were less committed to Calvinist orthodoxy. Such conflicts became particularly evident at the annual election of fellows in September; in 1656 they had resulted in some whimsical verses that included the lines:

> Because that Arminians we would not be thought,
> Our Election had its Predestination.[156]

Soon after Wilkins became master he had to arbitrate in the dispute between two candidates for a fellowship, one of whom accused the other (Robert Creighton) of blasphemy and of not attending private prayer meetings. Significantly, Creighton was the son of a chaplain of Charles I and a pupil of Duport. Wilkins ruled in favor of Creighton (thanks largely to Duport's favorable testimony), and he was duly elected as a fellow.[157]

It was probably Wilkins's attempts to overcome such divisions as the Creighton incident had revealed that prompted Gilbert Burnet to remark that "at Cambridge he joined with those who studied to propagate better thoughts, to take men off from being in parties, or from narrow notions, from superstitious conceits and fierceness about opinions."[158] Among those at Cambridge "who studied to propagate better thoughts" were Cudworth and More, whom Wilkins presumably came to know while master of Trinity and whom he later nominated as fellows of the Royal Society. Wilkins also appears to have become acquainted with Whichcote, for whom he later obtained the vicarage of St. Lawrence Jewry in London.[159]

Wilkins's irenic sympathies would have made Trinity a congenial place for someone like Barrow. Moreover, as at Wadham College, Oxford, which Wilkins had made a center of scientific activity while warden there from 1648 to 1659, Wilkins attempted to stimulate an interest in the "new philosophy" within Trinity and, according to Birch, "set up a like society [to the one at Wadham] there."[160] It was while at Trinity that Wilkins established his lifelong friendship with John Ray,[161] and he must also have come to know Barrow at the same time.

Wilkins, however, did not remain long at Trinity, since after the Restoration he was ejected to make way for Henry Ferne. The latter had a long-standing claim to the position, which was enhanced by a document signed by no less an authority than Charles I. Wilkins's departure prompted an unsuccessful petition from the fellows to Charles II imploring him to permit Wilkins to remain since by "his prudent government and sweet conversation [he has]... endeared himself to this Society"[162] – an indication of Wilkins's remarkable ability to win over people of widely varying views. The popularity of Wilkins meant that Ferne was regarded with some suspicion by a number of Trinity fellows – a suspicion heightened by the upheavals that accompanied the restoration of the discipline and rituals of the Established Church within Cambridge. John Ray, an admirer of Wilkins, described Ferne and the other restored Royalists as the "old gang," and in a letter written on 26 September 1660 to his friend Courthorpe, he commented that "they have brought all things here as they were in 1641: viz., services morning and evening, surplice Sundayes, and holydays, and their eves, organs, bowing, going bare [and] fasting nights."[163] Ray eventually refused to subscribe to the Act of Uniformity and was ejected in 1662.

Ray's experiences are indicative of the clash more generally between what he called the "old and new" university, which he thought would "never kindly mingle, or make one piece."[164] The conflict at Trinity over which liturgy to use was repeated elsewhere in the university. In a letter of 26 January 1661 George Dunte, a fellow of Corpus, reported to the secretary of state that there was a great deal of strife throughout Cambridge over the restoration of episcopal ceremonies.[165] At Emmanuel, formerly the preeminent Puritan college, Thomas Smith wrote to Sancroft, his former pupil, on 2 November 1660 to report, "In your college half the society are for the Liturgy, and half against it, so it is read one week and the Directory used another."[166] Ferne's determination to restore the Established Church to the position it enjoyed before the Civil War became an issue not only at Trinity but also within the university more generally, since he served as vice-chancellor in 1660 and 1661 and insisted that those intending to graduate as B.A.s should formally acknowledge their acceptance of the Book of Common Prayer and the Thirty-Nine Articles as well as the king's supremacy. This led to complaints from fifty would-be bachelors that he was exceeding his authority, since he was acting in a way that "was contrary to the King's declaration"[167] – a reference to the Declaration of Breda, which had preceded the Restoration and had spoken of a "liberty to tender consciences." As the Act of Uniformity of the following year made clear, however, Ferne's actions were in keeping with the character of the church settlement that finally emerged after the Restoration.

The Cambridge Platonists with their unenthusiastic attitude toward church discipline and rituals – More described such things as not "so good as to make men good, nor so bad as to make men bad"[168] – were a particular target for some of the high churchmen who returned to Cambridge determined that the university should act as a seminary for a revitalized Anglican clergy. Cudworth was accused of having been "notoriously disaffected to the royal cause,"[169] and in 1665 Henry More was publicly atacked by Joseph Beaumont, who, having been ejected from Peterhouse in 1644, had become master of the college in 1663. More's isolation within Cambridge is apparent in his outburst to Lady Conway following Beaumont's critique: "They push hard at the Latitude men as they call them, some in their pulpitts call them sons of Belial, others make the Devill a latitudinarian."[170]

Barrow was in the invidious position of having friends in both camps: His family connections and personal convictions meant that he would have shared the high churchmen's determination to rebuild the Church of England after its time of troubles, but he also had links with men like Ray, Wilkins, and More who, for a time at least, were under a cloud following the Restoration. We know little of what, if any, tensions such divided loyalties caused Barrow. A Latin epigram that Barrow addressed to Charles II soon after the Restoration[171] does suggest that he felt neglected by the king, which may be an indication that he was felt to lack the necessary zeal for immediate appointment to an influential position within the reconstituted Church of England, but the number of such complaints from disappointed Royalist clergy makes this highly conjectural. One office Barrow did gain soon after the Restoration was the Regius Professorship of Greek, which he had been refused in 1654. Widdrington, Barrow's former rival for the post, resigned the chair soon after the Restoration, perhaps fearing retribution for the way in which he had gained the position. Because of his close family connections with the government of the protector, Widdrington evidently felt under pressure to demonstrate his loyalty to the restored regime since he proceeded to accuse Cudworth and More of heresy and disaffection to the royal cause.[172]

Though in his inaugural lecture as Regius Professor of Greek Barrow referred – perhaps with some understandable pique – to Widdrington's infrequent lectures, Barrow himself had little success in attracting an audience, as Barrow himself testified in his *Oratio sarcasmica in schola graeca* delivered in 1661. In this speech Barrow also announced that he would take as the subject of his lectures in the next year the works of Aristotle, exhorting his students not to abandon Aristotle's philosophy because of the "breeze of novelty" or the "charm of a perverted fashion." Barrow soon made it clear that the dangerous philosophical fashion he had in mind was Cartesianism, since he contrasted Aristotelianism with "a new-fashioned one, which blunts the apprehension with too much 'Medidation'" and went on to praise Aristotle's natural philosophy as one that "is not based on arbitrary figments of the mind, does not resort to insensible causes, does not take refuge in absurd hypotheses, does not feed the mind with chimaeras nor vex it with tortures nor whirl its fancies into giddiness."[173]

Barrow's hostile attitude toward Descartes, which contrasts with the courteous, if critical, treatment of his work in the M.A. oration of 1652, reflects a more general antagonism toward Descartes's work within the university. This had become more marked in the period before and after the Restoration when Descartes's work was disseminated more widely. In 1656 William Dillingham, master of Emmanuel from 1653 to 1662 (when he was deprived for nonconformity), publicly asserted the need to combat the spread of Cartesian ideas and urged reverence for the Word as against a preoccupation with "naturall reason."[174] Cartesianism was particularly suspect to defenders of theological orthodoxy, whether of a Calvinist variety like Dillingham or of High Church Anglicanism like Duport. Thus, a life of John Wallis commented that round the time of the Restoration the "new philosophy" (a term often used for Cartesianism) "increased, and began to prevail very much at Cambridge" and that its followers were a "bug-bear not only to the sour and narrow soul'd Presbyterians, but also to the bigotted zealots of the Church of England."[175]

In his pamphlet of 1662 Simon Patrick remarked that during the Interregnum there had been attempts to forbid the study of the "new philosophy," with the result that it was "more eagerly studied and embraced."[176] After the Restoration, too, there were further attempts by the restored high churchmen to prevent the further spread of Cartesianism – though again to little effect. Roger North wrote that his study of Descartes while a student at Jesus from 1667 to 1669 led some to accuse him of having "impugned the very Gospel."[177] Elsewhere he comments that while his brother John was at the university after the Restoration "the new philosophy (as it was called) of Des Cartes entered full sail, and coming with strong credentialls from abroad, was greedily entertained by the yonger [sic] or more vigorous scollars; but the Doctors and graver sort adhered *mordicitus* to the old Qualitys of Aristotle." North added, "It appears that the good Dr. Barrow, was more inclined to follow after truth, than any authority beside it" and praised his 1652 M.A. oration for its "due censure of the philosofy" and "just encomium of the Author."[178] In 1668 the vice-chancellor issued an edict forbidding sophists and B.A.s from keeping acts based on Descartes's work, and it was stipulated that Aristotle's writings should be used instead (though the regulation appears to have had little effect).[179] This reaction reflects the fears of the newly restored high churchmen that Cartesianism would weaken the Christian orthodoxy, which they saw as their mission to instil within the university, a sentiment that Barrow also appears to have shared.

Barrow's change in attitude toward Descartes from qualified respect in 1652 to outright hostility in 1661 may also be explained by the growing influence of Hobbes's work, his *De corpore* having appeared in 1655 and his *De homine* in 1658. To Barrow, Hobbes's work may well have appeared to confirm his suspicions about the materialistic tendencies of Descartes's work. That such views were current within Restoration Cambridge is indicated by an M.A. oration of 1663 delivered by John Covel of Christ's, who, like Barrow, defended Aristotle from the criticisms of Gassendi and Descartes; he also praised the Stagirite's work as a bulwark against Hobbism.[180] As late as 1675–6 another member of Christ's felt the need to reassure a former teacher that his tutor was not too much of a "philo-Cartesius" since he forbade the reading of Hobbes.[181] For Barrow, as for other Cambridge opponents of Hobbes such as Thomas Tenison, Benjamin Laney, and Cudworth,[182] Hobbes became the whipping boy for much that they were afraid of in the "new philosophy." Throughout Barrow's work, then, there are frequent snipings at Hobbes: The *Mathematical Lectures* contain a long refutation of Hobbes's theory of proportionality, one of the cornerstones of his philosophy,[183] and in his sermon "The Being of God Proved from the Frame of the World" Barrow is particularly keen to cite Hobbes in support of the argument from design, since it "is the expression of another person well known among us, whom few do judge partial to this side, or suspicious of bearing a favourable prejudice to Religion."[184] Barrow's contempt for Hobbes's political thought is also evident in the following remark:

> It is therefore a monstrous paradox, crossing the common sense
> of men, which, in this loose and vain world, hath lately got
> such vogue, that all men naturally are enemies one to another:
> it pretendeth to be grounded on common observation and ex-
> perience; but it is only an observing the worst actions of the
> worst men.[185]

Barrow's hostility toward Hobbes may also account for his change of attitude toward atomism as well as Cartesianism. In his M.A. oration of 1652 Barrow had praised the work of Gassendi and Kenelm Digby, who had sought to cleanse atomism of its traditional antireligious connotations, but in a sermon preached after the Restoration Barrow plainly still regarded atomism with some suspicion: "He that can seriously ascribe" the workings of the universe, he writes, with perhaps Hobbes's natural philosophy in mind, "to an undisciplined and unconducted troop of atoms rumbling up and down confusedly through the field of indefinite space, what might he not as easily assert or admit?"[186]

This determination to combat the influence of Hobbes is an indication of Barrow's more general preoccupation with theological issues after the Restoration – his earlier interest in mathematics and science being at least in part the result of the natural tendency of someone with Barrow's Anglican beliefs to turn to disciplines other than theology during the troubled conditions of the Interregnum. Having helped to establish the Lucasian

Professorship on a firm foundation, it is not surprising that in 1669 Barrow resigned his mathematical chair in order to devote himself to the chief study of his profession: divinity.

But having resigned the Lucasian Professorship Barrow was now adrift in the stormy seas of ecclesiastical preferment. If he were not to remain becalmed as a mere fellow of Trinity, he needed patrons. In the first place he could call on his uncle, who, in 1669, was translated from Sodor and Man to St. Asaph; from him he received a small sinecure in Wales[187] and, given the lowly status of the see, could probably hope for little more – except, perhaps, through his uncle's influence with other, better endowed episcopal patrons. More promising was his association with John Wilkins, who, in 1668, was consecrated bishop of Chester – a remarkable feat for a brother-in-law of Cromwell. Wilkins owed his elevation to the dissolute court favorite, the duke of Buckingham, who approved of Wilkins's abortive efforts to win over the Dissenters by widening the doctrinal and liturgical boundaries of the Established Church. Shortly after Wilkins's appointment to Chester, Pepys noted in his diary that Wilkins was "a mighty rising man, as being a latitudinarian and the Duke of Buckingham his great friend."[188] It is possible, then, that Wilkins may have played a role in recommending Barrow to Buckingham, since it was the duke who was active in obtaining the mastership of Trinity for Barrow in 1672[189] – one year after Buckingham became chancellor of Cambridge, a post from which he was ejected at royal command in 1674 to be replaced by the royal bastard, the duke of Monmouth. However, Barrow also owed his mastership to the good offices of Gilbert Sheldon, the archbishop of Canterbury, who opposed Wilkins's and Buckingham's moves toward ecclesiastical comprehension and who had bitterly resented Wilkins's elevation to the episcopate.[190] It is an indication of Barrow's wide theological sympathies that he could attract support from two such differing quarters.

Another of Barrow's episcopal patrons was Seth Ward (bishop of Exeter, 1662–7, and of Salisbury, 1667–89). Ward had left Cambridge in 1644 after being ejected from his Sidney Sussex fellowship by the parliamentary visitors, and so his and Barrow's paths would not have crossed at the university. The two men, who shared a passion for mathematics, probably came to know each other through the Royal Society, perhaps after being introduced by Wilkins, a mutual friend. Ward evidently valued Barrow's friendship highly since, about the time that Barrow resigned his fellowship, the bishop invited Barrow "to live with him, not as a Chaplain, but rather as a Friend and Companion."[191] Barrow declined Ward's offer of an archdeaconry but later accepted a prebend's post from him, which he resigned after becoming master.[192] Perhaps Ward – who, unlike his friend Wilkins, approved of Sheldon's uncompromising policies toward Dissenters – played a part in recommending Barrow to the archbishop of Canterbury when the mastership of Trinity became vacant.

The diversity of Barrow's ecclesiastical patrons was of a piece with Barrow's eclectic theology. True to his upbringing, Barrow's theological works retain some of the most characteristic features of high church theology, notably an emphasis on the church fathers and church tradition. He also remained on friendly terms with some of the most vigorous defenders of the privileges of the Established Church. As vice-chancellor in 1676 he warmly praised the work of the Regius Professor of Divinity, Joseph Beaumont,[193] who was known for being an "incomparable Disputant against Schismatics."[194] Yet Barrow also had ties with latitudinarian Anglicans like Wilkins who wished to accommodate as many English Protestants as possible within the Established Church. Apart from Wilkins, Barrow's main link with the leading latitudinarian clergy, who, after the Restoration, were largely London-based (though generally graduates of Cambridge), was probably John Mapletoft. Having become interested in the biomedical sciences through the Nidd circle, Mapletoft had taken his M.D. from Cambridge in 1667 and practiced as a physician in London. He was a kinsman of Thomas Firmin, a philanthropic merchant whose house became a meeting place for latitudinarian clergy, whom Barrow came to know through Mapletoft.[195] Mapletoft had also gone to Westminster School with John Locke, who, thanks to Mapletoft, met prominent clergymen such as Tillotson and Barrow; on Barrow's death Locke spoke of him as one of his "very considerable friends."[196] Barrow's connections with the latitudinarians were strengthened by the fact that many of them were also associated with the Royal Society. Through his Royal Society connections Hooke mentions meeting Barrow in the company of men such as Whichcote, Firmin, and Tillotson in October 1673[197] (the last of whom was to act as Barrow's literary executor).

Barrow admired the way in which the latitudinarians had attempted to overcome the differences between the Established Church and the Dissenters, claiming, in a series of notes on the subject, that "it is evident that no sort of people have been more effectually diligent in promoting the real interest and honour of the Church than those who have been traduced by these odious names and characters" [i.e., terms like "Latitudinarians" or "Rationalists"].[198] Nevertheless, Barrow, unlike his latitudinarian colleagues, had suffered for his Royalist convictions during the Interregnum, and despite his praise for the latitudinarians' "temperate behaviour towards Dissenters," there are times when Barrow's own engrained hostility toward the heirs of the Puritans breaks through. In this same work he lists among the church's "advantages against Dissenters" "the gentle, sociable, discreet temper of the clergy in comparison to their adversaries' surly and morose way" and "the memory of those villainous pranks; those perjuries, cheats, rapines, giddinesses, and levities, &.c., whereof other parties have been guilty, exposed and condemned themselves."[199]

The same ambivalence in his relations with the latitudinarians can be seen in Barrow's theology. There is much in Barrow's theological work that resembles the preaching of the latitudinarians; like them he could emphasize the importance of reason, arguing, for example, that

> reason itself, well followed... [will] serve to produce faith. For that there is a God, reason from observation of appearances in nature and providence will collect; that goodness is one of his principal attributes, reason from the same grounds will infer... so hath reason led us to the door of faith, and being arrived thither, will (if our will be not averse) easily find entrance.[200]

Like the latitudinarians, too, Barrow decried theological speculation and its divisive consequences. Thus, he castigated those who departed "from the good ancient wholesome doctrine... who dote on curious empty speculations and idle questions, which engender strife, and yield no good fruit."[201]

However, Barrow's mention of "ancient wholesome doctrine" is an indication of the way that he placed rather more emphasis on doctrinal matters than did most latitudinarians. In particular, though agreeing with the latitudinarians about the importance of natural theology, Barrow gives much greater weight to the significance of revelation. Whereas Tillotson could claim that, "excepting in a very few particulars, they (natural law and Christianity) enjoin the very same things,"[202] Barrow, though stressing the importance of reason in religion, could also emphasize the gulf between human reason and divine revelation in a way that set him apart from the latitudinarians. "It is true," Barrow wrote, "some few sparks of this divine knowledge may possibly be driven out by rational consideration; philosophy may yield some twilight glimmerings thereof," but, he continued, true faith indicates "immediate influences from the fountain of life and wisdom, the divine Spirit."[203] The theological differences between Barrow and the latitudinarians reflect the differences in their situations: Barrow's outlook was still shaped largely by an institution concerned primarily with the training of clergy; the main task of the latitudinarians, by contrast, was to awaken some religious interest in congregations weary of theological strife.

Barrow's strongly clericalist conception of the university – a view that colored much of his preaching, since most of his sermons were delivered within the university – is apparent in his actions as master of Trinity from 1672 to 1677. In the year after he took up office he politely, but firmly, declined a request from the principal secretary of state to exempt Francis Aston from the college regulation requiring fellows of seven years' standing to take orders. Such a dispensation, argued Barrow, would result in "the subversion of the maine designe of our foundation, which is to breed

Divines."[204] To Barrow the life of a don was an essentially contemplative one, the primary object of which was to reflect on the eternal verities, or, as Barrow himself put it, "a calling, which doth not employ us in bodily toil, in worldly care, in pursuit of trivial affairs, in sordid drudgeries; but in those angelical operations of the soul, the contemplation of truth, and attainment of wisdom."[205]

Consistent with his fundamentally conservative view of the role of the university, Barrow was insistent that Cambridge's traditions and privileges should be protected. As master of Trinity he was permitted to marry but declined to do so since he thought it was "not agreeable with the statutes."[206] When, in 1675, the Royal College of Physicians attempted to challenge the traditional right of Cambridge M.D.s to practice medicine anywhere in the country, Barrow was determined to prevent "any thing in prejudice to our privilege," being willing, if necessary, to invoke the aid of the chancellor, the duke of Monmouth, who, commented Barrow, "we have found ready upon all occasions to protect our rights."[207] Barrow's close involvement in university politics, which this letter of February 1675 reflects, was, no doubt, partly prompted by the knowledge that in November 1675 he was to begin his term as vice-chancellor (a post that rotated among the heads of colleges), though to Barrow's great relief little of significance occurred during his one-year term.[208]

Barrow's praise for Monmouth as chancellor probably owed much to the fact that in 1674 the duke had obtained for the university a royal letter that offered Cambridge some protection against the tendency of the Crown to use the university's offices as a means of rewarding its more minor clients.[209] Despite this, however, the university had to remain vigilant to stem the flow of royal mandates, which, after the Restoration, had become a flood.[210] In October 1675, the month before Barrow took up the vice-chancellorship, the master of Jesus College vainly attempted to evade a royal mandate by pointing out that, thanks to the letter of 1674, the university "had a gracious liberty to refuse whatever letters should pretend to dispense with exercises or cautions."[211] Trinity, as a royal foundation, was a particularly vulnerable target for such royal pressure. However, John Pearson (Barrow's predecessor as master of Trinity from 1662 to 1672) had endeavored to prevent such royal meddling in the college's affairs. In 1671, for example, he sought the aid of the secretary of state to prevent, or at least reduce, such royal mandates.[212] Barrow's letter to the secretary of state of 1674 refusing a request to allow Aston a dispensation from the statute requiring him to take orders is an indication that he was determined to maintain his predecessor's firmness in resisting outside interference in the college's affairs. Part of the reason for the decline of Trinity after Barrow's death was that his successors, John North (1677–83) and John Montagu (1683–99), were more inclined to buckle under such royal pressure so that royal favor rather than academic merit became the path to advancement within the college.[213] But, though willing, if necessary, to prevent royal interference in the university's affairs Barrow remained a stalwart defender of the royal supremacy. The efforts of John Skinner, a fellow of Trinity, to edit the *State Letters* of John Milton in 1677 were met by Barrow with a letter curtly ordering him to refrain from publishing "any writing mischievous to the Church or State."[214]

Generally, however, Barrow appears to have had few disputes with Trinity's fellows, and like his friend John Wilkins, he succeeded in preserving harmony in a college prone to factionalism. Hill's remark that the "Senior Fellows so well understood and esteemed him, that with good-will and joy they received a Master much younger than any of themselves"[215] appears to have had some substance: In July 1673, more than a year after taking office, Barrow could report to his friend Mapletoft that "Trinity College is, God be thanked, in peace (I wish all Christendome were so well) and it is my duty, if I can, to keep uproars thence."[216] Very different was the experience of Barrow's successor, John North, whose tactlessness and irresolute attempts at discipline united the college against him. North's problems were compounded by the crisis in the college's finances caused by the building of the Wren Library, which was begun by Barrow in 1676.[217] With unaccustomed humor North left instructions that at his death he should be buried in the antechapel so that "the fellows might trample upon him dead as they had done living."[218] Trinity continued to drift under the mastership of the aristocratic Montagu, who, in 1700, was replaced by Richard Bentley, whose determined efforts to make the college the academic center of the university and himself the college's autocrat led to one of the longest and most tangled quarrels in the long annals of academic dispute.

Barrow's death in 1677, then, marked the beginning of a period of decline in the college's morale and internal harmony. It also came at a time when the college's enrollments began to drop markedly, a trend that was barely evident during Barrow's mastership. Under Ferne (master, 1660–2) the annual average enrollment was forty-five, under Pearson (master, 1662–72) it reached fifty, a figure that fell to forty-six under Barrow: Thereafter, however, the decline in numbers became more pronounced: During North's mastership (1677–83) the figure was thirty-five and a half, and under Montagu (1683–99) enrollment fell still further to twenty-nine[219] – figures that mirror the more general decline in the enrollments of both Oxford and Cambridge during this period and that continued until the mid-eighteenth century.

But such changes lay in the future when Barrow died in 1677. As master of Trinity he could reflect that his college, and the university more generally, had successfully weathered the storms of the mid-seventeenth century and was now playing an active part in consolidating the position of the reestablished episcopal church to which Barrow had been loyal in and out of season. Barrow was not faced with the conflict of loyalties that the Glorious Revolution of 1688 brought in its wake when churchmen raised on the principles of passive obedience had to choose between their loyalty to a Stuart king and the welfare of the Church of England, a dilemma that undermined much of the work to which Barrow's colleagues within Restoration Cambridge had devoted themselves.

Barrow died, too, at a time when it was becoming increasingly difficult to display that mastery of the major branches of learning that was one of the most remarkable features of his wide-ranging career. By the late seventeenth century such catholicity of learning was becoming increasingly rare for reasons that were reflected in the changes in the curriculum during Barrow's time at Cambridge. The view, associated with the "new philosophy," that knowledge was continually expanding led to the breakdown of the relatively static educational system that seventeenth-century Cambridge had inherited (with some modifications) from the Middle Ages. The traditional position that all knowledge, whatever its character, should be woven into a single synthesis came to be regarded as a hindrance to the growth of science, and consequently the various disciplines that had once shared a common set of assumptions grew farther apart from one another. To many this period was a time of great intellectual excitement, whereas for others (like Barrow's tutor, Duport) it meant a challenge to all that they held as basic, but few attempted to do justice to both the old and the new to the extent of Barrow, whose career illustrates the repercussions of the intellectual and political revolutions of the seventeenth century[220] – changes that were reflected, in microcosm, in the Cambridge of Isaac Barrow.

Notes

1. A. Napier (ed.), *The Theological Works of Isaac Barrow, D.D.,* 9 vols. (Cambridge, 1859), 9:xiv.
2. J. B. Mullinger, *The University of Cambridge,* 3 vols. (Cambridge, 1873-1911), 3:266.
3. J. D. Twigg, "The Parliamentary Visitation of the University of Cambridge, 1644-5," *English Historical Review,* 98 (1983): 522.
4. Mullinger, *University of Cambridge,* 3:369.
5. J. D. Twigg, "The University of Cambridge and the English Revolution, 1625-1688" (Ph.D. thesis, University of Cambridge, 1983), pp. 145-6.
6. C. H. Cooper, *Annals of Cambridge* (Cambridge, 1842-1908), 3:439, 447; and A. G. Matthews, *Calamy Revised...* (Oxford, 1934), p. xiii.
7. D. Hoyle, "A Commons Investigation of Arminianism and Popery in Cambridge on the Eve of the Civil War," *Historical Journal,* 29 (1986): 419-25; J. G. Hoffman, "The Puritan Revolution and the 'Beauty of Holiness' at Cambridge," *Proceedings of the Cambridge Antiquarian Society,* 72 (1982-3): 94-105; and N. Tyacke, *Anti-Calvinists: The Rise of English Arminianism c 1590-1640* (Oxford, 1987), p. 194.
8. H. Kearney, *Scholars and Gentlemen* (London, 1970), pp. 102-3; and G. B. Tatham, *The Puritans in Power* (Cambridge, 1913), p. 125.
9. Mullinger, *University of Cambridge,* 3:324.

10. H. C. Foxcroft (ed.), *Supplement to Burnet's "History of My Own Time"* (Oxford, 1902), p. 464.

11. M. H. Nicolson (ed.), *Conway Letters* (New Haven, Conn., 1930), p. 265; and M. H. Nicolson, "Christ's College and the Latitude-men," *Modern Philology, 27* (1929): 35–53.

12. C. Wordsworth, *Social Life at the English Universities in the Eighteenth Century* (Cambridge, 1874), p. 14.

13. J. Gascoigne, "Politics, Patronage and Newtonianism: The Cambridge Example," *Historical Journal, 27* (1984): 1–24; and idem, *Cambridge in the Age of the Enlightenment: Science, Religion and Politics from the Restoration to the French Revolution* (Cambridge, 1989), pp. 142–84.

14. L. Stone, "The Educational Revolution," *Past and Present, 28* (1964): 69.

15. L. Stone, "The Size and Composition of the Oxford Student Body, 1580–1909," in L. Stone (ed.), *The University in Society,* 2 vols. (Princeton, N.J., 1974), 1:92.

16. Ibid., p. 54.

17. R. S. Westfall, "Isaac Newton in Cambridge: The Restoration University and Scientific Creativity," in P. Zagorin (ed.), *Culture and Politics from Puritanism to the Enlightenment* (Berkeley and Los Angeles, 1980), pp. 142–3.

18. Cooper, *Annals,* 3:309.

19. Ibid., p. 288.

20. Mullinger, *University of Cambridge,* 3:283–6.

21. *Dictionary of National Biography,* s.v. "Isaac Barrow (1614–80)."

22. Napier (ed.), *Theological Works,* 1:xli.

23. R. S. Bosher, *The Making of the Restoration Settlement* (London, 1951), p. 36; and J. W. Packer, *The Transformation of Anglicanism, 1643–1660* (Manchester, 1969), passim.

24. Packer, *Transformation,* pp. 46–7.

25. Napier (ed.), *Theological Works,* 1:xxxix, 9:540.

26. Twigg, "Parliamentary Visitation," p. 522.

27. E. Carter, *The History of the University of Cambridge. . .* (London, 1753), pp. 339–48.

28. P. Hammond, "Dryden and Trinity," *Review of English Studies,* n.s., 36 (1985): 49.

29. Twigg, *Cambridge,* pp. 97–8.

30. Hammond, "Dryden," p. 49.

31. J. Heywood, *Cambridge University Transactions during the Puritan Controversies of the Sixteenth and Seventeenth Centuries,* 2 vols. (London, 1854), 2:531.

32. Hammond, "Dryden," p. 49.

33. J. H. Turner (ed.), *The Diaries of the Rev. Oliver Heywood,* 4 vols. (Brighouse, 1882–5), 1:162.

34. Twigg, *Cambridge,* p. 96.

35. Mullinger, *University of Cambridge,* 3:473–4.

36. Cooper, *Annals,* 3:417. See also ibid., 3:414–15, 418–19; and Twigg, *Cambridge,* pp. 117–19.

37. Figures derived from Matthews, *Calamy.*

38. A. Clark (ed.), *Aubrey's Brief Lives, 1669–1696,* 2 vols. (Oxford, 1898), 1:89.

39. The Walpoles behaved "with great circumspection during the Civil War," but Edward Walpole's Royalist credentials were sufficient for him to be elected to the Cavalier Parliament. J. H. Plumb, "The Walpoles: Father and Son," in his *Men and Places* (London, 1963), p. 123.

40. Cambridge University Library, Cholmondeley Mss., Correspondence No. 1.

41. J. H. Monk, "Memoir of James Duport," *Museum Criticum, 8* (1825): 684.

42. Napier (ed.), *Theological Works,* 1:xliii.

43. Ibid., 9:xxv; p. 141.

44. Tatham, *Puritans,* p. 122.

45. Napier (ed.), *Theological Works,* 1:xl.
46. Cambridge University Library, Mm.1.43 (Baker Ms.), fols. 396-8.
47. Trinity College, Cambridge, Ms. R.10.29.
48. J. Tulloch, *Rational Theology and Christian Philosophy in England in the Seventeenth Century,* 2 vols. (Edinburgh, 1872), 2:85.
49. Mullinger, *University of Cambridge,* 3:589.
50. Tulloch, *Rational Theology,* 2:52.
51. Ibid., 2:56.
52. T. Birch, *The Life of the Most Rev. John Tillotson* (London, 1752), pp. 399-400.
53. Love's argument that God is known by three books - "Scripture, Nature and Conscience" - also indicates some questioning of the orthodox Calvinist stress on *sola scriptura.* Kenneth Spencer Research Library, University of Kansas, Ms. A.41, fol. 49. (I am grateful to this library for supplying me with a microfilm of the manuscript, the reference for which was obtained from J. R. Jacob and M. C. Jacob, "The Anglican Origins of Modern Science: The Metaphysical Foundations of the Whig Constitution," *Isis,* 71 [1980]: 258.)
54. *Anti-Calvinists,* pp. 49, 57. See also H. R. Trevor Roper, *Catholics, Anglicans and Puritans: Seventeenth Century Essays* (London, 1987), pp. 81-90.
55. S. Patrick, *A Brief Account of the New Sect of Latitude-Men* (London, 1662), p. 5.
56. A. Taylor (ed.), *The Works of Symon Patrick,* 9 vols. (Oxford, 1858), 9:425.
57. Ibid., p. 419.
58. Cambridge University Library, Add. Ms. 84.
59. Hammond, "Dryden," p. 48.
60. Tulloch, *Rational Theology,* 2:55.
61. *Dictionary of National Biography,* s.v. "John Goodwin."
62. G. R. Cragg (ed.), *The Cambridge Platonists* (New York, 1968), p. 42.
63. A. Lichtenstein, *Henry More: The Rational Theology of a Cambridge Platonist* (Cambridge, Mass., 1962), p. 123.
64. Mullinger, *University of Cambridge,* 3:594.
65. A. Tuckney, *None but Christ, or a Sermon upon Acts 4.12. Preached at St. Maries in Cambridge July 4 1652* (London, 1654), p. 50.
66. Napier (ed.), *Theological Works,* 5:138. Barrow's library catalogue also indicates that he was widely read in Arminian theology. It includes three works by Arminius himself together with three by Grotius, who allied himself with the Arminians. Bodleian Library, Oxford, Ms. Rawlinson D.878, fols. 14, 33, 34.
67. G. R. Cragg, *From Puritanism to the Age of Reason* (London, 1960), p. 29.
68. Napier (ed.), *Theological Works,* 4:443. This was a distinctive feature of Arminianism. J. Hastings (ed.), *Encyclopedia of Religion and Ethics* (Edinburgh, 1908), s.v. "Arminianism." Barrow's links with the High Church tradition are evident in the way that he associates the doctrine of the Holy Spirit with his ecclesiology. He writes, for example, of the "donation of the Holy Spirit to the Christian Church and to all its members." Napier (ed.), *Theological Works,* 4:433.
69. J. Tulloch, *Rational Theology,* 1:35.
70. Napier (ed.), *Theological Works,* 5:126-7.
71. Ibid., p. 157.
72. Ibid., 4:296.
73. Ibid., p. 321.
74. J. Coolidge, *The Pauline Renaissance in England* (Oxford, 1970), p. 12.
75. Napier (ed.), *Theological Works,* 4:13.
76. Ibid., 5:153.
77. Ibid., 4:15.
78. Ibid., 7:[1].
79. Ibid., 6:509.

80. Ibid., p. 528.
81. J. Dillenberger, *Protestant Thought and Natural Science* (London, 1961), p. 14.
82. H. R. McAdoo, *The Spirit of Anglicanism: A Survey of Anglican Theological Method in the Seventeenth Century* (London, 1965), p. 384.
83. Napier (ed.), *Theological Works,* 5:167.
84. P. H. Osmond, *Isaac Barrow: His Life and Times* (London, 1944), p. 38; Latin original in Napier (ed.), *Theological Works,* 9:40.
85. Osmond, *Barrow,* pp. 26–7; Napier (ed.), *Theological Works,* 9:33.
86. For a summary of the kind of training in scholastic logic that Barrow would have received see W. J. Costello, *The Scholastic Curriculum at Early Seventeenth-Century Cambridge* (Cambridge, Mass., 1958), pp. 47–8.
87. Napier (ed.), *Theological Works,* 7:22.
88. Ibid., 9:161; Osmond, *Barrow,* p. 94.
89. In his "Rules to Be Observed by Young People and Schollars in the University," printed by G. M. Trevelyan under the title "Undergraduate Life under the Protectorate," *Cambridge Review,* 69 (1943): 330.
90. I. Barrow, *The Usefulness of Mathematical Learning,* trans. J. Kirby (London, 1734), p. 119.
91. W. Ong, *Ramus, Method and the Decay of Dialogue* (Cambridge, Mass., 1958), p. 217.
92. Kearney, *Scholars,* p. 61.
93. Trevelyan, "Undergraduate Life," p. 330.
94. Barrow, *Usefulness,* p. 404.
95. A. Gabbey, "Philosophia Cartesiana Triumphata: Henry More (1646–71)," in T. Lennon, J. M. Nicholas, and John W. Davis (eds.), *Problems of Cartesianism* (Kingston, Ontario, 1982), pp. 171–250.
96. D. B. Sailor, "Cudworth and Descartes," *Journal of the History of Ideas,* 23 (1962): 133; and J. E. Saveson, "Descartes's Influence on John Smith, Cambridge Platonist," ibid., 20 (1959), 258–63.
97. C. Webster, "Henry More and Descartes: Some New Sources," *British Journal for the History of Science,* 4 (1969): 361.
98. M. Nicolson, "The Early Stages of Cartesianism in England," *Studies in Philology,* 26 (1929): 360–1.
99. Webster, "Henry More," p. 360.
100. C. Webster, *The Great Instauration: Science, Medicine and Reform, 1626–1660* (London, 1975), p. 149n.
101. Nicolson, "Cartesianism," p. 358.
102. In a letter to Newton of 1687, written in response to the publication of the *Principia,* Clerke made the claim that he and Barrow "contributed neare 40 years since, as much or more than any two others, (to speake modestly) *in diebus illis,* to bring these things into place in ye university." H. W. Turnbull, J. F. Scott, A. R. Hall, and L. Tilling (eds.), *The Correspondence of Sir Isaac Newton,* 7 vols. (Cambridge, 1959–77), 2:493.
103. Nicolson, "Cartesianism," p. 368.
104. Osmond, *Barrow,* pp. 28–33; Napier (ed.), *Theological Works,* 9:79–104.
105. Tulloch, *Rational Theology,* 2:88.
106. R. Hurlbutt, *Hume, Newton and the Design Argument* (Omaha, Neb., 1965), p. 11; and J. T. Baker, *An Historical and Critical Examination of English Space and Time Theories from Henry More to Bishop Berkeley* (New York, 1930), p. 13.
107. Napier (ed.), *Theological Works,* 9:liv.
108. Ibid., 1:xliii.
109. In a letter of 14 February 1654 Worthington described Barrow to Hartlib as a man "of admirable parts." Ibid., 1:lxii.
110. B. J. T. Dobbs, *The Foundations of Newton's Alchemy* (Cambridge, 1975), p. 94.

111. C. E. Raven, *John Ray, Naturalist: His Life and Works* (Cambridge, 1950), p. 458.
112. H. R. McAdoo, *Anglicanism,* p. 148.
113. Turnbull, *Correspondence of Sir Isaac Newton,* 1:305.
114. Cragg, *Cambridge Platonists,* p. 27.
115. Napier (ed.), *Theological Works,* 5:371.
116. Sailor, "Cudworth and Descartes," p. 137.
117. Webster, *Great Instauration,* p. 146.
118. Cudworth to Boyle, 16 October 1684, in T. Birch, *The Works of Robert Boyle,* 6 vols. (London, 1772), 6:511.
119. A. Koyre, *From the Closed World to the Infinite Universe* (Baltimore, Md., 1970), pp. 110-54.
120. Bodleian Library, Oxford, Ms. Rawlinson D.878, fols. 41, 48v.
121. Barrow, *Usefulness,* pp. 170-1.
122. Ibid., pp. 167, 179.
123. Ibid., p. 171.
124. Ibid., p. 180.
125. R. A. Greene, "Henry More and Robert Boyle on the Spirit of Nature," *Journal of the History of Ideas,* 23 (1962): 451-74.
126. Baker, *Space and Time Theories,* p. 14.
127. I. Barrow, *The Geometrical Lectures,* trans. E. Stone (London, 1735), pp. 6-7.
128. J. A. Gunn, *The Problem of Time* (London, 1929), pp. 52-3.
129. Barrow, *Geometrical Lectures,* pp. 5-6.
130. E. A. Burtt, *The Metaphysical Foundations of Modern Physical Science* (London, 1964), pp. 152-3; and E. Strong, "Barrow and Newton," *Journal of the History of Philosophy,* 8 (1970): 155-72.
131. Napier (ed.), *Theological Works,* 5:277.
132. Ibid., 7:45-6.
133. Ibid., 1:63.
134. Ibid., 2:339.
135. Ibid., 4:482.
136. R. L. Armstrong, "Cambridge Platonists and Locke on Innate Ideas," *Journal of the History of Ideas,* 30 (1969): 188.
137. Napier (ed.), *Theological Works,* 4:476.
138. Armstrong, "Innate Ideas," p. 191.
139. Barrow, *Usefulness,* p. 107.
140. Ibid., p. 107.
141. Barrow, *Geometrical Lectures,* p. 7.
142. S. P. Lamprecht, "Innate Ideas in the Cambridge Platonists," *Philosophical Review,* 35 (1926): 569.
143. Barrow, *Usefulness,* pp. 70-1.
144. Ibid., pp. 73-4.
145. H. G. van Leeuwen, *The Problem of Certainty in English Thought, 1630-1690* (The Hague, 1963).
146. R. H. Kargon, *Atomism in England from Harriot to Newton* (Oxford, 1966), pp. 109-121; and idem, "Newton, Barrow and the Hypothetical Physics," *Centaurus,* 2 (1965): 46-56.
147. E. Cassirer, *The Platonist Renaissance in England,* trans. J. P. Pettegrove (Edinburgh, 1953), p. 4; and Armstrong, "Innate Ideas," p. 192.
148. Barrow, *Usefulness,* p. xxxi.
149. Tatham, *Puritans,* p. 140.
150. Bosher, *The Making,* p. 48.
151. Ibid., p. 38.
152. Napier (ed.), *Theological Works,* 9:xi-xiii.

153. R. W. T. Gunther, *Further Correspondence of John Ray* (London, 1928), p. 17.
154. B. J. Shapiro, *John Wilkins, 1614-1672* (Berkeley and Los Angeles, 1969), p. 146.
155. Ibid., p. 141.
156. Costello, *Scholastic Curriculum*, p. 128.
157. Shapiro, *Wilkins*, p. 142.
158. G. Burnet, *History of My Time*, ed. O. Airy, 2 vols. (Oxford, 1897-1900), 1:332-3.
159. Shapiro, *Wilkins*, p. 144.
160. Birch, *Tillotson*, p. 405.
161. Raven, *Ray*, p. 56.
162. Shapiro, *Wilkins*, p. 147.
163. Gunther, *Further Correspondence*, pp. 17-18.
164. Ibid., p. 29.
165. M. A. Everett Green, F. H. B. Daniell, and F. Bickley, *Calendar of State Papers, Domestic Series, of the Reign of Charles II* (1860-1947), 28 vols., 1660-1, p. 488.
166. Cambridge University Library, Ms. Mm.1.45; fol. 127.
167. Cooper, *Annals*, 3:491.
168. Foxcroft (ed.), *Supplement*, p. 463.
169. M. H. Nicolson, "Christ's College and the Latitude-men," *Modern Philology*, 27 (1929): 46.
170. M. H. Nicolson (ed.), *Conway Letters: The Correspondence of Anne, Viscountess Conway, Henry More, and Their Friends, 1642-1684* (New Haven, Conn., 1930), p. 243.
171. Which Whewell translated as "None more warmly had wished that Charles might return to his kingdom; / No one less has felt truly that Charles is returned." Napier (ed.), *Theological Works*, 9:xl.
172. Nicolson, "Christ's College," 46.
173. Osmond, *Barrow*, pp. 94-5; Napier (ed.), *Theological Works*, 9:165-6.
174. Twigg, *Cambridge*, p. 175.
175. British Library, Add. Ms. 32601, fol. 60.
176. Patrick, *New Sect*, p. 23.
177. R. North, *The Lives of the Norths*, ed. A. Jessopp, 3 vols. (London, 1890), 3:15.
178. P. T. Millard (ed.), "An Edition of the Life of Dr. John North by Roger North...," (D. Phil. thesis, University of Oxford, 1969), pp. 85-6.
179. Bodleian Library, Oxford, Rawlinson Ms. c. 146, fol. 37.
180. British Library, Add. Ms. 22910, fols. 13-15.
181. Northamptonshire Record Office, Isham Correspondence No. 963.
182. S. I. Mintz, *The Hunting of the Leviathan* (Cambridge, 1962), pp. 72-9, 125-33.
183. M. Carré, *Phases of Thought in England* (Oxford, 1949), p. 258.
184. Napier (ed.), *Theological Works*, 5:192. Barrow's anti-Hobbism is also evident from the fact that his library included about ten works (by Boyle, Bramhill, Cumberland, Echard, Templer, and Wallis) devoted to refuting Hobbes. Bodleian Library, Oxford, Ms. Rawlinson D.878, fols. 40v, 41, 41v, 43v, 44v, 49v.
185. Napier (ed.), *Theological Works*, 2:378.
186. Ibid., 5:203-4.
187. Ibid., 1:xlviii.
188. Shapiro, *Wilkins*, p. 185.
189. Napier (ed.), *Theological Works*, 1:xlix.
190. Birch, *Tillotson*, p. 37.
191. W. Pope, *The Life of Seth, Lord Bishop of Salisbury*, ed. J. B. Bamborough (Oxford, 1961), p. 152. Pope's sometimes unreliable account is here confirmed by a letter from Ward to Simon Patrick of 17 June 1671, which concludes, "Dr. Barrow who had done me the favour to accompany me in this place presents his hearty respect to you." Evidently, then, Barrow was well acquainted with Patrick, one of the most prominent of the latitudinarians. Queens' College, Cambridge, Ms. 73, fol. 25.

192. Pope, *Seth,* pp. 152–3.
193. Osmond, *Barrow,* p. 211.
194. P. Barwick, *The Life of the Reverend John Barwick* (London, 1724), p. 39.
195. Napier (ed.), *Theological Works,* 1:lxxii.
196. H. R. Fox-Bourne, *The Life of John Locke* (London, 1876), p. 310.
197. H. W. Robinson and W. Adams (eds.), *The Diary of Robert Hooke, 1672–80* (London, 1935), pp. 10, 11.
198. Napier (ed.), *Theological Works,* 9:584.
199. Ibid., 9:577.
200. Ibid., 5:50.
201. Ibid., 4:15–16.
202. McAdoo, *Anglicanism,* p. 175.
203. Napier (ed.), *Theological Works,* 4:451.
204. Ibid., 1:lxv.
205. Ibid., 3:437–8.
206. Ibid., 1:xlviii.
207. Ibid., 1:lxxiii.
208. Osmond, *Barrow,* pp. 210–11.
209. Cooper, *Annals,* 3:563.
210. E. F. Churchill, "The Dispensing Power of the Crown in Ecclesiastical Affairs," *Law Quarterly Review,* 38 (1922): 309–15.
211. Green et al., *State Papers,* 1675, p. 351.
212. Ibid., 1671, pp. 77–8.
213. F. E. Manuel, *A Portrait of Isaac Newton* (London, 1980), p. 103.
214. D. Masson, *The Life of Milton...* (Cambridge, 1859–94), 6:804. I owe this reference to Dr. H. Porter.
215. Napier (ed.), *Theological Works,* 1:xlix. In a letter to Collins of 10 December 1672, Newton remarked that "we are here very glad that we shall enjoy Dr. Barrow again, especially in the circumstances of Master, nor doth any rejoice at it more than [I]." S. P. Rigaud (ed.), *Correspondence of Scientific Men of the Seventeenth Century, Including Letters of Barrow, Flamsteed, Wallis and Newton,* 2 vols. (Oxford, 1841–62), 2:347.
216. Napier (ed.), *Theological Works,* 1:lxxii.
217. Westfall, *Never at Rest,* pp. 335–6.
218. Ibid., p. 337.
219. Figures derived from W. W. R. Ball and J. A. Venn, *Admissions to Trinity College Cambridge, 1546–1600,* 5 vols. (London, 1913–16), 1:10–11.
220. I am drawing here on the conclusion of my "Isaac Barrow, Theologian and Scientist: A Case Study of the Impact of the 'Scientific Revolution,'" *Flinders Journal of History and Politics,* 9 (1983): 45–55.

5

Barrow as a scholar

ANTHONY GRAFTON

The intellectual achievements of seventeenth-century England have hardly lacked attention. The political theories of rebels and rulers, the technical achievements of anatomists and chemists, and the crowning scientific glory of Newton's synthesis have always evoked interest and enthusiasm. But the widespread interest in these eminently modern aspects of English culture has had its problematic as well as its positive side. It has distracted attention from enterprises less attractive to modern tastes, less closely tied to modern disciplines, and not as well supplied with heroic figures. Classical scholarship of the form that Barrow practiced is a powerful case in point. Everyone knows that English classical scholarship really began with Richard Bentley's great books at the end of the century, the *Epistola ad Millium* and the *Dissertation on Phalaris*. Before these appeared, England produced one great edition of a Greek text, Stanley's Aeschylus, and harbored one great Hellenist from the Continent, Isaac Casaubon. Otherwise, it specialized only in the production of neat little textbook editions of school classics, like the popular editions of Roman poets put out by Farnaby the flogger. So even the most recent histories suggest.

In fact, this vision is at best a partial one, shaped as much by the limitations of modern scholars as by the breadth of the surviving historical documents. It assumes, for example, that the postclassical and the nonliterary aspects of antiquity are somehow peripheral. Thus, a great edition of a patristic Greek text – like the Eton Chrysostom – and a highly erudite edition of a nonliterary classical work – like John Selden's edition of and commentary on the *Marmor parium* – do not matter as much as the high-profile textual criticism of Latin poetry practiced on the Continent by Nicolaas Heinsius and J. F. Gronovius. It assumes that breadth of viewpoint cannot make up for lack of cutting edge. Thus, the broad and sympathetic re-creation of ancient social customs and material culture found in British antiquarian works like Dempster's edition of Rosinus or Potter's Greek antiquities is less important than the sharper and seemingly more modern critical writing of Continental critics like Madame Dacier. It assumes, in other words, that seventeenth-century scholars must be measured by a twentieth-century standard, in regard to both the tools they applied and the choice of objects to which they might apply them. And these assumptions, long abandoned in the parallel enterprise of the history of science, need scrutiny here too.[1]

Seventeenth-century English culture was nourished by several different forms or traditions of direct study of the classics. One of them – the best remembered now, thanks to scholars as diverse as A. E. Housman and L. Stone – was indeed pedagogical. English scholars like Thomas Farnaby proved unusually adept at explicating the Latin classics for schoolboys. Their commentaries, notable for concision, simplicity, and attention to grammatical and syntactical difficulties, were often reprinted in Holland and used in Continental schools. Their approach was hardly elaborate. It basically reflected a simple desire to move the student through the text, eliminating difficulties that might hinder a quick reading; it rarely or never involved the proposing of complex historical or allegorical explanations.[2]

A second tradition – restored to light above all by the work of the late Charles Schmitt – was Aristotelian. English scholars like John Case continued the long-lived tradition of basing university instruction in all fields so far as possible on Aristotle's texts and methods. Their own new textbooks, which combined Aristotelian approaches with more modern ones – like Case's *Lapis philosophicus* and *Sphaera civitatis* – won a considerable market. And as Schmitt conclusively showed, even some of the most innovative English students of the natural world, like Harvey, owed basic assumptions and methods not to their empirical investigations but to their equally intense and prolonged study of the Aristotelian corpus.[3]

A third tradition – studied in part but intensively by Ernst Cassirer long ago, and more broadly, in more recent times, by literary historians like Don Cameron Allen and S. K. Heninger, Jr., and by the great Warburg Institute scholar D. P. Walker – was encyclopedic. Englishmen like John Marsham and foreigners domiciled in England like Isaac Vossius made strenuous efforts to enfold in a single scheme of dates and eras all of early human history, biblical and classical, and to connect in a single developmental scheme all of early human culture. They tried to show the derivation of Greek mythology from Jewish history, to trace the settlers of Greece and Rome back to their ancestors on Noah's Ark, to assess the original intellectual achievements of Egyptian sages, Chaldean priests, and English druids as well as the derivative ones – as they thought – of Greek philosophers. Their works included Cudworth's *True Intellectual System of the Universe,* Stanley's *History of Philosophy,* and Marsham's

Canon Chronicus, all of which earned Continental reprints (in the first two cases in Latin translation, in the third in the Latin original) and provoked substantial replies and attacks. This encyclopedic brand of scholarship, with its efforts to preserve the belief in a *prisca theologia* and its desire to fuse Greek and Biblical traditions, long retained its stimulating power. Newton's strange chronology of the expedition of the Argonauts was only in part his original production and in part the last great flowering of the seventeenth-century encyclopedic tradition.[4]

A fourth tradition, finally, more specific and technical than the others, lay in an area now foreign to most classical scholars, but it was one that in the sixteenth and seventeenth centuries evoked widespread interest: the history of the sciences in antiquity. This tradition of study stretched back in England to the very dawn of Hellenism early in the sixteenth century, when Thomas Linacre translated the elementary textbook of astronomy then known as the *Sphere* of Proclus and worked on the Aldine edition of Galen.[5] It was brilliantly continued by John Caius, in his deft and meticulous efforts to collect and correct the writings of Galen. These combined expert knowledge of medical practice with a more than up-to-date mastery of philology; Caius, as Vivian Nutton has shown, engaged in such rare practices (for his time) as noting down even the erroneous readings from classical manuscripts, on the apparent ground that they might convey useful information, and systematically collecting fragments of lost works by Galen cited by other authors in works still extant.[6] This tradition was enormously enhanced in the 1560s and 1570s, when Henry Savile took fire from his reading of and meetings with Ramist intellectuals on the Continent. He accepted Ramus's view that the ancient sages of the Near East, from Adam on, had cultivated the mathematical sciences more proficiently than the Greeks whose works were preserved. He accepted also the Ramist program of detailed inquiry into the history of Greek science, which was meant to reveal what had gone wrong with mathematics and astronomy in Greek hands and to provide some means of restoring them to their primeval excellence. He presented Ramus's views – often in Ramus's words – to the audience of his Oxford lectures on Ptolemy in 1570.[7] And in one crucial respect he went beyond most contemporary Continental students of Greek mathematics. He embarked on a systematic effort to collect, study, and explicate the main ancient texts in the exact sciences, point by point and reading by reading. As his follower John Bainbridge put it in the preface to his edition of Ptolemy's *Planetary Hypotheses,* Savile

> thought it of no great value to have established two professorships in mathematics with liberal salaries unless he also restored the ancient – and the best – masters to their chairs, from which they were shamefully cast down by the barbarism and contempt

of earlier times. And he assembled various Greek manuscripts
of the ancients to that end.[8]

This enterprise proved extraordinarily successful. The Bodleian Library
became a central repository of Greek scientific manuscripts. Savile himself became one of the few Europeans to have a real mastery of Greek
astronomy, as his annotated copies of the *Almagest* and Theon's commentary on it, and his lectures on the text, clearly show.[9] He became
a recognized member of the European fraternity of mathematicians as
well. When Joseph Scaliger ventured his strange attack on mathematics
as a discipline, arguing that all its practitioners since ancient times had
gone wrong in their efforts to deal with the quadrature of the circle –
arguing, indeed, that some of the basic procedures they had used in their
works on the problem, like *reductio ad absurdum,* were formally illegiti-
mate – every luminary in the profession from Clavius to Viète ridiculed
him. He turned to Savile, whom he had met long before, in a desperate
search for sympathy and vindication – only to receive a very dusty answer
indeed, as Savile explained why *reductio ad absurdum* was a basic and
proper form of geometrical reasoning in the classical style. Savile's high
standing in the world of learning is clear from the uncharacteristic defer-
ence Scaliger showed him, sending him a long letter and a demonstration
elaborately written out. His reaction to Savile's reply is, unfortunately,
unrecorded.[10]

What Savile proposed, moreover, his professors disposed. They produced editions of technical Greek texts that can more than stand comparison with far later efforts. Bainbridge's edition of the *Planetary Hypotheses*
includes emendations made on technical grounds that were rejected by
the late-nineteenth-century editor of Ptolemy, J. L. Heiberg, only to be
forcefully vindicated in still more recent times by the greatest of all modern students of ancient astronomy, Otto Neugebauer.[11] The same work
includes an edition of the *Canon,* or dated list of Near Eastern, Greek,
and Roman rulers, that formed part of Theon's recension of Ptolemy's
Handy Tables. This crucial document for the history of the ancient world
was published several times after Joseph Scaliger first included it in his
Thesaurus temporum of 1606, but its status, its import, and its apparent
disagreements with the Bible puzzled some of its most competent early
readers, notably Johannes Kepler, who tried to emend the *Canon* to make
it match other sources of less precision.[12] Bainbridge saw at once that the
Canon was fundamentally correct as transmitted, that the regnal dates it
contained matched and provided a context for the chronological intervals
of the *Almagest.* His edition remains of fundamental interest to any student of the *Canon* – or, indeed, of ancient chronology.

No wonder, then, that Bainbridge also proposed, in a memorandum
still unpublished, a dazzling conjectural emendation of an eclipse dating
in the *Almagest.* Bainbridge's proposal may be unnecessary, but it deals

sensibly with a still puzzling passage of the text – so sensibly that it would be repeated two hundred years later by another great student of ancient astronomy and chronology, Ludwig Ideler, in total ignorance of Bainbridge's anticipation of him.[13] No wonder either that he produced perhaps the most original of all seventeenth-century works on ancient chronology, the *Canicularia,* which dealt with the risings and calendrical import of Sirius – as well as all poetic references to it – in the ancient world. This posthumously published and soon forgotten book displayed formidable knowledge of ancient literature as well as of astronomy and chronology, and demolished central portions of Scaliger's reconstruction of the Egyptian calendar.

Another Savilian professor, Bainbridge's friend Henry Briggs, edited Euclid's *Elements.* Yet another, John Wallis – a mathematician of great originality and proficiency – produced an edition of Ptolemy's *Harmonics* and related texts that remains of considerable interest and utility.[14] He thus connected scholarly inquiry into the past with some of the most advanced experimental work done in the Royal Society in his own day. England would not see such intensity of interest in or such proficiency in explication of ancient scientific texts again until the far later age of T. L. Heath and J. K. Fotheringham. Evidently Newton was far from being the only English scientist to combine an ample helping of humanistic and historical interests with his more technical pursuits. Evidently, too, a history of classical scholarship attuned to the interests of most seventeenth-century practitioners would find ample space for England and English scholars – and would offer sympathetic interest, rather than ridicule, for such now bizarre-seeming but then reasonable enterprises as Isaac Vossius's study of ancient prostitution and Edmund Dickinson's effort to show that the poems of Homer rewrote the Old Testament in dactylic hexameter.

Barrow's work is connected to all four of these traditions, in different ways and measures. His links to the first two, naturally, were official. Barrow's career included two posts teaching the humanities. He himself wrote a deliberately artistic and elaborate Latin prose and verse on many ceremonial occasions, and though he produced no neo-Latin best sellers, he knew his quantities and his Horace well enough to win a place in the great nineteenth-century anthology of Latin verse from Charterhouse, *Sertum Carthusianum.*[15] He argued powerfully to his Cambridge audiences for the unique value of the ancients as the basis of a coherent education, one that would prepare young men for all the intellectual and practical difficulties they would meet. And he insisted on Aristotle's unique pedagogical and intellectual value as the basis for university teaching and disputation. He was at once, in other words, a competent practitioner and a warm defender of the humanistic techniques and Aristotelian traditions that most histories would lead one to think died out in England after 1600.

Both the lengths to which Barrow went to prepare himself as a humanist and the flavor of the eloquence he polished and cherished may come as a surprise to the unprepared modern reader. Like any good university man a hundred years before, Barrow collected – as a later title to one of his notebooks tells us – "sentences...out of the old Greek tragedians and comedians." This carefully indexed florilegium dealt with a paralyzingly wide range of subjects:

> *Arbores*
> *Arbitri. Arbitrium*
> *Arcana v. Secreta*
> *Aristocratia*
> *Arithmetica*
> *Arma*
> *Architectura. Aedific.*[16]

So run a few headings under *A*. The collection itself consists of lines and short passages from the Greek dramatists, given without comment, though with occasional cross-references (and sometimes another sort of gloss; the excerpts under *Medicina,* all in Greek, end with the laconic, wicked remark "On voit plus de vieux yvrongs, que de vieux medecins.").[17] They are followed by longer extracts from the ancient historians. And the purpose of all this meticulous assembling of materials, this booty capitalism of the spirit, becomes clear as soon as one examines Barrow's Latin.

Barrow wrote a characteristic late humanist prose, ornate and rotund – so much so that it greatly offended Whewell, himself no Atticist and no stranger to seventeenth-century Latin. In particular, he missed no reasonable opportunity and invented many unreasonable ones to deploy one of the *sententiae* he had hunted and collected in order to make a passage more copious. The mechanical quality of his *bricolage* is clear from its repetitiveness. To take a simple example, no textbook offered more handy expressions to the young Latinist, more tags laden with meaning to prove their users' learning, than Erasmus's *Adages.* This book, long a best seller in early modern Cambridge, gave some thousands of pithy expressions (like *dulce bellum inexpertis* – "War seems great fun when you haven't tried it") in the variant forms in which they appeared in the classics, together with essays of various lengths about their import. Tags like these peppered the Latin of all good products of a seventeenth-century humanist education, providing an international language that – rather like geometry in the ancient Mediterranean or smoke signals in pre-Columbian North America – proved one's humanity and adherence to an internationally valid code of conduct and values.

One of the best-known humanist tags – one often useful in that standard humanist letter that instructed a friend to cheer up under adversity – was *Spartam nactus es, hanc orna,* "You're in Sparta now; make the best of it," a proverb that, according to Erasmus, "tells us that whatever province we happen to have made our own, we must fit ourselves to it, and suit our behaviour to its dignity."[18] No one ever found more ways to quote one maxim than Barrow. In his oration on assuming the post of Humanitatis praelector in 1654, he thanked his patrons for assigning him "*Spartam . . . curae omnis et periculi immunem, nec corpori nec animo ingratam,*" "a province free from any worry or danger, pleasant to both body and soul."[19] In his oration on assuming the chair of Greek in 1660, he lavishly praised his predecessors from Erasmus to Duport, "*qui hanc Spartam excolentes claruerunt,*" "who cultivated this province with special distinction."[20] And in the memorable "Sarcastic Oration" of 1661, in which he thanked his listeners for having stayed away from his first year's lectures on Sophocles, he explained why he had chosen Sophocles as his text when he accepted "*hanc Spartam pro virili mea ornandam.*"[21] One wonders if Barrow's repetitive rhetoric partially explains his lack of appeal as a professor. At all events, the entirely conventional nature of his eloquence is clear. All of Barrow's formal Latin prose is characterized by a similar coacervation of adages, allusions, references to history, mythology, and poetry. In the course of a mere two pages of complaint about his low enrollments, he compares himself to Prometheus chained to his rock, to Polyphemus frightening all comers away from his cave, and to an Ovidian lover singing "not to the hills or the woods, but to the walls and benches."[22] *Kunstprosa* evidently was alive and well in revolutionary and Restoration Cambridge; and even if we view that pursuit with less sympathy than Barrow himself clearly did, we must recognize its continued validity in its own time in England, as historians already do in the contemporary Holy Roman Empire.

Barrow's commitment to Aristotle, like that to the *studia humanitatis,* had both standard and original features. In defending Aristotle against all potential rivals from antiquity, Stoics, Skeptics, and Platonists, he took the normal position of a university lecturer on whom Aristotle's gifts of precise and formal argument to a clear end could not be lost. In defending Aristotle against modern rivals and detractors as well, he also shows no particular originality.[23] When he borrows Cicero's praise of Aristotle's *flumen orationis aureum* (golden stream of eloquence), for example, he merely repeats as hundreds of humanists had before him a *topos,* and an inappropriate one to boot. Cicero had in fact applied this phrase to Aristotle's lost literary dialogues rather than to the surviving corpus of technical works that Barrow actually read. But in switching his lectures – after a year's course on Sophocles' tragedies had drawn no listeners – to Aristotle's *Rhetoric,* Barrow showed a clear-eyed realism rare among pedagogues at any time. Greek courses often failed in early modern Europe, especially, perhaps, when they dealt with the greatest of Greek texts – Pindar, the tragedians, Thucydides – works that posed such terrifying linguistic tasks that they drove most students away. Courses succeeded when those who gave them chose realistic goals, construable texts, and subjects that connected readily with other segments of the curriculum. And here at least Barrow stood out. He explained that Aristotle should attract students not only by his brilliance and eloquence but by his practical value. One could not go on to study medicine, theology, or jurisprudence without knowing Aristotelian philosophy; nor could one hold any substantial job in the university. In offering this argument Barrow confirms the grip Aristotle retained over so much of the curriculum and reveals himself as able to temper his commitments and tailor his offerings neatly to contemporary tastes.[24] John Case was no more adept than Barrow at preserving the Aristotelian tradition.

The encyclopedic tradition also crops up visibly, but less frequently, in Barrow. His exposure to it probably began early – perhaps at the feet of his respected predecessor in the chair of Greek, James Duport. One of Barrow's notebooks, which bears the name "Duportus" on an early leaf, contains detailed notes on Homer that deploy a vast range of subsidiary information to make the text itself not only a mirror to the reader, but a window through which one could see into a richly reconstructed past. When traces of material culture appear – like Paris's hair in *Iliad* 11:385 or the bow with which he wounds Diomedes in 11:375 – the text becomes the pretext for detailed excursuses into classical material culture and social mores – matters that the lecturer describes as subsidiary, but into which he delves at such length as to bely his explicit statement.[25] When Diomedes says that Paris has wounded him, using the verb *epigrapsas* (11:388), the lecturer comments revealingly that "*graphein* and *epigraphein* are never found in Homer with the [later normal] sense of 'writing'"; he refers to a more detailed treatment elsewhere of the one apparent contradiction to this rule, the tale of the letter on a folding tablet that Bellerophon bore to his intended killer; and he explains that writing in the modern sense was devised long after Homer's time.[26] Other notes take off much farther from the text – in one case, into a detailed discussion of the question, much debated in the seventeenth century, of whether New Testament Greek had a particularly Hebraic flavor.[27] The notes are fragmentary, and their origin uncertain; but they do show Barrow in direct contact with the older, encyclopedic traditions that still lived in the Cambridge of Duport and Cudworth, and thus exposed to the broad-gauged reconstructive scholarship characteristic of the antiquarian tradition.

Barrow's main piece of philological work, however, was the influential edition of Archimedes, Apollonius, and Theodosius that he brought out in 1675. And here he employed a laconic and austere method, rather than the more expansive ones of his notebooks. His edition, as he himself recognized, could no longer use the large format and heavy commentary characteristic of influential earlier editions like Rivault's. Archimedes had now been surpassed by modern mathematicians; the old humanistic mathematics that tied most or all mathematical discoveries to the recovery or reconstruction of classical procedures had also become outdated. A new edition – and Barrow promised a "new method" on his title page – could not make a mark or find an audience by surrounding its island of text with a sea of obsolescent erudition. Rather, Barrow took care to emphasize the small size and low price of his edition as special attractions. Moreover, and more important, he emphasized that a modern edition of Archimedes et al. should aim not at the reconstruction of modern mathematics but at the preservation of a classical tradition that retained historical interest and pedagogical utility:

> To preserve ancient authors, the inventors of the sciences, from destruction seems an important task for their modern followers, who would otherwise earn a reputation for ingratitude. True, their contents can in large part be derived more rapidly or constructed more concisely by modern techniques; yet reading them retains its value. First, it seems pleasant to examine the foundations from which the sciences have been raised to their present height. Second, it will be of some interest to sample the sources from which virtually all the discoveries of the moderns are derived; for it was by studying or imitating the clever and subtle methods of the ancients that the industry of the latter reached its eminence. Furthermore, I think that there is no better source for learning a pure taste for and skill in demonstration than those whose skill and elegance became especially illustrious in their deduction of theorems.[28]

Barrow subordinated all decisions about format and editorial policy to the need for economy and clarity in presenting the mathematics of the texts. Like his brilliant younger contemporary Bentley, he treated his text as a rationalist should. What mattered was clarity; to provide it he frequently altered the Greek text he worked from, as his marginal notes show. Like Bentley too, Barrow often felt that his corrections needed no more justification than *lego*, "I read." But unlike Bentley – who reconstructed the histories of the texts he studied, dug energetically in libraries for new manuscript evidence, and qualified even his famous declaration that "I prefer reason and the nature of the case to a hundred manuscripts" with the revealing clause "especially since the old Vatican MS agrees with me" – Barrow devoted no space to philological problems or information.[29] His readers did not know which Greek text he had used as the basis for his translation, whether his emendations were manuscript variants or his own conjectures, or even what he thought about most problems of chronology and authenticity. One result of these laconic procedures has been that Barrow has lost credit for many of his alterations to the standard

Greek texts of his time, as they have since been found in manuscripts and are accordingly reported as variants, not conjectures, in modern critical apparatuses. For our purposes, however, ancestry supersedes progeny; the interest of Barrow's methods lies above all in their similarity to those of Savile and Bainbridge. In his ability to combine high technical skill in Greek and mathematics alike, his refusal to engage in the details of philological work, and his obsessive interest in producing a rigorous, clear, useful text, Barrow updated the English tradition of the late Renaissance to fit the needs of a less humanistically inclined age.

No segment of Barrow's edition is more revealing in this regard than his brief introduction to the *Lemmata* or *Liber assumptorum* – a short text that survived the Middle Ages not in Greek but in an Arabic translation by Thabit ibn Qurra.[30] This had been published twice in rapid succession: by S. Foster in London (1659) and G. A. Borelli in his Florentine edition of Apollonius's *Conics* (1661). Borelli in particular had found much to worry over in the book, which the Arabs had ascribed to Archimedes. On the one hand, he told his readers, neither Archimedes nor later Greek mathematicians had ascribed this work to Archimedes himself; moreover, Pappus quoted much similar material, which he did not ascribe to Archimedes. On the other hand, neither Pappus nor Archimedes had listed all of Archimedes' works; indeed, the Archimedean commentator Eutocius of Ascalon had discovered an anonymous geometrical work that he attributed, on the grounds of its Doric style as well as its method, to Archimedes. Could not Thabit have translated a still more corrupt version of this work? Borelli came down on the side of authenticity, but whispered his views as unassertively as possible: "Shall we hesitate," he asked the reader, "to ascribe this little work to Archimedes?"[31]

Barrow dispatched both this complex set of problems and Borelli's complex solution to them with contemptuous ease. "I am quite amazed," he wrote, "that this excellent man, who is clearly the second ornament [along with Archimedes] of Sicily, disputes so anxiously about the author of these *Lemmata*." To be sure, one or two propositions used Archimedean terms like *arbelos* (a figure containing three semicircles with coincident diameters, which resembled a semicircular cobbler's knife). But it also contained far more material also found in Pappus with no author's name attached. It was surely an "Opus...mixtum" containing a few ancient propositions, not a genuine piece by Archimedes.[32]

By modern standards, Barrow's solution is far superior to Borelli's. He is right about the massive correspondence between the *Lemmata* and Pappus, right about the small amount of genuine Archimedean material in the work, right, finally, that Archimedes wrote no work that used the format of the *Lemmata*. Yet historically the minimalism of his means is more impressive than the correctness of the end they lead to. What had stimulated Borelli to undertake a massive excursion into classical mathematics and ancient and modern methods of higher criticism became for Barrow a side issue, to be rapidly dispatched – a question no more interesting, in the end, than the origin of the excellent "Schemata" to the text that he drew from the manuscripts of Jacob Golius, even though he did not know whether Golius "happened on better MSS than Abraham Ecchellensis [the translator] and I saw, or rather used his genius and erudition to achieve these astonishing results."[33]

Barrow, then, was a master scholar by the best standard of his time. His command of classical languages and of the techniques of textual and historical criticism was clearly comparable to his command of classical and modern mathematics. But scholarship was for him a tool rather than an end, one implement among many; and he practiced it as a mathematician should, aiming at the absolute minimum of argument and exposition that could serve his needs.

Notes

1. For a learned and critical version of the traditional view, see C. O. Brink, *English Classical Scholarship* (New York, 1986); a more historical vision inspires J. Levine, *Humanism and History* (Ithaca, N.Y., 1987).

2. See *M. Annaei Lucani Belli civilis libri decem,* ed. A. E. Housman (Oxford, 1950), and, for the pedagogical context, L. Stone, *The Crisis of the Aristocracy, 1558–1641* (Oxford, 1965), pp. 684–7.

3. C. B. Schmitt, *John Case and Aristotelianism in Renaissance England* (Montreal, 1983).

4. E. Cassirer, *Die Platonische Renaissance in England und die Schule von Cambridge* (Leipzig, 1932); D. C. Allen, *Mysteriously Meant* (Baltimore, Md., 1970); S. K. Heninger, Jr., *Touches of Sweet Harmony* (San Marino, Calif., 1974); D. P. Walker, *The Ancient Theology* (London, 1972); F. E. Manuel, *Isaac Newton, Historian* (Cambridge, Mass., 1962).

5. See *Linacre Studies: Essays on the Life and Work of Thomas Linacre, c. 1460–1524,* ed. F. Maddison and C. Webster (Oxford, 1977).

6. V. Nutton, *John Caius and the Manuscripts of Galen* (Cambridge, 1987); see also his "'Prisci dissectionum professores': Greek Texts and Renaissance Anatomists," in *The Uses of Greek and Latin: Historical Essays,* ed. A. C. Dionisotti et al. (London, 1988), pp. 111–26.

7. See N. Jardine, *The Birth of History and Philosophy of Science* (Cambridge, 1984), pp. 268–9. A specimen of Savile's use of Ramus is presented in A. Grafton, "From *De die natali* to *De emendatione temporum:* The origins and setting of Scaliger's chronology," *Journal of the Warburg and Courtauld Institutes,* 48 (1985): 138–40.

8. *Procli Sphaera. Ptolemaei de hypothesibus planetarum liber singularis...,* ed. J. Bainbridge (London, 1620), epistola lectori.

9. Savile's Basel 1538 Ptolemy is Bodleian Library Savile W.14; for his work on the *Almagest* see Bodleian Library Mss. Savile 26–31, which await serious study.

10. O. Glucker, "An Autograph Letter of Joseph Scaliger to Sir Henry Savile," *Scientiarum Historia,* 8 (1966): 214–24.

11. O. Neugebauer, *A History of Ancient Mathematical Astronomy* (Heidelberg, 1975).

12. See Kepler to Scaliger, 27 October 1607, in Kepler, *Gesammelte Werke* (Munich, 1938-), 16:63–71.

13. Bodleian Library Ms. Rawl. C.936, fol. 15 recto–verso, on *Almagest* 4:9.
14. C. V. Palisca, "Scientific Empiricism and Musical Thought," in *Seventeenth-Century Science and the Arts,* ed. H. H. Rhys (Princeton, N.J., 1961), pp. 99–100.
15. L. Bradner, *Musae Anglicanae* (New York, 1940).
16. Trinity College Cambridge Ms. R.9.40, fol. 1 verso.
17. Ibid., fol. 29 verso.
18. M. M. Phillips, *Erasmus on His Times* (Cambridge, 1967), p. 100.
19. *The Theological Works of Isaac Barrow,* ed. Alexander Napier, 9 vols. (Cambridge, 1859), 9:130.
20. Ibid., p. 138.
21. Ibid., p. 158.
22. Ibid., pp. 156–7.
23. Ibid., pp. 164–9.
24. Ibid., p. 168. For the place of Aristotle at seventeenth-century Cambridge see, e.g., C. Wordsworth, *Scholae academicae* (Cambridge, 1877), pp. 22, 124–6, emphasizing that Aristotle retained preeminence longer at Oxford than at Cambridge; J. L. Mulder, *The Temple of the Mind* (New York, 1969).
25. Trinity College Cambridge Ms. R.9.36, esp. fol. 6 verso.
26. Ibid., fol. 10 verso.
27. Ibid., fol. 19 recto.
28. *Archimedis opera. . .,* ed. I. Barrow (London, 1675), epistola lectori.
29. For Bentley see S. Timpanaro, *La genesi del metodo del Lachmann,* new ed. (Padua, 1981), p. 14.
30. For a brief discussion see *Dictionary of Scientific Biography,* s.v., "Archimedes," by M. Clagett.
31. Apollonius Pergaeus, *Conicorum lib. v. vi. vii,* ed. G. A. Borelli (Florence, 1661), 379–83.
32. *Archimedis opera,* ed. Barrow, sig. B recto–verso.
33. Ibid., sig. B 2 recto.

6

The preacher

IRÈNE SIMON

When Barrow died on 4 May 1677, only one of his sermons, *The Duty and Reward of Bounty to the Poor,*[1] had appeared; he had also sent to the press, but did not live to see published, *Upon the Passion of Our Blessed Saviour.*[2] Both these sermons had been preached before the lord mayor and aldermen of the city, who had asked Barrow to print them; they were twice[3] reprinted before they were included in the collected edition. At his death Barrow's papers were left to his father, Thomas Barrow, who entrusted them to John Tillotson, then dean of Canterbury, and to Abraham Hill – both friends of Barrow's – "with a power to print such of them as they saw proper."[4] Not all these manuscripts have survived, but a large number of them are extant in the library of Trinity College, Cambridge,[5] and one (Mss. Landsdowne 356) is preserved in the British Library. From these Tillotson published *Sermons Preached upon Several Occasions* in 1678, 8vo (reprinted 1679), *Several Sermons against Evil-Speaking,* also in 1678, 8vo (reprinted 1678, 1682), and *Of the Love of God and Our Neighbour in Several Sermons* in 1680, 8vo. Tillotson also published in 1680 Barrow's *A Treatise of the Pope's Supremacy,* as well as *A Discourse Concerning the Unity of the Church,* and in 1681 *A Brief Exposition of the Lord's Prayer and the Decalogue; with The Doctrine of the Sacraments.* In 1683 Tillotson brought out the first volume of his collected edition: *The Works of the Learned Isaac Barrow, D.D.,*[6] late master of Trinity College in Cambridge, Published by the Reverend Dr Tillotson, Dean of Canterbury, London, 1683 (f°), Containing *Thirty-two Sermons upon Several Occasions* (i.e., the contents of the three octavo editions plus *Bounty to the Poor* and *Upon the Passion*), *An Exposition of the Lord's Prayer, etc.,* and a second edition of *A Treatise of the Pope's Supremacy,* together with the *Discourse Concerning the Unity of the Church.* The second volume of the *Works,* Containing (thirty-four) *Sermons and Expositions of the Articles in the Apostles Creed,* also came out in 1683. The third volume of the *Works,* Containing *Forty-five Sermons upon Several Occasions* and completing Barrow's English *Works,* appeared in 1686.

As Abraham Hill tells us in "Some Account of the Life of Dr Isaac Barrow" in a letter addressed to Tillotson and prefixed to the first volume of the *Works,* Barrow "was so exact as to write some of his sermons four or five times over."[7] From the manuscripts preserved in the library of Trinity College we can form some idea of the task Tillotson set himself, for several sermons indeed have survived in four or five different versions. Examination of the manuscripts reveals how Barrow proceeded as he prepared his sermons: Some are clearly drafts, though already written in Barrow's neat, legible hand; some are revised versions; and some appear to be fair copies probably to be used in the pulpit. The process of revision consisted mainly in starting afresh and expanding the previous version: Barrow would add further arguments in support of his theme, develop those he had propounded, amplify his sentences, and add further references in the margin. He hardly ever left out anything he had previously written, but sometimes erased a word and wrote another above the line. As far as we can judge – for all the manuscripts Tillotson used for his edition have not survived – the dean always chose what appears to be the final version (the papers he sent to the printer's bear the printer's marks either to the octavo or to the folio edition). Tillotson thus published 111 sermons by Barrow: thirty-two in Volume 1, numbering 488 folio pages; thirty-four in Volume 2, numbering 506 pages; and forty-five in Volume 3, numbering 535 pages.[8] Yet it appears that he overlooked one, *A Defence of the B. Trinity,* preached on Trinity Sunday 1663, which Brabazon Aylmer, who had published all of Barrow's previous works, brought out in 1697, that is, after Tillotson's death. As it says in "The Publisher's Advertisement," this oversight of the dean's is easy to understand "in so great a number" of papers.[9] Apart from *A Defence of the B. Trinity,* Barrow's sermons were thus known to the reading public through Tillotson's edition or through editions based on it,[10] until the publication in 1859 of Barrow's *Theological Works* by Alexander Napier (9 vols., Cambridge University Press).

Napier found much to criticize in Tillotson's edition: He blamed him for often substituting words in common usage for archaic or learned ones; for adding sentences here and there; for deleting some of Barrow's sentences; and for dividing continuous texts into several sermons.[11] His strictures have been repeated by later writers on pulpit oratory or on Barrow,[12] but closer examination of the manuscripts Tillotson sent to the printer reveals that his verbal revisions are few and that he generally has authority from the manuscript – either a larger blank between paragraphs or some mark in Barrow's hand – for printing some continuous versions as a series of sermons on the same text. On the other hand, Napier's edition is not free of blame: He often prints as long footnotes passages from duplicate manuscripts of Barrow's, ignoring the reordering of the arguments in the final version; in fact, what Napier prints is often a composite, not an authentic text.[13] We can therefore trust that the sermons of Barrow "which continued to be read with pleasure in the eighteenth century were really Barrow's not 'Tillotson's Barrow.'"[14]

In his preface to the first volume of the *Works* Tillotson tells us when and where some of the sermons in this volume were delivered. Three of them[15] "were preached upon solemn Occasions: The *First* of them upon the 29. of May, 1676, the Anniversary of His *Majesty's* happy *Restoration:* the *Second* upon the 5 of November, 1675,[16] in commemoration of the great deliverance from the *Powder-Treason;* both in the year of his Vice-Chancellourship.[17] The *Third* at the consecration of the Bishop of *Man*...his Uncle."[18] Tillotson further tells us that the first sermon Barrow ever preached was *The Pleasantness of Religion* (1:1; 1:3),[19] at St. Mary's Cambridge on 30 June, 1661. The "two excellent *Sermons* of *Thanksgiving* [1:8, 9; 1:10, 11] were the next" (the manuscripts[20] bear a note in Barrow's hand: on the first, "Jan. 17, 1662, St Mary's"; on the next, "July 19, 1663, St Mary's," and "Jan. 10, 1664, Gray's Inn)." The "fourth in order," that is, *The Reward of Honouring God,* says Tillotson, "was the first that he preached before the King's Majesty" (the manuscript has a note in Barrow's hand: "ad aulam"). This is, unfortunately, all he tells us about the time and place of delivery. From Evelyn's diary we learn that *Love of Our Neighbour* (either 1:25 or 1:26; 2:27 or 28) was preached to the Household, that is, at the Royal Chapel in Whitehall, on 25 April 1675. We also know the dates and places of the two sermons Barrow was asked to print.[21] The date of the sermon on the Trinity, Trinity Sunday 1663, appears opposite the title page. The manuscript of the sermon *That Jesus Was the True Messias*[22] (2:20; 6:20) has the date "1671" inscribed on it. We learn from Walter Pope that *The Folly of Slander* (1:17, 18; 2:19, 20) was preached in two parts at Westminster Abbey.[23] He also tells us that Barrow used to preach for Wilkins at St. Lawrence Jewry,[24] but he gives no further details. It is surprising that Tillotson, who was a close friend of Wilkins and himself lectured regularly at St. Lawrence Jewry, does not mention it in his Preface. Finally, there is a tradition that the sermon *Foolish Talking and Jesting* (1:14; 2:16) was preached at Court and aimed at Buckingham. Considering how meager our information is in this respect,[25] it is generally thought[26] that most of Barrow's sermons were delivered in the chapel of Trinity College, Cambridge (he was appointed college preacher in 1671). This circumstance, together with Barrow's scholarly bent, probably accounts for his manner and style of preaching.

Walter Pope thought Barrow's sermons were too long, "complete treatises rather than orations."[27] In our own century it has been said that they "were bound soon to be fatiguing,"[28] but there is no evidence that such was the response – or rather lack of – of the original audiences. The length of his sermons probably varied according to the place and occasion as well as to the subject treated. The longest of all, *Bounty to the Poor,* was preached at the spital before a city audience, who obviously could sit for hours listening to sermons; according to Walter Pope, it took Barrow three hours and a half to deliver it.[29] Yet the Court of Aldermen requested him to print it "with what farther he had prepared to deliver at that time."[30] This sermon *is* inordinately long: In the first collected edition it numbers forty-three folio pages, but we do not know how much of it was actually preached. The only other sermon Barrow saw through the press, *Upon the Passion,* is more than half as long (twenty-five folio pages), although the Court's request to print it does not refer to what Barrow had further prepared. Walter Pope, however, relates that one day when Barrow was to preach at Westminster Abbey he was asked to deliver only half of what he had prepared and that this part alone took one hour and a half to preach. Walter Pope adds, "This sermon [*The Folly of Slander*] is since published in two sermons as it was preached";[31] in the collected edition the first part numbers twelve folio pages and the second nine. The first part of *The Folly of Slander* is about half as long as *Upon the Passion;* if Pope's figures are correct the latter would have taken three hours to preach. From this we may infer that *Bounty to the Poor* is twice as long as what Barrow actually delivered at the spital.[32]

Barrow's sermons are certainly long compared with Tillotson's, but the majority of them are not longer than Stillingfleet's. Apart from the two he was asked to print, the longest are *On the King's Happy Return* (nineteen pages), *On the Gunpowder Treason* (twenty pages), and *A Consecration Sermon* (twenty pages). The solemnity of the occasion probably accounts for the length of these, while the matter treated justified a full treatment in the sermon *Of the Vertue and Reasonableness of Faith* (twenty-three pages), in the sermon on the article of the creed *Was Crucified* (twenty pages), and in *The Certainty and Circumstances of a Future Judgment* (twenty-three pages). A few number less than ten pages. Except for one, *He Descended into Hell,* those are parts of longer discourses on the same text that Barrow must have preached in a series of sermons. It was also his habit to preach several sermons on connected subjects, for instance, in defense of faith and of the Christian religion, against evil speaking, and so on. Tillotson recognized this when he selected some for his first editions of 1678 and 1680, irrespective of the times at which they had been delivered. The sermons on the Creed, however, must have been intended as a series and preached as successive lectures in the order in which they were printed. Again, several sermons in the third volume appear in the manuscript as a continuous development of the same text, although blanks indicate that each part was meant to be preached separately.

This is all that can be reconstructed about the sermons as preached. Of Barrow's mode of delivery we have no evidence, but we can infer from the court's order to print *Bounty to the Poor* as well as from Walter Pope's remark about *The Folly of Slander* that it was his habit to, and that he was expected to, prepare his sermons in full. To conclude from this, as some have done,[33] that Barrow *read* his sermons seems to be unwarranted; he certainly used his papers in the pulpit. The copies chosen by Tillotson are usually in a larger hand and appear to have been meant for aid in delivery; but we all know the difference between reading a paper and lecturing from a fully written out text. We can also infer that Charles II would not have chosen Barrow for one of his chaplains simply because he was "the best scholar in England";[34] nor would Barrow have been asked to preach the spital sermon if the lord mayor and aldermen had not expected him to persuade his audience to give money to aid the poor. If Walter Pope is correct in stating that Barrow sometimes preached at St. Lawrence Jewry for Wilkins, we may also infer that Wilkins knew his man to be not only a sound divine but an effective preacher.

Whatever qualities Barrow displayed in the pulpit, it is through his published sermons that he reached a large audience in the seventeenth century and after. We need only remember that Mr. Booth, the husband of Fielding's Amelia, owed his reformation to Barrow's "excellent sermons in proof of the Christian religion"[35] to realize that Barrow's convincing power did not depend on his mode of delivery. Boswell's praise[36] of the comprehensiveness and liveliness of *Against Foolish Talking and Jesting* is further evidence of the enduring value of his sermons (or at least some of them). Coleridge thought that in verbal imagination Barrow excelled "almost every other writer of prose."[37] Charles II was even nearer the mark when "in his facetious manner" he called Barrow "*an unfair preacher* because he exhausted every subject, and left no room for others to come after him."[38] The remark, like so many of the Merry Monarch's, is perhaps double-edged, but all the truer for being a sober assessment of Barrow as a preacher.

Barrow preached on both the *credenda* and the *agenda* of the Christian religion, but whatever subject he treated he drew from it applications to practice. He also preached on solemn occasions, such as the anniversary of the return of Charles II (Royal Oak Day, 29 May) and in commemoration of the discovery of the Gunpowder Plot (5 November); as far as we can tell from the manuscripts, however, he did not preach on the anniversary of the execution of Charles I (30 January), an occasion that often lent itself to attacks on the Puritans. Yet Barrow was a staunch defender of the Established Church: In his *Treatise of the Pope's Supremacy* he marshaled an impressive number of arguments against Rome, and in his *Discourse Concerning the Unity of the Church* he demonstrated the danger of separation and argued that unity in discipline is not a mere speculative point but has a direct bearing on the practice of duties. Further, he devoted four of his sermons to *Obedience to Our Spiritual Guides and Governours* in support of the episcopal system of government in the Church.

Whatever subject he treated, the appeal was always to the reasonableness or common sense of men, for he set out to prove that such was the right belief or course of action and that true faith must issue in right action. Some subjects, such as the Passion of Christ, lent themselves to appeal to the "affections" of his audience, but Barrow never attempted to move his hearers through sheer appeal to feelings: He invited them to *consider*. In his sermon *Bounty to the Poor,* for instance, he does not try to move his audience by drawing an affecting picture of the plight of the poor, but deals instead with the many reasons why Christians should give plentifully to their poorer brethren, as well as with the rewards of charity in this world and in the next.

It is certainly characteristic that the first sermon he is known to have preached was *The Pleasantness of Religion.* Clearly, he meant to win men over to religion or to strengthen their faith and encourage them in piety. For all the stress he lays on the need for men to consider their latter end and to submit to the will of god, Barrow's view of the Christian life is not one of sorrows and sufferings: He rather expatiates upon the goodness of God than on His wrath. He did believe in the efficacy of rewards and punishments, and he showed that men are sorely tried in this world, but he did not preach "hell-and-fire sermons." He chose instead to persuade his audience to "walk uprightly" in the sight of a good though just God. When dealing with faith, he did not ask men to assent blindly or to search their hearts to find out if they were among those promised salvation. Instead, he demonstrated that justifying faith is right belief issuing in the performance of the duties of religion, and he preached four sermons on *The Doctrine of Universal Redemption.* Surely, his familiarity with the Cambridge Platonists no less than his own scientific bent led him to reason men into truth and right living. Besides, in an age when skepticism and Socinianism were gaining ground, he saw that men had to be convinced if they were to hold fast to the truths of the Christian religion. When dealing with the mysteries of religion, however, he argued that some truths are beyond human reason but must be accepted on the authority of Scripture, once belief in God's word has been established on reasonable grounds.

All in all, his sermons teach men to lead the good life in imitation of Christ, not by destroying the old Adam but by leading them nearer to the second Adam. His sermons *Of the Love of God* and *Of Love of Our Neighbour* show that love of God is as strong an incitement to Christian virtue as is fear of judgment, and Barrow saw that the pleasantness of religion was the best argument in favor of Christianity. The virtues he extolled are no less characteristic of him: Charitableness, the right government of the tongue, a peaceable temper and carriage, contentment, patience, and so on. These were virtues to be especially recommended in an age when censuring, criticizing, and being malicious passed for wit; when acrimony in controversy and the spirit of faction were widespread; and when there was much loose living and idleness in the fashionable world. The fact that Barrow did not preach on the plea of tender conscience, but referred to it only occasionally, again reflects his equable temper and his love of peace.

Encouragements to religion

The Christian religion is "delectable," Barrow argues, because it is a revelation of truth, and truth is what all men thirst after.[39] Moreover, its teachings do not consist of barren theories but of such truths as will enable men to reach their true end in this world and the next. The Christian religion is a doctrine vouchsafed by Truth itself; it proposes to men objects that are attainable and enjoins duties that with the help of God it is in their power to perform. Compared with philosophy, which also teaches men to achieve content through government of the passions by directing them to a worthy end, religion shows what is men's *summum bonum* and promises assistance from above to those who truly seek it. The pious man will enjoy peace of conscience in this life and will be rewarded with blessings in the next. Moreover, piety is the best prop of government since it cements the parts of the body politic. By meditating on the goodness of God men will be prompted to pay homage to Him and to thank Him for His bountifulness to His creatures;[40] they will also be urged to love their neighbors in as high a degree as is possible since they are children of the same Father. Love of others is a natural tendency of human nature, and men are dependent on one another, so that it is only reasonable that they should be charitable in their dealings with one another.

Yet Barrow knew that although men bear the image of God, they are apt to think themselves their own masters; he therefore counted on the effects of rewards and punishments in the hereafter to engage good men, and to restrain bad ones, in their course of life. As he said in one sermon on future judgments,[41] "Hope and fear are the main springs, which set on work all the wheels of humane action" (2:453; 6:431). Belief in a future judgment is also a bulwark against irreligion, because the sorry state of the world inclines men to doubt or even deny the existence of God. Yet the impunity of bad men is no more than an index of the long-suffering patience of God, who gives them time to reform, while the afflictions that good men must bear only serve to try their faith. Contrary to what some Puritans had claimed, particularly during the troubled times, success in one's enterprise is no evidence of God's special favor for a particular cause or person: God's designs are unsearchable; He is truly a *deus absconditus* "suffering his goodness and justice to be under a cloud, that at length they may break out more gloriously" (2:463; 6:453).

A further incitement to religion and the good life is the belief in universal redemption through Christ's atonement, which makes all men *salvabiles* or *salvandos*. No man is preordained to eternal damnation; none is preordained to eternal salvation. Contrary to the doctrine of election and reprobation taught by some Puritans, Barrow asserts and demonstrates that Christ died for all men,[42] an article of faith that gives them hope and thereby encourages them to persevere in the right way. Taken all in all, then, Barrow's message to Christians is one of hope: He encourages them to walk as Christ did rather than frighten them by denouncing the torments reserved for the impious.

To confirm Christians in their faith and to convince others "in this Age of wavering and warping toward infidelity" (2:220; 5:457)[43] Barrow enlarges on the peculiar excellences of the Christian religion. Christianity describes God as "most amiable in his goodness, most terrible in his justice, most glorious and venerable in all his ways of providence" (2:221; 5:459). The Christians' God unites all the perfections that the philosophers have attributed to God, with nothing repugnant to reason, and He reveals things about His inscrutable ways that reason unaided cannot discover: Reason may be unable to fathom such mysteries, but neither can it disprove them since they are consistent with God's goodness. The Christian religion enlightens men about their origin, their present state, and their end, none of which could be learned from experience or from the history of mankind. Further, it teaches man the dignity of his nature, his lapse from happiness through disobedience, his consequent frailty, and his need of divine succor in order to be reinstated in that felicity, but it engages him to have confidence in God's mercy, provided that he complies with His will. The Christian rule of life is therefore "most congruous to reason, and sutable to [man's] nature"; it is conducive to each man's welfare, and it is "apt . . . to promote the public benefit of all." Finally, it shows man how to improve himself "into a resemblance of the divine nature" (2:223; 5:464–5).

Doctrine

Incitement to religion is all very well, provided that men believe in the Christian doctrine. To establish, or strengthen, this belief in his audience, Barrow had to prove the existence of God as revealed in Scripture and of the Christian religion as the full declaration of God's will. In four sermons on the article of the creed *I Believe in God*,[44] he first demonstrates the existence of God successively from the frame of the world, from the frame of human nature, from universal consent, and from supernatural effects. It is clear that in using the argument from design Barrow is out to counter the vogue of the new Epicureanism and the effect on some of the new interest in science. This argument was to be reinforced by Newton's formulation of laws governing the universe and was to be developed in the physico-theology of the late seventeenth and early eighteenth centuries. The argument from the twofold nature of man had long been used by theologians to demonstrate that the spiritual principle could not have sprung from matter and must therefore be of divine origin. By emphasizing that man was created in the image of God, Barrow takes his stand against the notion of the essential depravity of man: This image has been defaced, but men still have relics of God in them that enable them to come to the knowledge of Him; reason is God's light in man and, when aided by supernatural grace, can direct him in the right service of God. Universal consent had been used by Herbert of Cherbury and was to be invoked from the end of the seventeenth century in support of deism. Barrow, like many of his fellow Anglicans of the time, uses it to establish that natural religion is the foundation on which to build belief in the Christian religion. He also argues that, although God created an ordered whole, He has sometimes to interpose in order to manifest His "especial care over and love toward men, in suspending or thwarting his own established laws... for their sake" (2:129; 5:268). It has been objected that belief in such miraculous works is repugnant to reason, but men who regard as credible only such things as they can see or feel are inexcusable in view of the many interventions of God in the world after its creation. It may well seem that here Barrow's reasoning is circular since he invokes testimonies from Scripture before he has established that Scripture is the one true revelation of God.

Once belief in the existence of God has been proved, it is natural for reason to infer His main attributes: wisdom, goodness, and justice. Man's reason can next infer from the veracity, goodness, and justice of God that He cannot have left His creatures in the dark concerning their true end. He must therefore have revealed His truths to them in order to perfect the knowledge that men's natural but finite reason can discover unaided. Next, from comparing the various beliefs[45] that at one time or another have claimed to be grounded on a revelation from God, it appears that paganism was never a one uniform truth, but a bundle of inconsistent stories attended by foolish and cruel practices; that Mohammedanism, which proclaims itself a complete declaration of God's truth to all men, has not "stamped on it the genuine characters of a divine original and authority" (2:201; 5:419). The Jewish religion, on the contrary, is acknowledged by Christians to have originated from a revelation by God because it is true and good; yet it is defective in that it is not universal or suited to the nature of mankind. God did not speak to *all* men, nor did He disclose *all* His mind, and He declared that His full truth would be revealed only by His Messiah. Barrow is clearly treading on dangerous ground since the covenant of grace does not abrogate the covenant of law; he therefore explains that God spoke in the infancy of the world, when men were not ripe to receive a more perfect instruction. Only by degrees could – and did – He impart "further manifestations of light and grace" (2:212; 5:443), when prophets could raise men's minds to higher things and urge them to interior moral duties rather than external rites and ceremonies.

The next step is to prove that Jesus was the true Messiah who revealed to men "the mystical land of promise...flowing with milk and hony" (2:236; 5:491). He fulfilled all righteousness under the Law and completed it by leading the regenerate out of the bondage of sin "into the superior state of happy rest and joy" (2:236; 5:491) by vanquishing the devil, the world, and the flesh as Joshua had vanquished the Amorites. The fundamental article of the Christian faith is therefore that Jesus is the Christ and this comprises "all other doctrines of moment therein" (2:236; 5:492). Barrow demonstrates in four sermons that Jesus is the Christ[46] because He did in fact correspond to what had been foretold of the Messiah and because the religion He preached is perfect in all respects and enables man to reach his true end, felicity. Jesus achieved this by restoring men into God's favor through the covenant of grace, and the kingdom He erected is "spiritual in nature, universal in extent, and perpetual in duration" (2:257; 6:13), hence meets all the requirements of a true revelation by a good God. What He preached He practiced, thereby showing men that they, too, are able to perform the duties enjoined on them. His divine nature was attested when God interposed extraordinarily by producing "supernatural works," which by thwarting the course of nature declared clearly that He was sent by God.

Having grounded his argument in the reasonableness of faith, Barrow is at pains to show that miracles do not contradict reason, because some things are beyond the reach of reason. Besides, these miracles were performed in public and were open to the inspection of witnesses whose word there is no reason to doubt. God also imparted to the Apostles the power to work miracles. Both of these are further proof that Jesus was indeed the true Messiah. Yet God did not rely solely on miracles to show that He was the Christ: Jesus was endowed with preeminent virtues that enabled Him to fulfill His mission on earth; moreover, the sublimity of His doctrine proves that He was the true Messiah, and the success of the propagation of the Gospel as well as its manner are further evidence of God's approbation. In two sermons[47] Barrow deals with future judgments. He first demonstrates the soundness of the inference of Ecclesiastes, who concluded from the contemplation of the world that there must be a future judgment, and therefore a future life, since everything in this world is vanity and vexation of spirit. Among the many considerations Barrow offers in proof of the point the most conclusive is that man has been given free choice and power over his actions. It is therefore just that a distinction should be made between men according to their behavior.

In his four sermons on *The Doctrine of Universal Redemption*,[48] Barrow proves that Christ atoned for *all* men by grounding the assertion on texts from Scripture and by demonstrating that this is consonant with God's intention: As the sin of Adam provoked God's anger against all his descendants, so the righteousness and sacrifice of Christ were meant to suspend the fatal sentence for all. Some men fail to take advantage of this new covenant, yet their ruin is not to be imputed to God's will nor to any incapacity of man as he was first created, but to be blamed entirely on these men's neglect of the means of instruction put at their disposal by God. Barrow also meets the objection that only those to whom this revelation was conveyed could benefit from redemption by adhering to Christ. We need only consider the lives of holy men in the Old Testament, Barrow says, to see that mankind was never left destitute of help from above, though in a measure different from that which is vouchsafed to Christians. Thanks to that help, men in all times and places were enabled to seek God and serve Him in *some* measure, a measure "answerable to such light and strength: no more doth God require" (3:449; 4:297).

Barrow's demonstration of the reasonableness of the Christian religion is at the same time a demonstration *a contrario* of the unreasonableness of infidelity.[49] He attributes unbelief to men's neglect of instruction as a consequence of their pursuit of things of this world. Although he several times invokes the example of the Israelites' resistance to Christ's message, it appears that the sermon is a tract for the times. In this perspective it is only natural that he should argue that men resist the truths of religion because Christ requires men to fight against their lusts and that their vicious inclinations drive away grace, "which is requisite to the production and preservation of faith" (2:12; 5:30). The unbelievers whom he here reproves are men who do not recognize any spiritual principle in man, "so turning him into a beast, or into a puppet, a whirligig of fate or chance" (2:12; 5:30). They revile mankind as void of all goodness; they ascribe all actions and events to necessity, explode all difference between good and evil, and discard conscience as a bugbear. This covers a broad spectrum of opinions, but of such as were current at the time and in some way or other were connected with the new Epicureanism, which was anathema to all believers. Such men acknowledge no rule of life but brutish sensuality and thereby turn the world into chaos where no government is possible. But, Barrow adds, the wickedness of these men in no way argues that man as such is wholly corrupt.

Faith is grounded in reason, but men must be ready and willing to give their assent in order for the argument to have power to persuade them and for God's grace to work on them. Barrow therefore defines faith[50] as "a hearty and firm persuasion concerning the principal doctrines of our Religion, from divine revelation taught by our Lord and his Apostles" (2:16; 5:36). Men are neither irresistibly subdued by the force of arguments nor overpowered by grace, either of which would negate their free choice in giving assent. Faith is superior to human knowledge such as philosophers have taught because natural light or reason can take men only so far as it can reach. As Barrow said in the sermon on universal redemption, through faith in Christ men are loosed from the burden of sin and made *salvabiles,* provided that they are ready to comply "with those reasonable conditions which by Gods wisdom are required" (3:454; 4:303), for faith without answerable works is a dead faith. A man having received baptism is thereby made acceptable to God through Christ's merit, but he will be saved only when Christ declares him so, that is, when he appears before Him as sovereign judge: No man can be pronounced saved or damned in this life. The grace conferred by baptism will continue to help a man so long as he obeys the commands of Christ; if he sins, God will always be ready to receive him into His favor again, provided that he truly repents and amends his ways. Moreover, God has provided "competent aids" for men to subdue their evil inclinations and to withstand temptation by dispensing His Holy Spirit on them as members of the visible Church. This gift enables them to perform their duties *if they choose to avail themselves of it.* Barrow thus denies the doctrine of absolute decrees sanctifying some men only; he even asserts that God's grace is not confined to Christians since He is no less ready "to afford help to his poor creatures *wherever it is needful or opportune*" (3:462, my italics; 4:320).

Faith requires that man bend his reason and will to accept the mysteries revealed by Christ: He must "submit his understanding, and resign his judgement to God . . . being ever ready to believe whatever God declareth however to his seeming unintelligible and incredible" (2:30; 5:66).[51] In his exposition of the article of the creed *I believe* Barrow thus moves from plain truths to mysteries of faith, from man's natural response to God's goodness to the need for him to thwart his natural inclinations. Yet his demonstration is so cogent that there is no contradiction between the two aspects of Christianity he puts before his audiences.

From faith flow natural effects[52] on a man who embraces Christianity and obeys its commands: charity, meekness, sincerity of heart, humility, temperance, contentedness, peace of conscience, courage to resist temptation, and patience to bear trials. Besides such natural effects, faith also procures recompense from divine bounty, that is, the supernatural effects that are referred to in Scripture as justification. This having been a highly controversial point since the Reformation, Barrow defines its meaning according to the Church of England. To have faith means that reason is persuaded of all the truths, explicit or implicit, declared by God, and that the affections and the will are ready to comply. It thus involves the whole man. First and foremost, faith signifies what the object of faith, Christ, implies; but it also means that one is prepared to do what Christ commands. Faith especially implies belief in God's promise to restore men thanks to the oblation of Christ, belief that Christ is men's lord and will be their judge, as well as the corollaries from this belief. Barrow is yet careful to add that these propositions are general and therefore, once again, that no man has the right to consider that his own sins are forgiven; nor should a man constantly search his conscience to discover if he himself is justified: Presumption and doubt are equally dangerous. The right attitude is that of the centurion who felt himself unworthy but "declared his persuasion that *if Christ should only speak a word, his child should be healed*" (2:64; 5:133). Without adverting to the Puritans – he merely mentions Calvin in the margin – Barrow thus rejects their teaching of assurance and election. This clearly defines the position of the Church of England versus the sects, which admitted into their fold only those who were assured through some supernatural experience that they were regenerated. Contrary to the practice of gathered churches Barrow and his fellow Anglicans believe in a visible church in which all those who believe faithfully in the promises of God and live accordingly *may* be saved. He also rejects a newer interpretation of justifying faith – by such men as Ames – as a direct experience of the *person* of Christ: To him these are meaningless notions that only obfuscate the mind and distort Scripture. Barrow equally rejects Bellarmine's interpretation of justification as an infusion of righteousness and the interpretation of the Council of Trent that justification means infusion of a habit of grace or charity. "The Spirit of God as a principle of righteousness," he says, is bestowed by baptism on every sound believer, but each man must contribute to his own righteousness by obeying Christ's commands, although works as such cannot justify him. The term "justification," or "men justified," Barrow says, is applied in Scripture to various acts of God: Whether one applies it to any or to all these acts does not signify and certainly does not deserve to raise controversies. Justification is in any case "an act of judgment, performed by God...in acceptance of a competent satisfaction offered to him" (2:78; 5:162), which results from man's redemption by Christ.

God is the Father Almighty[53] whose providence governs all created things and beings. He is the maker of heaven and earth,[54] a point that was denied by the Manicheans of old and that was qualified by such philosophers as Aristotle, who believed that the work of God consisted in framing and ordering a passive primeval principle, matter. Some of the old philosophers even contended that matter was the only principle of all things, an opinion that has been "revived and embraced by some persons in our days" (2:176; 5:367), Barrow says. All these opinions are false and run counter to what Scripture asserts, namely, that God created the universe out of nothing by His mere will and command. In elaborating this point Barrow is clearly aiming at the new materialism; appeal to Scripture is therefore not sufficient. Barrow proves the point by arguing that conclusions drawn from the observation of effects in the natural world are not applicable to matters outside the compass of man's observation. Besides, no idea innate in man allows him to assert that creation out of nothing is impossible for God. Finally, to posit the preexistence of matter is to erect a second principle equal to God. In order to account for what is imperfect in the world, Barrow resorts to the principle of plenitude; as to real imperfections and evils, these, he says, are caused by the "wilfulness and impotency" of creatures. Pain and grief are either caused by men or consequent upon their misdemeanor, that is, inflicted by God for their faults. Another false opinion Barrow refutes is that God created the world "as a natural or necessary emanation or result; as heat from fire" (2:183; 5:384). The error is dangerous since it implies that God was "fatally determined," whereas He is a free agent who created the world out of His natural bounty because goodness is "freely diffusive and communicative of it self" (2:184; 5:387).

God's ways with men are, however, unsearchable.[55] Ancient philosophers were led to doubt God's providence when reflecting upon the course of human affairs. Barrow's answer to the objection is that, whereas things in nature obey necessity, men are self-moving agents so that in human affairs it is not always easy to distinguish between human and divine agency since God acts not for the good of particular persons but for the whole world, both present and future. Consequently, men should repress all wanton curiosity touching the ways of Providence; they should guard against both conceitedness and discontent; above all they should guard against being led to unbelief through their bafflement in front of God's unsearchable purposes.

Faith takes over where reason stops, and belief in Christ implies that men accept the mysteries of religion without trying to probe into them. Barrow does rely on reason to prove the existence of God, but he can only appeal to Scripture for belief in the Trinity, in the Incarnation of Christ, His Resurrection and Ascension. The Messiah is God's only begotten Son,[56] to whom the Father gave dignity and authority next to Himself. The Son existed from all eternity: He is one with God, having His divine essence by "communication" from the Father. Further, the divine essence is "communicated" from the Father and the Son to the Holy Ghost. The three persons or hypostases are coessential and coeternal yet distinct from each other, and they have different offices. The Son[57] is the lord of men and their savior, their master, owner, teacher, and leader; He has dominion over all things because He was appointed voluntarily the Son of Man and because of His capacity and power to govern the world, to make and preserve men. The Holy Ghost[58] is distinct from both the Father and the Son, yet like them subsisting as a hypostasis, not an accident; He "proceeds" jointly from the Father and the Son "in a manner incomprehensible and ineffable" (2:502; 6:528), but so that one divine essence is common to all three. It is through the operation of the Holy Ghost that the Son was made man;[59] besides, the Holy Ghost ratified the covenant of grace by descending on the Apostles. God's blessed Spirit is also the food of men's thoughts, and God has imparted it "as a continual guide and assistant" (3:524; 4:436) to all those who believe in the promise of mercy through Christ's satisfaction. It is through the Spirit that the frame of men's minds is repaired, with His assistance that Christians can follow the dictates of Christ, however "abominable to [men's] carnal humour" these may be (3:530; 4:449). Man's reason is no capable judge of such propositions as the Trinity, and Barrow warns men not to try to demonstrate the mysteries of faith, for this is as presumptuous as to deny them.[60] The danger of trying to give a rational explanation of the Trinity was to appear most clearly in the last decade of the century, when defenders of the Church's position arguing against the Anti-Trinitarians proved to be as obnoxious to right belief as the Anti-Trinitarians (see, e.g., William Sherlock's *A Vindication of the Doctrine of the Trinity*, 1690, which was attacked by South as an assertion of "Tritheism").

The Incarnation of the Son[61] is equally incomprehensible to men. As son of man Jesus was like unto men in all respects *before man's will was abused and lapsed into sin;* as Son of God, He remained coeternal and coessential with the Father. How this union of the divine and the human natures could be effected is a mystery, yet men can easily grasp why God should assume human nature out of mercy to them. As man, the Son could satisfy God; as man also, He was sensible of men's needs, could declare God's truth to the capacity of men, and could set before them an exact pattern of righteousness. The incarnation of the Son thus raises men's minds to the sense of the dignity of man and brings them comfort and joy. The efficient cause of the conception of Christ was the Holy Ghost.[62] All one can assert about this entirely mysterious process is that Jesus was not conceived "seminally" but "operatively," not of the substance of the Holy Ghost, but "by virtue of it." Barrow's use of the received terms sheds no more light on this mystery than does the term "procession" for the generation of the Holy Ghost. All he asserts is that the generation of Jesus resembles man's spiritual regeneration, both being effected "by the operation of the same divine spirit" (2:348; 6:248). He is yet careful to add that although Mary deserves respect she should not be the object of divine worship (*latria* or *hyperduleia*), as she is in the Church of Rome: Such practices are blasphemous because they ascribe to her attributes proper to God alone. Christians should rejoice, he says in another sermon,[63] in the good news that the birth of Christ brought to all men: The Nativity altered the world by abolishing the distinction between Jews and Gentiles and brought to all men a promise of redemption.

Christ's death was witnessed by both believers and unbelievers.[64] His resurrection from the dead can be doubted only on two grounds: either because it seems impossible or because the testimonies of it are insufficient.[65] Barrow demonstrates that such a miracle is not beyond the power, wisdom, and goodness of God and that it involves nothing repugnant to reason if one believes in the immortality of the soul. Next, he proves that the witnesses were credible and could not have been cheating, for the deceit would soon have been exposed. The Resurrection was necessary to illustrate the veracity, power, and goodness of God by making good what the ancient Scriptures had presignified and foretold.[66] As the natural body of Christ was raised from the dead, so His mystical body, the Church, was rescued from bondage. The ascension of Christ,[67] another mystery, was confirmed by the testimony of the Apostles and also agreed with the presignifications and prophecies concerning the Messiah. Christ's ascension to sit at the right hand of God revealed Him in His office as king, in His priestly office, and in His office as prophet since He announced the coming of the Holy Ghost. Barrow warns his audience against errors such as that of the Lutheran consubstantialists and of the Roman transubstantialists, who believe that the body of Christ is present wherever the host is kept. Christ, Barrow asserts, is nowhere "corporeally" present on earth though everywhere present in His divinity.

Three of Barrow's sermons deal with the Passion. Two are expositions of the Creed;[68] the other is *Upon the Passion*.[69] Given the subject, it is not surprising that all three should be more "pathetic" than is usual with Barrow. They are meditations on the Crucifixion, its manner and circumstances, and Barrow invites Christians to consider the trial and death of Christ under several aspects, each of which manifests God's purpose and is particularly instructive to men. As meditations on the Passion, they sometimes remind us of Donne, though Barrow lays greater stress on the instruction to be derived from the Crucifixion. Christ's sufferings were appointed by God[70] and undergone by Christ's own choice; they were such as to make clear the necessity of a harsh punishment for men's sins, as well as the meekness and fortitude of Christ in submitting to His Father's will. The accusation by the Jews and the final appeal to the Roman governor show that Christ died for *all* men. By undergoing these sufferings without murmur Christ proved that the kingdom He claimed to erect is not of this world. "The Cross," Barrow says, "was a Throne whereon humility and patience did sit in high state and glorious Majesty, advanced above all worldly pride and insolence" (2:370; 6:284–5). By looking on the Cross men will be reminded of how their salvation was bought, whence it proceeds, and from whom they may expect help; they will learn, too, how the flesh is to be dealt with since the death of Christ "shadows" men's death to sin, just as the bitter pain of Christ "shadows" the bitterness of the repentance they ought to feel; finally, they must learn that if they are truly virtuous they must expect to be exposed to the envy and hatred of men. Barrow draws no less than fourteen applications to practice, to conclude that "the willing susception and chearful sustenance of the Cross" is the chief characteristic of the Christian; that what was scandal to the Jews and foolish conceit to the heathens, is a great comfort to those who are "meek of heart and enlightened by divine grace" (2:357; 6:263). One point Barrow makes by way of demonstration is that all judgment is of God and is administered in His name, that magistrates are His officers so that what is done in a formal way at a public trial is to be regarded as executed by God. Barrow thus implicitly rejects the plea of tender consciences, all the more clearly as he adverts to this when considering that Christ received His doom from the governor Pontius Pilate, "as it were by God's own mouth" (1:470; 1:112).

Practice

Men have been given a perfect pattern of life by Christ, which they must try to imitate even though they cannot achieve such a high degree of goodness.[71] True works of piety, however, can be performed only through the divine power imparted to men and through reliance on God's direction and assistance;[72] but if men really trust in God He will help them with His grace as He did the holy men in the Old Testament; through His aid they will be able to overcome their evil dispositions.[73] God does not, however, overpower them through His grace, for the whole worth of obedience consists in the free submission of man to God: He can be served only by "volunteers."[74]

Man's first duty is to love and honour God through prayer and thanksgiving;[75] but his second duty, love of his neighbor, is like unto the first and can manifest itself in innumerable ways,[76] notably through a peaceful temper and carriage toward others.[77] Barrow stresses that the duty of charity applies to all men whatever their nation, sect, or religion; he condemns forthrightly the wrangles that divided his contemporaries and advises men not to meddle with dangerous controversies touching the public weal for fear of sapping the basis of "sacred authority."[78] The most impressive sermon Barrow preached on charity is no doubt *Bounty to the Poor*.[79] Almsgiving, Barrow argues, is an attempt to restore the primeval equality of God's creatures, who were endowed equally until sin introduced degree by forging "these pestilent words *meum* and *tuum*" (1:443; 1:52). Barrow is not advocating redistribution of wealth since he adds that degree was ordained by God as a consequence of man's sin; what he does advocate is the doctrine of stewardship.

Some of Barrow's sermons are more specifically concerned with right interior dispositions and with conduct. One is *Rejoice Evermore*,[80] which recommends joy as a duty enjoined by God. It is scandalous, he says, to consider religion as prohibiting all delight, mirth, and good humor; on the contrary, the Christian religion is a source of true, pure, and steady joy since God in His goodness has made all delight to be man's duty. The Gospel is the good tidings declaring God to be a god of love, hope, peace, and consolation, who vouchsafes men the gift of the blessed comforter, the Holy Ghost. To rejoice evermore is thus the same as to exercise piety through prayer and thanksgiving. Barrow yet advises Christians to watch over their hearts with all diligence,[81] for knowledge of one's inmost self is necessary if one is to direct one's life to its proper end. In connection with this, Barrow recommends a true *contemptus mundi*[82] and urges men to meditate upon the transitoriness of all worldly goods. Barrow also advises men not to delay in repentance,[83] because reformation is a slow process and because those who persevere in evil will find their hearts hardened against grace or will discover that God's patience with them is exhausted and that He has withdrawn His grace from them. Besides rejoicing evermore Barrow counsels contentment[84] with whatever station in life on the ground that the latter is determined – or at least allowed – by God. Men should see that whatever happens to them is good and fit, that according to God's design everything tends to man's welfare, and that everyone's condition is best for him. This is not the only place where Barrow harps on the theme that whatever is, is right. The example of Christ's submission to the Father's will, which Barrow invokes in the sermon *Patience*,[85] is probably a better inducement to men to bear their crosses than any general considerations he offers elsewhere. This is no wonder if one remembers that most attempts at a theodicy are unsatisfactory and strike a note of complacency in their account of evil.

Although men should not be unduly concerned with things of this world, they should apply themselves with industry to whatever task their calling involves. In a series of sermons on *Industry*[86] Barrow argues that every man has some peculiar business to prosecute and that everyone is expected to use his talents for his own benefit and for the good of others. Moreover, industry is a "fence to innocence and vertue" (3:220; 3:379) and contributes to the welfare of the commonwealth by perfecting "those arts whereby humane life is civilized" (3:223; 3:388). In the sermon on industry *In Their Particular Calling: As Scholars* Barrow extols the scholar's search for knowledge as a worthy exercise of men's reason and sweetest entertainment of their minds. He particularly praises the study of divinity, yet his praise in no way detracts from the value of secular learning.

Among the evils to be shunned, excessive self-love has a prominent place[87] since all other vices derive from it. Love of self is, however, not culpable as such: It was implanted in man by God, and reason allows man to pursue what appears to him good provided that it does not hurt others. Barrow also denounces what he regarded as a widespread vice of his age, namely, men's failure to manifest their piety in public.[88] He blames these new Pharisees, who, out of bashfulness or desire to be "civil," fear the censure of scoffers and of dissolute men and are therefore content to be pious in private. Not that he recommends assuming a formal garb of singular virtue like the Pharisees of old and, we may add, like some of his more self-righteous contemporaries; but he insists that some duties must be performed in public, such as joining in public worship, practicing charity, defending truth, and denouncing vice and irreligion.

Barrow also preached a series of sermons against evil speaking. The way men speak, Barrow argues, is apt to influence other people's thoughts and behavior,[89] and blasphemous speaking, as well as aspersing piety, casts a slur on religion. Men who do not believe in God should forbear to vent their opinions, because these run counter to men's common sense. The second sermon, *Against Foolish Talking and Jesting,*[90] is probably the best known because in it Barrow distinguishes between innocent facetiousness and culpable raillery. This distinction is the more necessary, he says, "in this our age (this pleasant and jocular age) which is so infinitely addicted to this sort of speaking" and in which all reputation appears "to vail and stoop to that of being a wit" (1:194; 2:3). There follows the by now famous definition of wit as that "versatile and multiform thing . . ., so variously apprehended by several eyes and judgments, that it seemeth no less hard to settle a clear and certain notion thereof, than to make a pourtraict of *Proteus,* or to define the figure of the fleeting air" (1:195; 2:4). Still, facetiousness is allowable when it is used for harmless divertisement, for the Christian religion is not so harsh as to forbid all pleasure. It is also permitted and even useful when it exposes vile things to contempt, reproves vices, or corrects errors, and it is often the best way of defense against unjust reproaches. On the contrary, to jest about holy things, to use scurrilous language, or to be facetious about obscene and smutty matters as a vain ostentation, is to be condemned most firmly. In the sermon *Against Rash and Vain Swearing*[91] Barrow explains that oaths are part of religious worship and are also administered between private persons and between nations as a guarantee to bind the faith of men. The careless use of oaths is not only uncivil, but detracts from their value and is apt to encourage perjury, hence to sap the very foundation of society.

Of Evil-Speaking in General[92] best reveals Barrow's gentle temper in that he condemns all harsh speaking by private persons as well as all use of words that may inflame men's passions against each other. Yet Barrow regards evil speaking as allowable to men in authority in the prosecution of justice and to God's ministers in order to inveigh against sin and vice. What he does recommend, however, is moderation, meekness, fraternal reproof of faults, and fair language in defense of truth since "a modest and friendly style do suit truth" (1:232; 2:84). In *The Folly of Slander*[93] Barrow denounces "this vile and common sin," which, he says,

rages especially "in our age and country" because "conceits newly coined...
are levelled at the disparagement of Piety, Charity and Justice, substituting interest in the room of Conscience" (1:242; 2:103). He ascribes the
current practice of slander to the implacable dissensions among his contemporaries, to the bitter zeal that derives from these, as well as to the
affectation of seeming wise and witty by any means. All sectarian spirit
is evil, while meekness and charity should be observed toward dissenters
in opinion, even toward enemies. Detraction too[94] is a dangerous evil
because it discourages the practice of goodness. Barrow similarly condemns *Rash Censuring and Judging,*[95] a practice, he says, that is now
very rife. It is taken as an index of wit and wisdom, and it is further encouraged by the many dissensions in opinions and by men's addiction to
parties. Men should not meddle unduly with their neighbors' private actions, nor should they be inquisitive into the proceedings and purposes of
their superiors, or even of their equals and of those not subject to their
charge and care.

This series of sermons ends with two on *Quietness and Doing Our Own
Business,*[96] which once again testify to the price Barrow set on amicable
relations among men, particularly on order and quiet in civil societies.
He warns men especially against meddling with affairs of state and with
religious controversies: Private persons should not presume to advise their
superiors, for this is to trespass against the bounds of their calling and station in life. Barrow firmly condemns "this licentious age" (1:293; 2:212),
in which so many men arrogate to themselves the right to be counselors
of state. These men work mischief against public tranquillity by forget-
ting that magistrates are accountable only to God, that there is a kind of
sacredness in the mysteries of state, and that princes' judgments are more
than human since these men are representatives of God on earth. In sum,
what Barrow recommends is passive obedience and acceptance of the status quo. In *Obedience to Our Spiritual Guides and Governours,*[97] he argues that the Church is an *acies ordinata* whose captain general is Christ.
Such an order or subordination is required by reason and appointed clearly
in Scripture. Barrow sums up the main claims put forward by the defenders of episcopal discipline and warns his audience to guard against
seducers who are "Ministers of *Satan,* the Pests of Christendom, the Enemies and Murderers of Souls" (3:278; 4:19). However peaceable his temper, Barrow does not hesitate to brand these men with opprobrious terms
that clearly advert to the Nonconformists, especially to the extreme sects.
Barrow thus clearly takes his stand against the plea of tender conscience,
ignoring – or genuinely failing to understand – what it was grounded on.
According to him disobedience from the laws of the Church makes a man
an outlaw from the kingdom of Christ and can derive only from "the
worst fruits of the flesh and corrupt nature" (3:286; 4:36). That a moderate man like Barrow should attribute the plea of conscience to such low

motives speaks volumes for the spirit of the age. He further argues that for Christians to follow the advice of their guides does not clash with following their conscience since in that case their conscience is enlightened by the advice of men more skillful in interpreting the divine law. In the *Consecration Sermon*[98] Barrow argues that pastors must have authority to reprove people whose life does not conform to God's commands. This, he says, they cannot do if they appear in despicable garb. Contempt of God's ministers has too often led to contempt of religion itself. Considering their employment in the Church and their condition in the world, pastors need to be provided for in order to enjoy a free and safe condition of life. Barrow's sermon is thus another defense of episcopal discipline, in particular of tithes.

As vice-chancellor of the university in 1675–76 Barrow was appointed to preach the sermon *On the King's Happy Return*.[99] The Apostles, he explains, commanded Christians to give thanks for kings and for all in authority. Christians should pray even for bad kings as Scripture enjoins, because these too are representatives of God on earth. Should the king be a bad ruler, the only redress men may seek is that through prayer; they should neither criticize his actions nor murmur against him. Although he refers to the late times of anarchy, his tone throughout is moderate yet firm. He is more outspoken in the sermon on the commemoration of the Gunpowder treason.[100] If men would consider wisely, he argues, they would see that the plot was discovered through God's hand, and he seizes the opportunity to condemn "the brood of Epicureans" (1:144; 1:450) who deny the interference of God in the affairs of men. Barrow is yet careful not to encourage belief in the intervention of God on any and every occasion; but having recognized God's help in their deliverance from a great evil, men should rejoice and trust in God to go on assisting them. It is remarkable that on an occasion that lent itself to violent attacks against the Church of Rome Barrow should steer clear of casting any aspersions on the authors of the plot, but rather dwell on the happy deliverance through the hand of God.

Clearly, Barrow was not the man to meddle in political strife; he was aware of the feuds of his time, but usually alluded to them in fairly general terms in order to warn Christians against dissensions and to guard against the threat of anarchy. In this no less than in the tenor of all his sermons – *A Treatise of the Pope's Supremacy* is a very different affair – Barrow reveals himself to be a true apostle of meekness. He was a man of peace, and he preserved his equanimity in an age that, as he repeatedly pointed out, was given to animosity and censuring. This is probably the reason why he has often been called a Latitudinarian; yet though he preached no sermon on "the mischief of Separation," he clearly rejected the plea of tender consciences and would hardly have been one to seek accommodation with the Dissenters as did his friend and editor Tillotson.

Manner and style

It will have appeared from the preceding account of Barrow's sermons that his mode of preaching conforms to the model recommended by Wilkins in his *Ecclesiastes* (1646) and favored by Anglican divines after the Restoration, namely, using a threefold division: *explication* of the text, *confirmation,* and *application.* In order to make clear the meaning of his text Barrow usually *explicates* the words and phrases rather than the context and not seldom has recourse to the corresponding Greek words (sometimes even a Hebrew word). For *confirmation* he adduces a number of considerations under which to view his subject, thereby providing proof of what he is arguing. What characterizes his method is the fertility of his thought in amplifying his theme and the wide compass from which he draws his arguments. By the time he concludes he has accumulated so many considerations that the topic may seem to be exhausted. He seldom orders his considerations so as to reach a climax, but his "Lastly" or "Finally" usually introduces the consideration that such or such a way of acting will gain God's favor or that such or such a truth is consonant with God's purpose for men's welfare. In some cases he resorts to close reasoning, as in the proof of the existence of God but even then each proof – for instance, the harmonious order and variety of the universe – is developed under several aspects. The convincing force of this mode of arguing derives from the cumulative power of the "considerations," some of which are explicated under several heads. The conclusion itself is generally a long list of *applications,* a list Barrow sometimes cuts short because "time will not allow."

His proofs, as we saw, are from the reason of the thing and from Scripture: Appeal to reason and experience is constant even when the purpose is to argue the need for reason to bow before the word of God; Scripture is appealed to either in support of the reasons adduced or as sole confirmation of the theme, as in the case of prefigurations of Christ and prophecies declaring the coming of the Messiah. The more usual mode of arguing for Barrow is to show that "so reason dictateth, and Holy Scripture more plainly declareth." As we saw, he occasionally refutes some interpretations that he regards as dangerous errors, but he never engages in controversy or oversubtle arguments. He eschews both the witty mode of preaching of the Caroline divines and the excessive subdivisions of some Puritan preachers. In spite of his many references to the fathers of the Church or to pagan philosophers, he is never guilty of ostentation of learning, even when he quotes those in the original Latin or Greek. Besides, in some cases – probably before a university audience – the interpretation of the texts demands that he have recourse to the Greek version. It is characteristic of Barrow that, although he draws attention to types of Christ in the Old Testament, he does not resort to typological

or anagogical arguments. Above all he avoids drawing mystical signifi-
cances or inferences from his texts; for him the plain meaning of Scripture
is the true meaning.

The structure of his sermons no less than the tenor of his thought marks
him as belonging to the later-seventeenth-century Anglican orators. Yet
he is not a plain preacher, if only because of the number and amplitude
of his "considerations." As J. E. Kempe remarked, every sermon by Barrow "is extensive in the sense of being a comprehensive discussion of all
the component parts of his subject. He goes through them all, one by
one, step by step, and places each in its right position. The process, it
must be owned, is sometimes tedious, but it must also be allowed that the
result in the hands of a strong and laborious workman like Barrow is
vastly impressive."[101] The result is indeed impressive, the more so when
the various considerations are presented in parallel syntactical structures
that emphasize the weight of the argument. For instance, in *The Pleasantness of Religion* he adduces fifteen considerations to show that religion is "delectable and satisfactory," because it implies a revelation of
truth, assures us that we take the best course, begets a hope of success,
prevents discouragement, makes all grief bearable, always has a good
conscience attending it, and so on; each point is amplified with a wealth
of words developing the contents of each consideration.

No less amazing than the wide compass of his thought is his command
of words. His verbal imagination manifests itself in the copiousness of
his style, yet the amplifications are not mere tautologies. They serve to
illustrate the point he is arguing and make it clearer through concrete
comparisons. For instance, in *The Profitableness of Religion* he argues
that men need a sure guide of conscience and he explains, among other
considerations, how wild their course is apt to be for lack of such a guide:

> There is scarce in nature any thing so wild, so untractable, so
> unintelligible as a man who hath no bridle of Conscience to
> guide or check him. A profane man is like a Ship, without
> Anchor to stay him, or Rudder to steer him, or Compass to
> guide him; so that he is tost with any wind, and driven with any
> wave none knoweth whither; whither bodily temper doth sway
> him, or example leadeth him, or company inveigleth and haleth
> him, or humour transporteth him; whither any such variable
> and unaccountable causes determine him, or divers of them
> together distract him: whence he so rambleth and hovereth,
> that he can seldom himself tell what in any case he should doe,
> nor another guess it; so that you cannot at any time know
> where to find him, or how to deal with him: you cannot with
> reason ever rely upon him, *so unstable he is in all his ways.* He
> is in effect a mere Child, all humour and giddiness, sometimes

> worse than a Beast, which, following the instinct of its nature,
> is constant and regular, and thence tractable; or at least so
> untractable that no man will be deceived in meddling with him.
> Nothing therefore can be more unmanly than such a person,
> nothing can be more unpleasant than to have to doe with him.
> (1:27; 1:204)

The point he is making is certainly not difficult to grasp, but the amplification makes it more vivid as well as more precise, for no two words he
uses are really synonyms (e.g, "wild," "untractable," "unintelligible"). In
such cases – and this is his method throughout – the amplification is at
bottom an explication, like the analysis and paraphrases through which
he explains his texts. It will also appear from the above quotation that his
sentences, though compound, are not very complex, since he repeats the
same syntactical pattern through apposition. This is more striking still
in long sentences – some of these cover a whole folio page – in which a
number of subordinates of the same type precede the main clause. For
instance, the Passion sermon opens with a series of when-clauses that
make for a stately movement giving weight to the aspects submitted for
consideration:

> When, in consequence of the original apostacy from God,
> which did banish us from Paradise, and ...
>   When poor man, having deserted his natural Lord and
> Protector ...
>   When, according to an eternal rule of justice, that sin de-
> serveth punishment ...
>   When, according to St Paul's expressions, *all the world was
> become guilty before God* ...
>   When, for us, being plunged into so wretched a condition ...
>   When, this, I say, was our forlorn and desperate case, then
> Almighty God, out of his infinite goodness ... did conclude to
> restore us. (1:464–5; 1:99–101)

Anaphora is in fact the most frequent rhetorical scheme Barrow uses,
and the resulting rhythm contributes to the impression of deep serious-
ness and to the weightiness of his arguments. Such a development of his
considerations, together with the slow march of his sentences, may indeed become monotonous, but it is relieved both by shorter sentences and
by the imagery. However, Barrow does not seem to have had regard for
variety in the rhythm of his discourse, for he has long passages of weighty
aggregate sentences or of rhetorical questions and at times equally long
passages in shorter, expository style. Modulation from one kind of style
to another is fairly rare, so that the reader might be tempted to skip were
it not for the vigorous mind that shines through. Besides, the care with

which Barrow defines and distinguishes is a pleasure that amply makes up for the length of his amplifications. The copiousness of words is astounding as such, but it is not an end in itself since it draws nice distinctions that serve to describe, inter alia, the diversity of men's behavior or actions. In the sermon *Not to Offend in Words,* for instance, a sentence containing a number of appositions catches fine nuances in the effects of ill-governed speech:

> Whereas most of the enormities, the mischiefs and the troubles whereby the Souls of men are defiled, their minds discomposed, and their lives disquieted, are the fruits of ill-governed Speech; it being that chiefly which perverteth justice, which soweth dissensions, which raiseth all bad passions and animosities, which embroileth the world in seditions and factions, by which men wrong and abuse, deceive and seduce, defame and disgrace one another, whereby consequently innumerable vexations and disturbances are created among men. (1:185, 535)

No reader would call this style elegant, especially since Barrow is fond of the cumbersome "-ing" forms, either with a subject to form a subordinate, as a participial form, or as a nominal form. The modern reader may find this all the more troublesome because Barrow often uses an archaic, Latinate, word order, as in "The Will of men may be so depraved... that good men by very bad men for doing well may be envied and hated" (1:28; 1:207). In addition Barrow uses a good many words that were already becoming obsolete in his time or that he simply coined from the Latin, for example, "defailance," "discoast," "fraternal correption," "a loose and slattering life," "how much a flamme this apology is," "like slumme and poison." His editor, Tillotson, crossed out a number of such words and substituted for them more current ones, such as "thrust out" for Barrow's "extrude," "fitness" for "idoneity," "being anointed" for "inunction," and so on.[102] He yet left quite a few that may make the modern reader pause but that also give a special flavor to Barrow's language, for he does not smell of the lamp as some earlier seventeenth-century writers did; in many cases the Latinate words were simply unavoidable, for instance, when he was expounding theological notions that had been defined by the fathers or the schoolmen. What is perhaps offensive – it did shock Coleridge – is the abrupt shift in some of Barrow's discourses from Latinisms to colloquial language or even slang. Clearly, unlike Dryden, Barrow did not seek his model for style in the conversation of gentlemen: He was a man of the old school. Yet he was abreast of the latest scientific developments in his time, although he may not have approved – and certainly did not strive to attain – the Royal Society's ideal for expository style. He had been a friend of Dr. Hammond, and he was a friend of Wilkins and of Tillotson, both of whom preferred a plainer style yet highly esteemed Barrow as a preacher. Coleridge was not far wrong when he said that Barrow closed "the first great period in the English language."[103] His style as well as his numerous quotations from Scripture and from the fathers in support of his arguments from reason and experience – his margins are cluttered with references – set him off from his near contemporaries Tillotson, Stillingfleet, and South. In his thought, however, he is close to other Restoration divines, and his care for words reflects his care for right thinking; for him as for them, false notions of religion result from abuse of words. No wonder Barrow rejects all mysterious interpretations of Scripture as meaningless. Besides, the riches of his verbal imagination enabled him to give new vigor to old truths. This appears best, perhaps, in *Bounty to the Poor,* in which the variety of observations and considerations he submits to his public somehow "images the liberal bounty which [the sermon] recommends."[104]

His imagery at times recalls the witty conceits of a former age, for instance, when preaching *The Duty of Thanksgiving* he says, "No holocaust is acceptable to God, as the Heart enflamed with the sense of his Goodness" (1:96; 1:350); or when on the article of the creed *Was Crucified* he draws instruction from the manner of Christ's punishment: "His posture on the Cross might represent unto us that large and comprehensive charity which he bare in his heart toward us, stretching his armes of kindness, pity and mercy, with them as it were to embrace the world, receiving all mankind under the wings of his gracious protection" (2:370; 6:283). On the whole, however, his comparisons are drawn from ordinary experience; in *The Profitableness of Godliness,* for instance, he asks, "Were a man designed onely, like a flie, to buzz about here for a time, sucking in the air, and licking the dew, then to vanish back into nothing, or to be transformed into worms; how sorry and despicable a thing were he?" (1:36, 224). Or, again, when urging Christ's example over precepts in *Of Being Imitatours of Christ* he says, "A systeme of precepts, though exquisitely compacted, is in comparison but a *Skeleton,* a dry, meagre, lifeless bulk, exhibiting nothing of person, place, time, manner, degree, wherein chiefly the flesh and bloud, the colours and graces, the life and soul of things do consist" (3:18; 2:500). In *The Consideration of Our Latter End* he argues that state and grandeur are mere shadows, that a moment of time will extinguish their luster and make all men equal in death: "as a drop and a pint bottle in compare with the Ocean are in a sort equal, that is, both altogether inconsiderable" (3:160; 3:254). Or, again, in *The Danger and Mischief of Delaying Repentance* he argues that the work of repentance requires time since "vertue is not a mushroom, that springeth up of it self in one night, when we are asleep and regard it not: but a delicate plant, that groweth slowly and tenderly, needing much pains to cultivate it, much care to guard it, much time to mature it, in our untoward soil, in this World's unkindly weather" (3:185; 3:307).

All of these – and there are many more – are homely images that speak to the capacity of men and that Barrow develops leisurely so that the object can be viewed in its several aspects. Yet these homely images do not strike a strange note in the context because whatever rare words Barrow may use these refer mostly to the kind of experience ordinary men have of the world or express ideas that common sense can grasp.

However learned Barrow was, however extensive his verbal imagination – he greatly admired Chrysostom, whom he read in Constantinople – his sermons evince a sense of the common life of men, of their experience in the world, of their qualities, and of their failings. Barrow never played down to his public; rather he exploited his own experience and his learning to bring home to his congregation truths they could understand and apply to their own lives. In this respect, too, Barrow exemplifies the new trend in pulpit oratory that prevailed among Anglican divines after the Restoration.

Notes

References to Barrow's sermons, in the text and in the notes, are to John Tillotson's edition of *The Works of the Learned Isaac Barrow, D.D.* (London, 1683–6), 3 vols. f°, followed by those to Alexander Napier's edition of Barrow's *Theological Works* (Cambridge, 1859), 9 vols. References are given by volume number and sermon number.

1. Isaac Barrow, *The Duty and Reward of Bounty to the Poor: In a Sermon Preached at the Spittal Upon Wednesday in Easter Week, 1671* (London: Brabazon Aylmer, 1671, 4to).
2. Isaac Barrow, *A Sermon Upon the Passion of Our Blessed Saviour: Preached at the Guildhall Chappell on Good Friday, the 13th day of April, 1677.* (London: Brabazon Aylmer, 1677, 4to).
3. Isaac Barrow, *Bounty to the Poor* (London: Brabazon Aylmer, 1671, 1680); idem, *Upon the Passion* (London: Brabazon Aylmer, 1678, 1682).
4. John Ward, *The Lives of the Professors of Gresham College* (London, 1740), p. 164.
5. For the manuscripts used by Tillotson see Irène Simon, *Three Restoration Divines: Barrow, South, and Tillotson: Selected Sermons,* vol. 1 (Paris, 1967), p. 308.
6. John Tillotson, ed. *The Works of the Learned Isaac Barrow, D.D.* (London, 1683–6); hereafter referred to as *Works.*
7. Ibid., vol. 1, sig. c.
8. It is reported that Brabazon Aylmer paid Thomas Barrow £470 for his son's sermons. See P. H. Osmond, *Isaac Barrow: His Life and Times* (London, 1944), p. 146.
9. Isaac Barrow, *The Defence of the B. Trinity* (London: Brabazon Aylmer, 1697), sig. A3.
10. The *Works* were later reprinted in 1700, 1716, 1722, 1741; further collections appeared in 1751, 1818, 1830–1, and 1841–2.
11. Alexander Napier, ed. *Theological Works of Isaac Barrow* (Cambridge, 1859), 1:xiv; hereafter referred to as Napier.
12. See W. Fraser Mitchell, *English Pulpit Oratory from Andrewes to Tillotson* (London, 1932), p. 29. See also Osmond, *Isaac Barrow,* Preliminary Note.
13. For Napier's strictures on Tillotson's tampering see Irène Simon, "Tillotson's Barrow," *English Studies* 45 (1964), 193–211, 274–88. See also idem, *Three Restoration Divines,* 1:307–10.

14. Simon, "Tillotson's Barrow," p. 288.
15. *Works,* 1:10–12; Napier, 1:12–14.
16. In the text of the *Works* the date given in the margin is 1673; this must be a printer's error since Barrow was elected vice-chancellor in 1675.
17. Presumably before the university.
18. *Works,* Vol. 1, sig. 3vo; Napier, 1:lxxxii.
19. Except for the two sermons Barrow himself sent to the press, the titles are Tillotson's.
20. Trinity College Cambridge (T.C.C.) Ms. R.10.26.
21. See notes 1 and 2.
22. T.C.C. Ms. R.10.21.
23. Walter Pope, *The Life of Seth Ward, Lord of Salisbury* (London, 1697), p. 147.
24. Ibid., p. 148.
25. The scant information we have may be summed up in the following table:

Works, 1:1; Napier, 1:3, *The Pleasantness of Religion,* St. Mary's, Cambridge, 30 June 1661

Works, 1:4; Napier, 1:6, *The Reward of Honouring God,* at court, August 1670

Works, 1:8; Napier, 1:10, *Of the Duty of Thanksgiving,* St. Mary's, Cambridge, 17 January 1662

Works, 1:9; Napier, 1:11, St. Mary's, Cambridge, 19 July, 1663; Gray's Inn, 10 January 1664

Works, 1:10; Napier, 1:12, *On the King's Happy Return,* before the university, 29 May 1676

Works, 1:11; Napier, 1:13, *On the Gunpowder-Treason,* ibid., 5 November 1675

Works, 1:12; Napier, 1:14, *A Consecration Sermon,* Westminster Abbey, 4 July 1663

Works, 1:25 or 26; Napier, 2:27 or 28, *The Love of Our Neighbour,* Whitehall, 25 April 1675

Works, 1:31; Napier, 1:1, *Bounty to the Poor,* St. Mary's Spital, Wednesday, Easter Week, 1671

Works, 1:32; Napier, 1:2, *Upon the Passion,* Guildhall Chapel, 13 April 1677.

Works, 1:17 and 18; Napier, 2:19, 20, *The Folly of Slander,* Westminster Abbey (n.d.)

Works, 1:14; Napier, 2:16, *Against Foolish Talking and Jesting,* probably at court (n.d.)

Works, 2:20; Napier, 5:20, *That Jesus Is the True Messias* (place unknown), 1671

The Defence of the B. Trinity (place unknown), Trinity Sunday, 1663 (Napier 4:492–523)

---

26. It has been suggested that some of Barrow's sermons were never preached (Caroline F. Richardson, *English Preachers and Preaching* [New York, 1928], p. 86). I have found no evidence of this.
27. Pope, *Life of Seth Ward,* p. 104.
28. Mitchell, *English Pulpit Oratory,* p. 104.
29. Pope, *Life of Seth Ward,* p. 104.
30. See the court's request in the first (1671) edition, opposite the title page.
31. *Works,* 1:17, 18; Napier, 2:18, 19.
32. It is well known that the printed sermon was often an expanded form of what had been delivered in the pulpit. A case in point is Tillotson's *On the Wisdom of Being Religious,* which was recast and expanded from the first printed version (1664) so that in the collected edition it numbers thirty folio pages (1696).
33. Mitchell, *English Pulpit Oratory,* p. 59, quoting Napier, *Theological Works,* 1:xv.
34. J. Ward, *Lives,* p. 162.
35. Henry Fielding, *Amelia,* book 12, chap. 5.

36. *Boswell's Life of Johnson,* ed. G. Birkbeck Hill, rev. L. F. Powell (Oxford, 1934–50), pp. 105–6.
37. *Anima Poetae,* ed. E. H. Coleridge (London, 1895), 25 November 1802.
38. Ward, *Lives,* p. 162.
39. *Works,* 1:1; Napier, 1:3; *Works,* 1:2 and 3; Napier, 1:4 and 5; *Works,* 1:5; Napier, 1:7.
40. *Works,* 1:23 and 24; Napier, 2:25 and 26; *Works,* 1:6 and 7; Napier, 1:8 and 9; *Works,* 1:8 and 9; Napier, 1:10 and 11; see also *Works,* 3:36 and 37; Napier, 3:49.
41. *Works,* 2:32; Napier, 6:31.
42. *Works,* 3:39–42; Napier, 4:58–60.
43. *Works,* 2:16; Napier, 5:16.
44. *Works,* 2:6–9; Napier, 5:6–9.
45. *Works,* 2:13–16; Napier, 5:13–16.
46. *Works,* 2:17–20; Napier, 5:17–20.
47. *Works,* 2:32 and 33; Napier, 5:31 and 32.
48. *Works,* 3:39–42; Napier, 4:58–60.
49. *Works,* 2:1; Napier, 5:1.
50. Ibid.
51. Ibid.
52. *Works,* 2:2 and 3; Napier, 5:2 and 3.
53. *Works,* 2:10 and 11; Napier, 5:10 and 11.
54. *Works,* 2:12; Napier, 5:12.
55. *Works,* 3:23; Napier, 3:48.
56. *Works,* 2:21; Napier, 6:21.
57. *Works,* 2:22; Napier, 6:22.
58. *Works,* 2:34; Napier, 6:33.
59. *Works,* 3:45; Napier, 4:63.
60. *A Defence of the B. Trinity,* Napier, 4:492–523.
61. *Works,* 2:23; Napier, 6:24.
62. *Works,* 2:24; Napier, 6:25.
63. *Works,* 3:43; Napier, 4:61.
64. *Works,* 2:27; Napier, 6:27.
65. *Works,* 2:29; Napier, 6:28.
66. *Works,* 2:30; Napier, 6:29.
67. *Works,* 2:31; Napier, 6:30.
68. *Works,* 2:25 and 26; Napier, 6:26.
69. *Works,* 1:32; Napier, 1:2.
70. See also *The Sufferings of Christ Foretold in the Old Testament, Works,* 3:44; Napier, 4:62.
71. *Works,* 3:3; Napier, 2:35.
72. *Works,* 3:1; Napier, 2:33.
73. *Works,* 3:2 and 4; Napier, 2:34 and 36.
74. *Works,* 3:4; Napier, 2:36.
75. *Works,* 1:6 and 7; Napier, 1:8 and 9; *Works,* 1:8 and 9, Napier, 1:10 and 11.
76. *Works,* 1:27 and 28; Napier, 2:29 and 30.
77. *Works,* 1:29 and 30; Napier, 2:31 and 32.
78. *Works,* 1:25 and 26; Napier, 2:27 and 28.
79. *Works,* 1:31; Napier, 1:1.
80. *Works,* 3:11; Napier, 3:39.
81. *Works,* 3:12 and 13; Napier, 3:40.
82. *Works,* 3:14 and 15; Napier, 3:41 and 42.
83. *Works,* 3:16 and 17; Napier, 3:43.
84. *Works,* 3:5–9; Napier, 3:37; see also *Works,* 3:38; Napier, 4:57.
85. *Works,* 3:10; Napier, 3:38.

86. *Works,* 3:18-22; Napier, 3:44-7.
87. *Works,* 3:28-31; Napier, 4:51 and 52.
88. *Works,* 3:32; Napier, 4:53 and 54; *Works,* 3:34 and 35; Napier, 4:55 and 56.
89. *Works,* 1:13; Napier, 1:15.
90. *Works,* 1:14; Napier, 2:16.
91. *Works,* 1:15; Napier, 2:17.
92. *Works,* 1:16; Napier, 2:18.
93. *Works,* 1:17 and 18; Napier, 2:19 and 20.
94. *Works,* 1:19; Napier, 2:21.
95. *Works,* 1:20; Napier, 2:22.
96. *Works,* 1:21 and 22; Napier, 2:23 and 24.
97. *Works,* 3:24-7; Napier, 4:50.
98. *Works,* 1:12; Napier, 1:14.
99. *Works,* 1:10; Napier, 1:12.
100. *Works,* 1:11; Napier, 1:13.
101. *The Classic Preachers of the English Church* (London, 1877), pp. 38-9.
102. For further examples see Simon, "Tillotson's Barrow," p. 7.
103. S. T. Coleridge, *Table Talk,* ed. H. N. Coleridge (London, 1884), July 1834, p. 294.
104. Simon, *Three Restoration Divines,* 1:226.

7

Isaac Barrow's library

MORDECHAI FEINGOLD

Introduction

The "estate" left by Barrow, wrote Abraham Hill, Barrow's first biographer, "was books."[1] Hill's statement, however, should not be interpreted solely as a monetary assessment. Most educated men (and not just scholars) in the early modern period regarded their libraries as among their most cherished possessions. For such people books were a chief pleasure, a solace in times of difficulty. And they pursued good books with an eagerness that has little counterpart today, even among the most unreformed book lovers.

Certainly, Barrow is an excellent specimen of a seventeenth-century bibliophile. He apparently garnered few, if any, books from either his father or his uncle the bishop, whose deaths succeeded his own, and consequently his collection represents a labor of love. To John Collins, Barrow articulated this passion for books, which extended even to those with debatable merit, and his willingness to countenance virtually any extravagance or expense in its name:

> I love to have by me divers books, which I do not much esteem,
> upon which score you need not scruple at your discretion to
> send me any book that I have not. I never matter the point
> of money in this case, and shall take any willingly and thank-
> fully from you: 'tis hard if there be not one thing at least to be
> learned out of any new book, and that satisfies me more than
> the expense of a few shillings can displease me.[2]

At the time of his death, Barrow possessed a private library consisting of 990 titles in some 1,100 volumes. This total can be supplemented by the 71 books (in 69 volumes) Barrow had donated during the previous few years to Trinity College Library. Although not a very large library compared with other major collections of the late seventeenth century, it was, despite Barrow's modest disclaimers, a "choice one"; indeed, according to Hill, the books Barrow bought were "so well chosen as to be sold for more than they cost."[3]

Be this as it may, the manuscript catalogue of Barrow's library that was prepared after his death presents us with a problem of assessment. As noted in Chapter 1, Barrow purportedly sold his books in 1655 in order to finance his Continental tour. Hence, no attempt has been made to maintain that any of the books Barrow owned in 1677 had been in his possession during his student days. Nevertheless, it remains a distinct possibility that upon returning to Cambridge after more than three years abroad, Barrow was able to reclaim at least a portion of the library he had sold. A not uncommon practice during the 1640s and 1650s was the exchange of books in lieu of money among friends, and perhaps someone, such as Duport or Worthington, offered to "purchase" Barrow's library in 1655 only to allow his friend to redeem it after the Restoration. For his part, Barrow never alluded to a "sale" of his library. On the contrary, he once said, "I have parted with few bookes, though I have been so unwise as to purchase very many."[4] And the fact that the library included not only all the traditional textbooks, but also the contemporary mathematical and philosophical literature he is known to have read while a student, adds likelihood to such a speculation.

The books themselves are almost evenly distributed among the three areas studied by Barrow. Some 340 are theological books, including Bibles, editions of the church fathers, and English controversial divinity; 330 books – nearly half of which are mathematical – are devoted to science, medicine, and philosophy; and slightly more than 300 works represent the humanities, classical literature, history and antiquities, biography, chronology, and modern literature (the last comprising a modest section of 10 books). The remaining books consist of a small collection of law (some 20 works), as well as a smattering of music books, university and college statutes, and a few volumes of academic verse. Predictably for the period, more than 80 percent of the books are in Greek and Latin. Of the rest, 155 are in English, 20 in French, 15 in Italian, and 15 in Hebrew and Arabic.

Barrow's surviving correspondence gives some interesting insight into his book-purchasing practices. For most of the scientific books he acquired during the 1660s, Barrow was indebted to John Collins, his good friend and future publisher who always "desired and delight[ed] to furnish...friends...and others of the Royal Society...with books they cannot meet with in stationers' shops."[5] Barrow was one of Collins's best clients. Delighted to receive whatever books Collins sent on to Cambridge, Barrow told him that in the matter of choosing books for his library, "I refer it wholly to your discretion, and shall be glad to have what you shall think good." Barrow was "troubled" only by Collins's recurrent practice of conferring books as gifts, a habit that elicited Barrow's frequent protests. On one occasion he complained, "Your excessive goodness doth continue to plunge me into such depth of obligation, which I shall never be able to get out of." On another he pleaded, "But I pray put it upon my account, for it is too much in conscience to put you upon so much trouble and expense too."[6] Collins's extremely good relations with London booksellers – as well as his network of friends abroad, which facilitated his procurement of new as well as rare books – enabled Barrow to enrich his library, either by purchase or gift, with works by Apollonius, Snell, Bettinus, Mengolius, Pascal, Huygens, Cavalieri, Alhazen, Fabri, and Sluse. In return, Collins often solicited Barrow's judgment concerning diverse theorems or his opinion about various mathematical books. Information thus received was freely circulated among Collins's correspondents. Certainly, Collins's responsibility for this flow of opinions and information made him the most important intelligencer of mathematical knowledge in England in the two decades following the Restoration.

Other books Barrow acquired either by himself, in bookshops in Cambridge or London, or through friends and relatives. In 1664, for example, Barrow invested £6 to buy various classical texts from John Mapletoft, who had forsaken academic life and was residing with the earl of Northumberland. On another occasion, Barrow instructed Collins to reimburse himself for books he had purchased for him by applying to Barrow père. This may suggest that even Barrow's father, who had been quite resourceful in purchasing books for Henry Oxinden before the Civil War,[7] may have assisted in his son's collecting activities.[8]

Not a few books in the library were presentation copies conferred upon Barrow by friends. At least some of Robert Boyle's works may have been gifts of their author. On 14 September 1673, for example, Henry Oldenburg sent presentation copies of *Several Tracts of the strange Subtility...of Effluviums* – on Boyle's behalf – to Barrow, More, and Newton. Similarly, two of the four works by Edward Stillingfleet in Barrow's library – subsequently to be acquired by Newton – were presentation copies.[9] It is also safe to assume that Barrow's past and present friends at Trinity conferred upon him copies of their works. John Ray is an obvious case. In addition to four of his own books, Ray most likely presented Barrow with a copy of his edition of Francis Willughby's *Ornithologia*. Another member of the "scientific club" at Trinity College, John Templer, may also have presented a copy of his *Idea theologiae Leviathanis* to Barrow in 1673. Thomas Gale and John North, more recent friends at Trinity, undoubtedly conferred their respective publications upon Barrow. Finally, of the few sermons in his library, the three by Barrow's former Trinity colleague, Samuel Rolle, probably represent gifts. In a like manner, a sermon published by Daniel Whitby, a chaplain to Seth Ward, in

1671 – during which time Barrow was residing with the bishop – can safely be regarded as the author's gift. Ward's own volume of sermons, published the following year, must also have been a presentation copy.[10]

Clearly, then, the gift of presentation copies to friends was an important aspect of the exchange of books. And this, together with the willingness to open one's private library to friends, was an important part of early modern book culture. In Barrow's case, such gifts to colleagues became almost an indulgence; his main concern about the delays in the publication of his geometrical and optical lectures, it may be remembered, was his inability to present copies of the book to his friends. According to Collins's count, some eighty copies were designated by Barrow for just this purpose. Isaac Newton, for example, received his copy on 7 July 1670, and other recipients – whose copies have since been identified – included Seth Ward, Edward Stillingfleet, and John Locke.[11]

Barrow appears to have been equally gratified to bestow books by other scholars on friends and deserving – or needy – students. Indeed, such a practice on the part of tutors and heads of house was common at the time, and Barrow himself had almost certainly benefited in this way. Not surprisingly, like most beneficiaries, Barrow perpetuated the system. Thus, he told John Mapletoft that one of the volumes he had purchased from him, "a Greeke testament, Gr. and Lat. in octavo (with some notes of beza)," he had rebound and gave "away to a friend, who beginninge to study Greeke, desired such a one."[12] At times Barrow acquired duplicate copies of certain works specifically in order to give them away. In 1668, for example, he asked Collins to set aside for him some copies of Sluse's *Mesolabium,* "which [he] would dispose of to friends."[13] It is interesting that at least forty books in Barrow's library, mostly mathematical books, were duplicates. Some he gave to Trinity College Library, whereas others, obviously not recorded in the catalogue, were likely to have been given to students.

Barrow was equally liberal in lending books to friends. To Collins he once sent the catalogue of his mathematical books and assured him that he would "be glad if any of them may ever accommodate" him.[14] Undoubtedly, Newton had had free access to Barrow's library ever since his undergraduate days. All those mathematical books that Newton borrowed and read during 1664–5 – Schooten's edition of Descartes's *Geometria* as well as Schooten's own *Exercitationum mathematicarum;* Viète's *Opera mathematica;* and John Wallis's *Operum mathematicorum, Tractatus duo,* and *Commercium epistolicum* – were most likely read in, or borrowed from, Barrow's library. Documentary evidence for this use, however, comes only in 1670. Collins then referred Newton to works by Huygens, Lalovera, and Dulaurens, "which bookes Dr Barrow can shew you," a suggestion that elicited from Newton the reply, "I have hitherto deferred writing to you, waiting for Dr Barrows returne from London that I might consult his Library about what you propounded in your last letter."[15]

Barrow died unexpectedly and intestate on 4 May 1677. In the absence of a will his property was made over to his old father, who, in financial straits and with no use for the books, put them up for sale. The act of dispersal appears to have been immediate. The catalogue of the library, the preparation of which was supervised by Isaac Newton, is dated 14 July, but apparently by that date some books had already been discarded. For example, Gilbert Atkinson, a scholar at Trinity (M.A. 1679), had acquired Barrow's copy of Sleidan's *Commentarii de statu religionis* as early as 4 June 1677, and four other books six days later.[16] However, the six volumes bought by Trinity College Library – consisting of Descartes's works and Regius's *Philosophia naturalis* – were entered into the library's records under the date 10 December.[17] It is possible, therefore, that the library, or at least part of it, was not sent up for sale in London, as is generally assumed. Rather it was dismantled at Cambridge, and only the catalogue was sent to "S.S.," perhaps at the request of a friend or a client who was interested in purchasing certain volumes. Be this as it may, other volumes from Barrow's library eventually made their way into various public and private libraries in ways not recorded. Thus, a copy of Marcus Aurelius's *De rebus suis* was to be found in the library of Jesus College, Oxford, before the end of the seventeenth century; at least one volume, Boulliau's *De lineis spiralibus,* was acquired by Thomas Tenison and became part of the collection the future archbishop donated to the public library he erected in London in 1684.[18]

Isaac Newton appears to have been a major beneficiary of the dispersal of the library, just as previously he had been a beneficiary of its contents. In his study of Newton's library, Harrison speculated that Newton may have obtained some books from Barrow's library "as part of the 'fee' for going through the library and despatching it for sale." However, Harrison made little effort to evaluate the extent of such gain, referring in his introduction only to two volumes in Newton's library bearing a Barrovian provenance: the 1653 Greek edition of the Bible and John Cosin's *Historia transubstantiationis paplis.*[19] I have already speculated that the former may actually have been a gift to Newton as early as 1663. But Cosin's book is not the only one to bear Barrow's provenance. Two of Edward Stillingfleet's books in Newton's library were undoubtedly presentation copies by the author to Barrow and were purchased by Newton in 1677 for 2s 6d apiece. Newton's copy of St. Gregory of Nazianzus also suggests a Barrovian lineage, for it had previously belonged to James Duport, Barrow's mentor and friend, who had most likely given it to Barrow, from whose library Newton purchased it for £2 10s.[20]

In fact, I would like to suggest that Newton's acquisition may possibly have been larger still, for a close comparison of the catalogues of the two libraries uncovers some remarkable correlations of holdings. Unfortunately, the whereabouts of most such volumes eluded Harrison's search, and consequently it is difficult to verify the hypothesis. Unquestionably,

at least some similarity of holdings can be attributed to common interests of contemporary scholars, such as Fromondus's *Meteorologicorum
libri sex,* Aelianus's *Variae historiae,* or some classical texts. But other
instances are more intriguing. For example, Barrow owned only two single issues of the *Philosophical Transactions,* numbers 38 and 43 for August 1668 and January 1668/9, respectively. These two issues, dating from
the period before Newton had become a fellow of the Royal Society, are
also the only two single issues of the *Transactions* in Newton's library.
(Newton owned a complete run of the *Transactions,* which he acquired
later.) Moreover, these two issues are bound together in Newton's library
with a few other works that may also have been acquired via Barrow:
Mercator's *Logarithmo-technia* and three of James Gregory's works.[21]
Finally, Newton may well have acquired from Barrow's library the copies
of some of the very books he had used years earlier. Schooten's two-
volume edition of Descartes's *Geometria,* which Newton owned at the
time of his death, for example, was not in his possession when he first
studied the work in 1664–5, and the annotations he made in the book
were entered in the late 1670s, when he composed his "Errores Cartesij
geometriae." Similarly, the notes taken by Newton on Schooten's *Exercitationum mathematicarum* seem to have preceded the date of acquisition
of the book, which is clean of annotations save for a single sheet containing a note taken by Newton in late 1664, later inserted loosely into the
book.[22]

Having had free access to Barrow's library, Newton was not compelled,
I would suggest, to acquire a large number of books. And although the
size of Newton's library in 1677 is not known, only 41 percent of the books
he owned were published before 1677, and at least some of these were
acquired, no doubt, in the years after Barrow's death. At any rate, it is
remarkable that the largest category of books known to have been purchased by Newton before 1677 were alchemical books, a subject conspicuously absent from Barrow's library catalogue.[23] It is not improbable,
therefore, that Barrow's death, depriving Newton of an immediate and
convenient source of books he needed on a regular basis, prompted him
to begin seriously building his own collection. And what more fortuitous
way to begin than to acquire some of Barrow's books that had been particularly useful or known to him? For such a reason, it seems to me, a
list of Newton's books that may possibly have belonged to Barrow – ex-
cluding classical texts and school textbooks – is not amiss, and this is
given here in an appendix to Barrow's library catalogue.

The present modernized rendition of Barrow's library catalogue[24] aims
to provide the reader with intelligible information about the books that
Barrow owned at the time of his death, together with the books that Barrow gave to Trinity College Library in the 1670s.[25] The catalogue appears

to have been compiled, on the whole, in great haste and, judging by the multiple samples of handwritings, by some three or four men mobilized to complete the project. As a result, only a small portion of the entries contain full bibliographical descriptions of the books: author, title or date, and place of publication. Many of the entries consist of only the author's name and an abbreviated, not always precise title. Other entries contain even fewer details, merely relating a book title or an author's name. Identification of some books – and many editions – is therefore tentative. In some cases, when no sufficient bibliographical information was given, and when more than one edition of a work exists, it was possible to determine the probable edition that Barrow owned according to the size of the book indicated in the catalogue. Also, especially in the case of classical and patristic texts, tentative identification was possible when the name of an editor or compiler was given. In all other cases, when it was not possible to pinpoint an exact, or even probable, edition owned by Barrow, I employed the method of indicating the first post-Reformation edition of the book as it appears in the British Library Catalogue of Printed Books.[26] The British Library Catalogue was also the chief source for the spelling of authors and titles, although all Latinized forms of place of publication have been rendered into the English.

Notes

1. *The Theological Works of Isaac Barrow,* ed. Alexander Napier, 9 vols. (Cambridge, 1859), 1:li.
2. *Correspondence of Scientific Men of the Seventeenth Century,* ed. Stephen J. Rigaud, 2 vols. (Oxford, 1841), 2:41. The remark was made a propos Barrow's somewhat slighting opinion concerning the obscurity of Mengoli's *Geometria speciosa.*
3. Napier, ed., *Theological Works,* 1:li.
4. Ibid., 1:lxix.
5. Rigaud, ed., *Correspondence of Scientific Men,* 1:141.
6. Ibid., 2:33, 32, 35.
7. See, e.g., *The Oxinden Letters,* ed. Dorothy Gardiner (London, 1933), p. 176.
8. Rigaud, ed., *Correspondence of Scientific Men,* 2:37; Napier, ed., *Theological Works,* 1:lxix.
9. *The Correspondence of Isaac Newton,* ed. H. W. Turnbull, J. F. Scott, A. R. Hall, and Laura Trilling, 7 vols. (Cambridge, 1959–77), 1:305; John Harrison, *The Library of Isaac Newton* (Cambridge, 1978), p. 244, nos. 1561–2.
10. Harrison found it hard to believe that Barrow's library included no "pamphlets, single sermons or other small items" and concluded that whatever their fate, such "ephemeral" literature was simply not included in the library catalogue. *The Library of Isaac Newton,* p. 61.
11. Ibid., p. 94 no. 122; Ward's copies of the optical and geometrical lectures are in the Salisbury Cathedral Library; Locke's copy of the geometrical lectures is at Harvard; and that of Tillotson, at Cambridge University Library. Barrow's presentation copy to Edmund Matthews, fellow of Sidney Sussex College, of his edition of Archimedes, Apollonius, and Theodosius was subsequently acquired by Newton. See ibid., p. 89, no. 76.

12. Napier, ed., *Theological Works,* 1:lxix. The "friend" may well have been Newton. His library includes a copy of the 1653 London edition of the Greek Old and New Testaments, with Barrow's name and a few notes in them. Harrison, *The Library of Isaac Newton,* p. 102, nos. 196, 200.

13. Rigaud, ed., *Correspondence of Scientific Men,* 2:68.

14. Ibid., 2:41.

15. *The Mathematical Papers of Isaac Newton,* ed. D. T. Whiteside, 8 vols. (Cambridge, 1967–81), 1:7–8, 19–24; Turnbull et al., eds., *The Correspondence of Isaac Newton,* 1:37–8, 42. For a similar, if more moderate, view concerning Newton's free access to Barrow's library, see Harrison, *The Library of Isaac Newton,* pp. 6–7, 60.

16. Edwin Wolf II, *The Library of James Logan of Philadelphia, 1674–1751* (Philadelphia, 1974), p. 454, no. 1855 (catalogue nos. 800–87); Atkinson's second purchase is bound together in a quarto volume at Cambridge University Library (sig. M.5.42), and the volume includes works by Zucchi, Broscius, Remmelinus, and Mulerius (catalogue nos. 1,052–193, 794, 667). See Harrison, *The Library of Isaac Newton,* p. 64 n 3. In the absence of any indication that Atkinson paid for the books, it is possible that he was one of the scholars employed to compile the catalogue and the books represent the reward for his trouble.

17. Trinity College Library, Ms. a.106, p. 45.

18. C. J. Fordyce and T. M. Knox, *The Library of Jesus College, Oxford,* Proceedings and Papers of the Oxford Bibliographical Society, 5 (1937), p. 18; Sotheby's 1861 sale catalogue of Tenison's library, no. 367 (catalogue nos. 45, 168, respectively).

19. Harrison, *The Library of Isaac Newton,* p. 11.

20. Ibid., pp. 244, 153 (nos. 1561–2, 704, respectively). Interestingly, unlike the case of Stillingfleet's and Gregory's books, there is no indication of the price paid, if any, for Cosin's book.

21. Harrison, *The Library of Isaac Newton,* pp. 216, 191, 154 (nos. 1305–6, 1073, 711, 712, 174).

22. Ibid., p. 61; Whiteside, ed., *Mathematical Papers,* 1:46–57, 4:336–45, 7:194 n 46.

23. Harrison, *The Library of Isaac Newton,* p. 78. Harrison has also speculated that Newton may have "abstracted some alchemical items for his own library," although he hastened to add "that there is nothing – inscriptions or notes – in the surviving books themselves which begin to substantiate this hypothesis." Ibid., p. 64. Yet Barrow had little interest in chemistry – and less in alchemy – beyond his student days. Dobbs's attempts to establish Barrow as the person who may have directed newton to the study of alchemy, and whose sympathy for Neoplatonism and Hermeticism was imparted to Newton, are based on an unwarranted reading of the documents. B. J. T. Dobbs, *The Foundations of Newton's Alchemy* (Cambridge, 1975), pp. 95–102.

24. Reconstructed from Bodleian Library manuscript Rawl. D878, fols. 33–59.

25. Extracted from Trinity College Library, Add. Ms. a.101 pp. 35, 64; Add. Ms. a.106 pp. 19, 33–4.

26. *British Museum Catalogue of Printed Books* (London, 1966–75). Additional information was derived from the following sources: H. M. Adams, *Catalogue of Books Printed on the Continent of Europe, 1501–1600, in Cambridge Libraries,* 2 vols. (Cambridge, 1967); A. W. Pollard and G. R. Redgrave, *A Short-Title Catalogue of Books Printed in England...1475–1640,* 2d ed., revised and enlarged by W. A. Jackson, F. S. Ferguson and K. F. Pantzer, 2 vols. (London, 1976–86); D. G. Wing, *Short-Title Catalogue of Books Printed in England...1641–1700,* 3 vols. (New York, 1945–51); *Catalogue Général des Livres Imprimés de la Bibliothéque Nationale* (Paris, 1902–81).

A Catalogue of the bookes of Dr Isaac Barrow sent to S.S. by Mr Isaac Newton, Fellow of Trin: Coll: Cambs. July 14. 1677

1. Acontius, Jacobus. Stratagemata Satanae. 12⁰, Basle, 1565.
2. – [Another copy/edition.]
3. Adamus, Melchior. Vitae Germanorum theologorum. 8⁰, Heidelberg, 1620.
4. Addison, Lancelot. West Barbary. 8⁰, Oxford, 1671.
5. Aelianus, **Tacticus**. De militaribus ordinibus instituendis more graecorum. 4⁰, Leyden, 1613.
6. – [Another copy/edition.]
7. Aelianus, Claudius. De animalium natura. (**Greek & Latin**.) 16⁰, Geneva, 1611.
8. – [Another copy/edition.]
9. – Variae historiae. (**Greek & Latin**.) 8⁰, Strassburg, 1662.
10. – [Another edition.] 12⁰, Leyden, 1616.
11. Aeschylus. Tragoedia. (**Greek**.) 8⁰, Paris, 1552.
12. – [Another edition.] 4⁰, Geneva, 1557.
13. Aesop. Fabulae. (**Greek**.) 8⁰, Basle, 1550.
14. – Fabulae. (**Greek & Latin**.) 8⁰, Venice, 1561.
15. Agobard, **Saint, Abp of Lyons**. Opera, Ed. Etienne Baluze. 8⁰, Paris, 1666.
16. Agricola, Georgius. De re metallica. Folio, Basle, 1546.
17. – [Another edition.] Folio, Basle, 1621.
18. Agricola, Rudolphus. De inventione dialectica. 8⁰, Cologne, 1557.
19. Aguiloni, Franciscus. Opticorum libri sex. Folio, Antwerp, 1613.
20. Ainsworth, Henry. Annotations upon...Genesis. Folio, London, 1627.
21. Alagona, Petrus. Sancti Thomae Aquinatis theologicae summae compendium. Folio, Rome, 1619.
22. Alciatus, Andreas. Emblematum liber. 8⁰, Augsburg, 1531.
23. Alexander, **Trallianus**. Alexandri Tralliani medici libri duodecim. (**Greek & Latin**.) 8⁰, Basle, 1556.
24. Alexandro, Alexander ab. Genialium dierum libri sex. 8⁰, Paris, 1570.
25. Alhazen (Ibn al-Haytham.) Opticae Thesaurus Alhazeni Arabis libri septem ...Item Vitellonis Thuringopoloni libri X. (**Greek & Latin**.) 8⁰, Basle, 1572.
26. Allestree, Richard. The Causes of the Decay of Christian Piety. 8⁰, London, 1667.
27. – The Whole Duty of Man. 8⁰, London, 1658.
28. Alsted, Johann Heinrich. Methodus admirandorum mathematicorum. 12⁰, Herborn, 1613.
29. – Thesaurus chronologiae. 8⁰, Herborn, 1650.
30. Ambrose, **Saint, Bp of Milan**. Opera. 3 Vols., Folio, Basle, 1567.
31. Ames, William. Opera. 5 Vols. 16⁰, Amsterdam, 1658.
32. Amphilochius, **Saint, Bp of Iconium**. Methodii...et Andreae Cretensis opera omnia. 2 Vols. (**Greek & Latin**.) Folio, Paris, 1644.
33. Anderson, Robert. Gaging Promoted. 8⁰, London, 1669.
34. Andrewes, Lancelot, **Bp of Winchester**. Responsio ad apologiam cardinalis Bellarmini. 4⁰, London, 1610.
35. – Tortura Torti. 4⁰, London, 1609.

36. Angeli, Stephano degli. Accessionis ad steriometriam, et mecanicam, pars prima. 4⁰, Venice, 1662.
37. – De infinitarum cochlearum mensuris, ac centris gravitatis. 4⁰, Venice, 1661.
38. – De infinitis parabolis, de infinitisque solidis. 4⁰, Venice, 1659.
39. – De infinitorum spiralium spatiorum mensura, opuscula geometricum. 4⁰, Venice, 1660.
40. – Miscellaneum geometricum in quatuor partes divisum. 4⁰, Venice, 1660.
41. – Miscellaneum hyperbolicum et parabolicum. 4⁰, Venice, 1659.
42. – Opusculum geometricum de linea sinuum et cycloide. 4⁰, Rome, 1659.
43. – Problemata geometrica sexaginta. 4⁰, Venice, 1658.
44. – De superficie ungulae. 4⁰, Venice, 1661.
45. Antoninus, Marcus Aurelius. De rebus suis. 4⁰, Cambridge, 1652.
46. Apollodorus, **Atheniensis**. Bibliotheces, sive de deorum origine. (**Greek & Latin**.) 8⁰, Rome, 1555.
47. Apollonius, **Pergaeus**. Conicorum libri IV. Folio, Bologna, 1566.
48. – [Another copy.]
49. – Conicorum libri IV, Folio, Antwerp, 1655.
50. – Conicorum libri V, VI, VII. Folio, Florence, 1661.
51. – [Another copy.]
52. Apollonius, **Rhodius**. Argonauticorum libri IV. 8⁰, Leyden, 1641.
53. Appian, **of Alexandria**. Romanae historiae. (**Greek & Latin**.) Folio, Geneva, 1592.
54. – [Another copy/edition.]
55. Apuleius, Lucius. Metamorphoseos libri XI. 8⁰, Gouda, 1650.
56. Aratus, **of Soli**. Phaenomena. (**Greek**.) 4⁰, Paris, 1559.
57. Archimedes. Opera. (**Greek**.) Folio, Basle, 1544.
58. – Opera quae extant. Ed. David Rivault. (**Greek & Latin**.) Folio, Paris, 1615.
59. – [Another copy.]
60. – Archimedis Syracusani Arenarius, et dimensio circuli. Ed. John Wallis. (**Greek & Latin**.) 8⁰, Oxford, 1676.
61. – De insidentibus Aquae. 4⁰, Venice, 1565.
62. – De iis quae vehuntur in aqua libri duo. 4⁰, Bologna, 1565.
63. – Paschasii Hamellii...Commentarius in Archimedis...librum de numero arenae. 8⁰, Paris, 1557.
64. Ariosto, Lodovico. Orlando Furioso. 4⁰, Venice, 1603.
65. Aristides, Aelius. Orationum tomi III. [2 Vols.] (**Greek & Latin**.) 8⁰, Geneva, 1604.
66. Aristophanes. Comoediae XI. (**Greek**.) 12⁰, Leyden, 1600.
67. – [Another edition.] Folio, Geneva, 1607.
68. Aristotle. Opera. (**Greek & Latin**.), 6 Vols., 8⁰, Venice, 1551–53.
69. – Opera. 2 Vols. [of 4?] (**Greek & Latin**.) 8⁰, Geneva, 1597.
70. – Mechanica. (**Greek & Latin**.) 4⁰ Paris, 1599.
71. – [Another copy.]
72. – Organon. (**Greek & Latin**.), 4⁰, Frankfurt, 1597.
73. – Problemata. (**Greek**.) 4⁰, Frankfurt, 1585.
74. Arminius, Jacob. Amica cum D. Francisco Junio de praedestinatione. 8⁰, Leyden, 1613.

75. – Examen modestum libelli quem D. Gulielmus Perkinsius...edidit...de praedestinationis modo & ordine. 8⁰, Leyden, 1612.

76. – Disputationes. 8⁰, Leyden, 1610.

77. – Orationes. 8⁰, Leyden, 1611.

78. Arnauld, Antoine. De la Fréquente Communion. 4⁰, Paris, 1643.

79. – Logique. 8⁰, London, 1664.

80. Arrianus, Flavius. Expeditio Alexandri. **(Greek & Latin.)** Folio, Geneva, 1575.

81. Asconius Pedianus, Quintus. Commentationes in aliquot M. T. Ciceronis orationes. 8⁰, Leyden, 1644.

82. Athanasius, **Saint, Patriarch of Alexandria**. Opera. **(Greek & Latin.)** Folio, Heidelberg, 1600.

83. Athenaeus, **Naucratita**. Deipnosophistarum libri XV. Ed. Isaac Casaubon, **(Greek & Latin.)** Folio, Leyden, 1612.

84. Augsburg Pharmacopoeia. Pharmacopoeia Augustana reformata. et eius mantissa. cum animadversionibus Johannis Zwelferi. 8⁰, Gouda, 1653.

85. Augustan History. Historiae Augustae scriptores sex. 8⁰, Leyden, 1661.

86. Augustine, **Saint**. Confessionum libri tredecim. 12⁰, Cologne, 1619.

87. – De Doctrina Christiana. 8⁰, Cologne, 1527.

88. – [Another copy/edition.]

89. – Opera. Folio, Paris, 1531.

90. – Saint Austins care for the dead. 12⁰, London, 1651.

91. Ausonius, Decimus Magnus. Opera. Ed. Joseph Scaliger, 8⁰, Heidelberg, 1588.

92. – [Another edition.] 8⁰, Geneva, 1595.

93. Bacon, Francis. Opera omnia. 5 Vols.(?) Folio, Frankfurt, 1665.

94. – Opuscula varia posthuma. 8⁰, London, 1656.

95. – Sylva sylvarum: or a Naturall Historie. Folio, London, 1628.

96. Baker, Augustine. Sancta Sophia. 8⁰, Douai, 1657.

97. Baluz, Etienne. Concilia Galliae Narbonensis. 8⁰, Paris, 1668.

98. Barclay, John. Paraenesis ad sectarios. 8⁰, Cologne, 1617.

99. Barlow, William. Magneticall Advertisements. 4⁰, London, 1616.

100. – [Another copy.]

101. Barrough, Philip. The Method of Phisicke. 4⁰, London, 1590.

102. Barrow, Isaac. Lectiones opticae & geometricae. 4⁰, London, 1669.

103. Bartholinus, Erasmus. Dioristice seu aequationum determinationes duabus methodis propositae. 4⁰, Copenhagen, 1663.

104. – [Another copy.]

105. Bartsch, Jacob. Planisphaerium stellatum. 4⁰, Nuremberg, 1661.

106. – [Another copy.]

107. Basil, **Saint, Abp of Caesaria in Cappodocia**. Opera omnia. 3 Vols. **(Greek & Latin.)** Folio, Paris, 1618.

108. Bates, William. Consideration of the existence of God. 8⁰, London, 1676.

109. – The harmony of the divine attributes. 4⁰, London, 1674.

110. Baudius, Dominicus. Epistolarum centuriae tres. 12⁰, Amsterdam, 1642.

111. Baxter, Richard. The Poor Man's Family Book. 8⁰, London, 1674.

112. Beaumont Joseph. Some observations upon the apologie of H. More. 4⁰, Cambridge, 1665.

113. Bellarmine, Robert. Roberti Bellarmini pro responsione sua ad librum Iacobi Magnae Britanniae regis. 8°, Rome, 1610.

114. Benedetti, Giovanni Battista. Diversarum speculationum mathematicarum & physicarum liber. Folio, Turin, 1585.

115. Bentivoglio, Guido. Della guerra di Fiandra. 8°, Cologne, 1635.

116. – Raccolta di lettre scritte del Cardinal Bentivoglio. 4°, Cologne, 1631.

117. Bernard, **Saint**. Opera omnia. Folio, Paris, 1640.

118. Bettinus, Marius. Aerarium philosophiae mathematicae. 3 Vols. Folio, Bologna, 1648.

119. – Apiaria universae philosophiae mathematicae. 3 Vols. Folio, Bologna, 1645–55.

120. Beveridge, William. Institutionum chronologicarum libri II. 4°, London, 1669.

121. Beverwyck, Johan van. Idea medicinae veterum. 8°, Leyden, 1637.

122. – [Another copy.]

123. Beyer, Johannes Hartmannus. Stereometriae inanium nova et facilis ratio geometricis demonstrationibus confirmata. 4°, Frankfurt, 1603.

124. – [Another copy.]

125. Bible, **Polyglot**. Biblia Sacra Polyglotta. 6 Vols. Folio, London, 1655–7.

126. – **Greek**. Divinae scripturae, nempe veteris ac novi testamenti, omnia. Folio, Frankfurt, 1597.

127. – **Latin**. Testamenti veteris Biblia sacra. Ed. Immanuel Tremellius. 4°, London, 1580.

128. – Biblia sacra. Ed. B. Arias Montanus. 8°, Antwerp, 1574.

129. – **English**. The Holy Bible. 8°, London, 1611. (Interleaved in 6 Vols.)

130. – The Holy Bible. 4°, London, 1613.

131. – [Another copy.]

132. – **Old Testament, Hebrew**. 8°, Amsterdam, 1639.

133. – **Greek**. Vetus Testamentum Graecorum ex versione Septuaginta interpretum. 3 Vols. 8°, Cambridge, 1655.

134. – **Psalms, Hebrew**. [Edition not identified.]

135. – **English**. The Psalmes of David. 8°, London, 1567.

136. – **New Testament, Polyglot**. Novum Testamentum. (Annotated by Theodore de Beze) Folio, London, 1642.

137. – [Another edition.] Ed. Immanuel Tremellius. (**Greek, Latin & Syriac**.) Folio, Geneva, 1569.

138. – **Greek**. [Edition not identified.]

139. – **Syriac**. Variae lectiones ex Novi Testamenti Syrici manuscripto codice collectae. 8°, Antwerp, 1572.

140. – **Concordance**. Concordantiae Testamenti Novi Graecolatinae. Ed. Henricus Stephanus. Folio, Geneva, 1600.

141. Biddle, John. Brevis disquisitio. 8°, London, 1653.

142. Billingsley, Robert. An Idea of Arithmetick. 8°, Cambridge, 1655.

143. Billy, Jacques de. Nova geometriae clavis algebra. 4°, Paris, 1643.

144. Bilson, Thomas. **Bp of Winchester**. The true difference betweene christian subjection and unchristian rebellion. 4°, Oxford, 1585.

145. Binius, Severinus. Concilia generalia, et provincialia, quotquot reperiri potuerunt. 9 Vols. Folio, Paris, 1636.

146. Blaeu, Willem. Institutio astronomica de usu globorum. 8⁰, Amsterdam, 1634.
147. Blagrave, John. The mathematical jewel. Folio, London, 1585.
148. Blancanus, Josephus. Aristotelis loca mathematica ex universis ipsius operibus collecta, & explicata. 4⁰, Bologna, 1615.
149. – [Another copy.]
150. – Sphaera mundi. 4⁰, Bologna. 1620.
151. – [Another copy.]
152. Blondel, David. Pseudo–Isidorus et Turrianus vapulantes. 4⁰, Geneva, 1628.
153. Blount, **Sir** Henry. A voyage into the Levant. 4⁰, London, 1636.
154. Boate, Gerard. Irelands Naturall History. 8⁰, London, 1652.
155. Boccaccio, Giovanni. Il Decamerone. 4⁰, Venice, 1612.
156. Bochart, Samuel. Epistola qua respondetur ad tres quaestiones. 12⁰, Amsterdam, 1650.
157. Bodin, Jean. Methodus historica. 8⁰, Basle, 1576.
158. – [Another copy/edition.]
159. – Les six livres de la République. 8⁰, Paris, 1580.
160. Boethius, Anicius Manlius Torquatus Severinus. De consolatione philosophiae. 8⁰, Venice, 1515.
161. Boileau–Despréaux, Nicholas. Oeuvres diverses. 4⁰, Paris, 1674.
162. Bonaventura, **Saint**. Divi Francisci vita. 8⁰, Florence, 1509.
163. Borelli, Giovanni Alfonso. Euclides restitutus. 4⁰, Pisa, 1658.
164. – Historia et meteorologia incendii Aetnaei anni 1669...accessit. 4⁰, Bologna, 1670.
165. Boreman, Robert. An antidote against swearing. 8⁰, London, 1662.
166. Boulenger, Pierre. Institutionum Christianorum. 8⁰, Paris, 1560.
167. Boulliau, Ismael. Astronomia Philolaica. Folio, Paris, 1645.
168. – De lineis spiralibus demonstrationes novae. 4⁰, Paris, 1657.
169. – Philolai, sive dissertationis de vero systemate mundi, libri IV. 4⁰, Amsterdam, 1639.
170. – [Another copy.]
171. Boxhorn, Marc Zuerius. Historia universalis. 4⁰, Frankfurt, 1675.
172. Boyle, Robert. Essays of the Strange Subtility...of Effluviums. 8⁰, London, 1673.
173. – Experiments and Considerations Touching Colours. 8⁰, London, 1664.
174. – Hydrostatical Paradoxes. 8⁰, Oxford, 1666.
175. – New Experiments and Observations Touching Cold. 8⁰, London, 1665.
176. – New Experiments Physico-Mechanical, Touching the Air. 2nd Ed. 4⁰, Oxford, 1662.
177. – Some Considerations about the Reconcileableness of Reason and Religion. 8⁰, London, 1675.
178. – Some Considerations Touching the Style of the H. Scriptures. 8⁰, London, 1661.
179. – Some Physico-Theological considerations about the possibility of the Resurrection. 8⁰, London, 1675.
180. – Some Considerations Touching the Usefulnesse of Experimental Naturall Philosophy. [Vol. 1 only?] 4⁰, Oxford, 1663-71.
181. – Some motives and Incentives to the Love of God. 8⁰, London, 1659.

182. – Tracts...about the Cosmicall Qualities of things. 8⁰, Oxford, 1670.
183. – Tracts...containing New Experiments touching the Relation betwixt Flame and Air. 8⁰, London, 1672.
184. Boys, John. Works. Folio, London, 1629.
185. Brahe, Tycho. Astronomiae instauratae progymnasmata. 2 Vols. 4⁰, Frankfurt, 1610.
186. Bramhall, John, **successively Bp of Derry and Abp of Armagh**. Castigations of Mr. Hobbes. 8⁰, London, 1657.
187. – A just vindication of the Church of England. 8⁰, London, 1654.
188. Brerewood, Edward. Enquiries touching the Diversity of Languages, and Religions. 4⁰, London, 1635.
189. Briet, Philippe. Annales mundi. 3 Vols. (of 4) 8⁰, Paris, 1662–63.
190. Briggs, Henry. Arithmetica Logarithmica. Folio, London, 1624.
191. Briggs, William. Ophthalmo–graphia, sive Oculi ejusque partium descriptio anatomica. 8⁰, London, 1676.
192. Brisson, Barnabe. Ritu nuptiarum & Jure connubiorum. 12⁰, Leyden, 1641.
193. Broscius, Johannes. Apologia pro Aristotle et Euclide contra Petrum Ramum et alios. 4⁰, Danzig, 1652.
194. – De numeris perfectis disceptatio. 4⁰, Amsterdam, 1638.
195. Browne, **Sir** Thomas. Hydriotaphia, Urne–buriall, or, a Discourse of the Sepulchrall urnes lately found in Norfolk. 8⁰, London, 1658.
196. – Pseudodoxia Epidemica. 4th Ed., 4⁰, London, 1658.
197. – Religio Medici. 8⁰, London, 1643.
198. Buchanan, George. Rerum Scoticarum Historia. 8⁰, Frankfurt, 1584.
199. Burgersdijck, Franco. Institutionum logicarum libri duo. 8⁰, Cambridge, 1637.
200. Burthogge, Richard. Divine goodness explicated and vindicated. 8⁰, London, 1672.
201. Butler, Charles. The Principles of Musik. 4⁰, London, 1636.
202. Butler, Samuel. Hudibras. 8⁰, London, 1663.
203. Buxtorf, Johann. Lexicon Hebraicum et Chaldaicum. 8⁰, Basle, 1615.
204. – Synagoga Judaica. 8⁰, Basle, 1641.
205. – Thesaurus grammaticus linguae sanctae hebraeae. 8⁰, Basle, 1651.
206. Caesar, Caius Julius. Commentariorum de bello Gallico libri VIII. 8⁰, Venice, 1564.
207. Caesarius, **Saint, Abp of Arles**. Homiliae XIV. Ed. Etienne Baluze. 8⁰, Paris, 1669.
208. Caesarius, **Brother of St. Gregory Nazianzus**. Quaestiones theologicae et philosophicae. 4⁰, Augsburg, 1626.
209. Callimachus. Hymni. 4⁰, Geneva. 1577.
210. Calvert, James. Naphtali. 4⁰, London, 1672.
211. Calvin, Jean. Epistolae et responsa. Folio, Geneva, 1575.
212. – Institutio Christianae religionis. 8⁰, Basle, 1536.
213. Cambridge University. Statutes. (Manuscript.)
214. – Trinity College. Statutes. (Manuscript.)
215. – Cambridge & Oxford verses. 3 Vols. 4⁰.
216. Camden, William. Britannia. 8⁰, London, 1586.
217. – Remaines. 4⁰, London, 1605.

218. Cameron, John. Myrothecium evangelicum. 4⁰, Geneva, 1632.
219. Campian, Edmund. Rationes decem. 8⁰, Antwerp, 1582.
220. Caninius, Angelus. Hellenismus. (**Greek.**) 8⁰, London, 1613.
221. Cano, Francisco Melchior. Opera. 8⁰, Cologne, 1605.
222. Cardano, Girolamo. De subtilitate libri XXI. 8⁰, Basle, 1560.
223. Carleton, George, **Bp of Chichester.** A thankfull remembrance of Gods mercy. 4⁰, London, 1624.
224. Carranza, Bartholome. Summa conciliorum. 8⁰, Paris, 1550.
225. Carthage, **Conference of.** Gesta collationis Cartagini habitae. 8⁰, Paris, 1588.
226. Casaubon, Isaac. Animadversionum in Athenaei Dipnosophistas. 2 Vols. Folio, Leyden, 1600.
227. – Epistolae. 4⁰, Magdeburg et Helmstadt, 1656.
228. Casaubon, Meric. De quatuor linguis commentationis pars prior. 8⁰, London, 1650.
229. Cassianus, Joannes, Eremita. Opera. 8⁰, Antwerp, 1578.
230. Cassiodorus Senator (Flavius Magnus Aurelius.) Opera omnia. 8⁰, Geneva, 1622.
231. Castell, Edmund. Lexicon Heptaglotton. 2 Vols. Folio, London, 1669.
232. – Oratio in scholis theologicis. 4⁰, London, 1667.
233. Castellanus, Petrus. De festis Graecorum syntagma. 8⁰, Antwerp, 1617.
234. Castelli, Benedetto. Della misura dell'acque correnti. 4⁰, Rome, 1628.
235. Castro, Alphonsus à. De potestate legis poenalis libri duo. 8⁰, Antwerp, 1568.
236. Cato, Marcus Porcius. Disticha de moribus ad filium. Ed. Joseph Scaliger. 8⁰, Edinburgh, 1620.
237. Catullus, Caius Valerius. Catulli, Tibulli, Propertii. Ed. J. J. Scaliger. 8⁰, Paris, 1577.
238. – [Another copy/edition.]
239. Cavalieri, Bonaventura. Directorium generale uranometricum. 4⁰, Bologna, 1632.
240. – Exercitationes geometricae sex. 4⁰, Bologna, 1647.
241. – [Another copy.]
242. – Geometria indivisibilibus. 4⁰, Bologna, 1635.
243. – Lo Specchio ustoria. 4⁰, Bologna, 1632.
244. – Trigonometria plana et sphaerica. 4⁰, Bologna, 1643.
245. Celsus, Aulus Cornelius. De re medica. 8⁰, Leyden, 1549.
246. Chamberlayne, Edward. Anglia notitia. 12⁰, London, 1669.
247. – [Another copy/edition.]
248. Chamier, Daniel. Panstratiae Catholicae, sive Controversiarum de religione adversus Pontificios corpus. 4 Vols. Folio, Geneva, 1626–30.
249. Charles V, **Emperor.** De capitalibus iudiciis constitutio. Folio, Basle, 1543.
250. Charles I, **King of Great Britain.** The Workes of King Charles the Martyr. 8⁰, London, 1648.
251. Charleton, Walter. Oeconomia animalis. 12⁰, Cambridge, 1669.
252. Charron, Pierre. De la sagesse trois livres. 8⁰, Paris, 1613.
253. Cheke, **Sir** John. De pronuntiatione Graecae. 8⁰, Basle, 1555.
254. Chemnitz, Martinus. Examinis Concilii Tridentini opus integrum. Folio, Frankfurt, 1573.

255. Chillingworth, William. The Religion of Protestants. Folio, London, 1638.
256. Cicero, Marcus Tullius. Opera. 2 Vols. 4°, Lyon, 1588.
257. Claudianus, Claudius. Opera. 8°, Hanau, 1612.
258. Clavius, Christophorus. Astrolabium. 4°, Rome, 1593.
259. - Epitome arithmeticae practicae. 8°, Rome, 1585.
260. - Gnomonices libri octo. Folio, Rome, 1581.
261. - In Sphaeram Iohannis de Sacro Bosco commentarius. 4°, Rome, 1581.
262. - Opera mathematica. (One of 5 Volumes?) Folio, Mainz, 1611–12.
263. Clement, of Alexandria. Opera. Folio, Florence, 1550.
264. Clement I, Pope, Saint. Epistolae ad Corinthios. (Greek & Latin.) 4°, Oxford, 1633.
265. Clifford, Martin. A Treatise of Humane Reason. 12°, London, 1674.
266. Cluverius, Philippus. Introductio in universam geographiam. 4°, Leyden, 1624.
267. Collins, John. An Introduction to merchants accounts. Folio, London, 1656.
268. - The doctrine of decimal arithmetic. 8°, London, 1665.
269. Cologne. Council of. Cannones concilii provincialis Coloniensis anno... MDXXXVI. 12°, Paris, 1548.
270. Comines, Philippe de. Mémoires. 8°, Paris, 1539.
271. Commandino, Federigo. Liber de centro gravitatis solidorum. 4°, Bologna, 1565.
272. - [Another copy.]
273. Constantinus, Robertus. Lexicon Graecolatinum. Folio, Lyon, 1637.
274. Conti, Natalis. Historiarum sui temporis libri X. 4°, Venice, 1572.
275. - Mythologiae. 8°, Frankfurt, 1581.
276. - [Another copy/edition.]
277. Copernicus, Nicolaus. Astronomia instaurata, libris sex comprehensa. 4°, Amsterdam, 1617.
278. Cosin, John, Bp of Durham. Historia transubstantiationis papalis. 8°, London, 1675.
279. Costerus, Franciscus. Enchiridion controversiarum. 8°, Cologne, 1589.
280. - [Another copy.]
281. Council of Trent. Catechismus Romanus. 8°, Dillingen, 1567.
282. Courcelles, Estienne de. Opera theologica. Folio, Amsterdam, 1675.
283. Courcier, Petrus. Opusculum de sectione superficiei sphaericae. 4°, Paris, 1663.
284. Cowell, John. Institutiones Iuris Anglicani. 12°, Oxford, 1664.
285. - The interpreter. 4°, Cambridge, 1607.
286. Croft, Herbert. The Naked Truth. 4°, London, 1675.
287. Croone, William. De ratione motus musculorum. 4°, London, 1664.
288. Crueger, Petrus. Tetragonismus circuli per lineas. Leipzig, 1607.
289. Cucchus, Marcus Antonius. Institutionum juris canonici libri quatuor. 8°, Cologne, 1564.
290. Cumberland, Richard. De legibus naturae disquisitio philosophica. 4°, London, 1672.
291. Cureau de la Chambre, Maris. The character of the Passions. 8°, London, 1650.

292. Curio, Coelius Secundus. De amplitudine beati regni Dei. 8°, Basle, 1554.
293. Curio, Hieronymus. Lexicon Graecolatinum. Folio, Paris, 1530.
294. Curtius Rufus, Quintus. Historiae Alexandri magni. 8°, Leyden, 1649.
295. Cyprian, **Bp of Carthage**. Opera. Ed. I. Pamelius. Folio, Paris, 1574.
296. Cyril, **Saint, Patriarch of Alexandria**. Opera. 7 Vols. **(Greek & Latin.)** Folio, Paris, 1638.
297. Cyril, **Saint, Patriarch of Jerusalem**. Opera. **(Greek & Latin.)** Folio, Paris, 1631.
298. Daillé, Jean. Apologia pro ecclesiis reformatis. 8°, Amsterdam, 1652.
299. – De duobus Latinorum ex unctione sacramentis confirmatione et extrema ut vocant unctione. 4°, Geneva, 1659.
300. – De la créance des pères, sur la fait des images. 8°, Geneva, 1641.
301. – De cultibus religiosis Latinorum. 4°, Geneva, 1671.
302. – De imaginibus libri IV. 12°, Geneva, 1641.
303. – De jejuniis et quadragesima liber. 8°, Daventer, 1654.
304. – De poenis et satisfactionibus humanis libri VII. 4°, Amsterdam, 1649.
305. – De pseudepigraphis apostolicis. 8°, Harderwijk, 1653.
306. – De sacramentali sive auriculari Latinorum confessione disputatio. 4°, Geneva, 1661.
307. Dary, Michael. The general doctrine of equation. 8°, London, 1664.
308. Davenant, John, **Bp of Salisbury**. Ad fraternam communionem inter evangelicas ecclesiaes restaurandam adhoratio. 12°, Cambridge, 1640.
309. – Animadversions upon a treatise intitled Gods love to mankind. 8°, Cambridge, 1641.
310. Day, William. A paraphrase and commentary upon the Epistles of Saint Paul to the Romans. Folio, London, 1661.
311. Dee, John. Parallaticae commentationis praxeosque nucleus quidam. 4°, London, 1573.
312. – [Another copy.]
313. Demetrius, **Phalereus**. De elocutione. 8°, Florence, 1542.
314. Demosthenes. Orationes. **(Greek.)** Folio, Paris, 1570.
315. – Olynthiacae orationes tres, et Philippicae quatuor. Ed. Nicholas Carr. 4°, London, 1571.
316. – [Edition not identified.]
317. Descartes, René. Geometria. Ed. Franciscus van Schooten. 2 Vols. 4°, Amsterdam, 1659–61.
318. – De homine. 4°, Leyden, 1662.
319. – Lettres de Mr Descartes. 2 Vols. [of 3] 4°, Paris, 1657–67.
320. – Meditationes de prima philosophia. 4°, Amsterdam, 1654.
321. – Principia philosophiae. 4°, Amsterdam, 1644.
322. Digges, Thomas. Alae seu scalae mathematicae. 4°, London, 1573.
323. – [Another copy.]
324. Diodorus, Siculus. Bibliothecae historicae. **(Greek.)** Folio, Geneva, 1559.
325. Dion Cassius. Historiae Romanae. **(Greek & Latin.)** 8°, Frankfurt, 1592.
326. Dionysius, **Saint (the Areopagite)**. Opera. **(Greek & Latin.)** 2 Vols. Folio, Antwerp, 1634.
327. Dionysius, **of Halicarnassus**. Antiquitatum Romanarum libri X. **(Greek.)** Folio, Paris, 1546.

328. – Scripta quae extant omnia. 2 Vols. 8⁰, Hanau, 1615.
329. Diophantus. Arithmeticorum libri sex. Folio, Paris, 1621.
330. Dioscorides, Pedacius. Opera quae extant omnia. Folio, Frankfurt, 1598.
331. Dodwell, Henry. Some considerations of present concernment; how far the romanists may be trusted by princes of another persuasion. 8⁰, London, 1675.
332. – Two Letters of advice. 8⁰, London, 1672.
333. Doughty, John. Analecta Sacra. 8⁰, London, 1658.
334. – Velitationes polemicae. 8⁰, London, 1651.
335. Downes, Andrew. Praelectiones in Philippicam de pace Demosthenis. **(Greek & Latin.)** 8⁰, London, 1621.
336. Drexelius, Hieremias. The considerations of Drexelius upon eternitie. 12⁰, Cambridge, 1639.
337. – Opuscula. 8 Vols. [Edition not identified.]
338. Drusius, Joannes. Interpretum veterum Graecorum quae extant in totum Vetus Testamentum. 4⁰, Franecker, 1619.
339. – De sectis Judaicis commentarii. 4⁰, Arnheim, 1619.
340. Du Chesne, André. Les antiquitez des villes et places de France. 8⁰, Paris, 1609.
341. Dulaurens, Franciscus. Specimina mathematica. 4⁰, Paris, 1667.
342. Duport, James. Homeri...gnomologia. 4⁰, Cambridge, 1660.
343. – Metaphrasis libri psalmorum graecis versibus contexta. 4⁰, Cambridge, 1666.
344. – Musae subsecivae. 8⁰, Cambridge, 1676.
345. Du Praissac, Sieur. The Art of Warre. 8⁰, Cambridge, 1639.
346. Durandus, Gulielmus. Rationale divinorum officiorum. 8⁰, Venice, 1568.
347. Eachard, John. Mr. Hobbs's state of nature considered. 8⁰, London, 1672.
348. Emmius, Ubbo. Vetus Graecia illustrata. 8⁰, Leyden, 1626.
349. – Graecorum respublicae. 16⁰, Leyden, 1632.
350. England, **Church of**. Certaine sermons or Homilies appoynted to be read in Churches. Folio, London, 1635.
351. Ennodius, **Saint, Bp of Pavia**. Opera. Ed. J. Sirmond. 8⁰, Paris, 1611.
352. Epictetus. Arriani de Epicteti dissertationibus libri III. **(Greek & Latin.)** 4⁰, Basle, 1554.
353. – Enchiridion. **(Greek & Latin.)** 8⁰, Basle, 1563.
354. – [Another edition.] 8⁰, Geneva, 1594.
355. – [Another edition.] 8⁰, Cambridge, 1655.
356. Epiphanius, **Saint**. Opera. **(Greek.)** Folio, Basle, 1544.
357. Erasmus, Desiderius. Adagia. Ed. Johann Jacob Grynaeus. Folio, Frankfurt, 1643.
358. – Apophthegmatum opus. 8⁰, Leyden, 1534.
359. – Colloquiorum familiarium opus. 12⁰, Leyden, 1550.
360. – De duplici copia verborum. 12⁰, Amsterdam, 1645.
361. – Paraphrases in libros Novi Testamenti. Folio, Basle, 1540.
362. – [Another edition.] 3 Vols. 8⁰, Paris, 1540.
363. Erpenius, Thomas. Grammatica Arabica. 4⁰, Leyden, 1636.
364. – Orationes tres de linguarum Ebraeae atque Arabicae dignitate. 12⁰, Leyden, 1621.

365. Eschinardi, Francisco. Centuria problematum opticorum. 4⁰, Rome, 1666, 1668.

366. Euclid. Elementorum libri XV. Ed. Christophorus Clavius. Folio, Maintz, 1612.

367. – Elementorum libri XV. Ed. Isaac Barrow. 12⁰, Cambridge, 1656.

368. – C. Dibvadii in geometriam Euclidis prioribus sex elementorum libris comprehensam demonstratio numeralis. 4⁰, Leyden, 1603.

369. – [Another copy.]

370. – Euclidis data. Ed. Claude Hardy. (**Greek & Latin**.) 4⁰, Paris, 1625.

371. – Euclidis optica et catoptrica. 4⁰, Paris, 1557.

372. – Procli Diadochi in primum Euclidis elementorum librum commentaria. Folio, Padua, 1560.

373. Eudaemon–Johannes, Andreas. Parallelus Torti ac Tortoris eius L. Cicestrensio. 8⁰, Cologne, 1611.

374. Euripides. Tragoedia XIX. 2 Vols. (**Greek & Latin**.) 8⁰, Heidelberg, 1597.

375. Eusebius, **Bp of Emesa**. Homeliae. 8⁰, Paris, 1547.

376. – [Another copy/edition.]

377. Eusebius, **Pamphili**. Chronicon. Ed. J. J. Scaliger. Folio, Leyden, 1606.

378. – Ecclesiasticae historiae. 2 Vols. (**Greek & Latin**.) Folio, Paris, 1544.

379. Eustache de Saint–Paul. Ethica. 8⁰, Cambridge, 1654.

380. – Summa philosophiae. 8⁰, Cambridge, 1648.

381. Evelyn, John. A character of England. 12⁰, London, 1659.

382. – Sylva, or a discourse of forest–trees. Folio, London, 1664.

383. Fabri, Honoré. Synopsis geometrica. 8⁰, Lyon, 1669.

384. – Synopsis optica. 4⁰, Lyon, 1667.

385. – [Another copy.]

386. Fabricius, Joannes Seobaldus. De unitate ecclesiae Britannicae meditationes sacrae. 8⁰, Oxford, 1676.

387. Farnaby, Thomas. Index poeticus. 12⁰, London, 1638.

388. – Index Rhetoricus. 12⁰, London, 1634.

389. Field, Richard. Of the Church. 4⁰, London, 1606.

390. Finé, Oronce. Liber de geometria practica. 4⁰, Strassburg, 1544.

391. Fitz–Herbert, Thomas. An sit utilis in scelere, vel de inplicitate principis Macchiavelliani. 8⁰, Rome, 1610.

392. Florus, Publius Annius. Historiae Romanae. 4⁰, Paris, 1532.

393. – [Another edition.] 12⁰, Cambridge, 1664.

394. Forbes, John. Instructiones historico–theologicae de Doctrina Christiana. Folio, Amsterdam, 1645

395. Fournier, George. Geographica orbis notitia. 12⁰, Paris, 1648.

396. Fragmenta vetustissimorum autorum. 4⁰, Basle, 1530.

397. Franciscus, **a Sancta Clara** [Christopher Davenport.] Apologia episcoporum. Folio, Douai, 1667.

398. Franckenberger, Reinhold. Chronologiae Scaligero–Petavianae breve compendium. 4⁰, Wittenberg, 1661.

399. Francklin, Richard. Tractatus de tonis in lingua Graecanica. 8⁰, London, 1630.

400. Fromondus, Libertus. Meteorologicorum libri sex. 4⁰, Antwerp, 1627.

401. Frontinus, Sextus Julius. Quae extant. 8⁰, Amsterdam, 1661.
402. Fulbert, **Bp of Chartres**. Opera varia. 8⁰, Paris, 1608.
403. Fuller, Nicholas. Miscellaneorum theologicorum...libri tres. 4⁰, Oxford, 1616.
404. Gale, Thomas. Opuscula mythologica. 8⁰, Cambridge, 1671.
405. Galilei, Galileo. Discorso...intorno alle cose che stanno in su l'acqua, o che in quella si nuovono. 4⁰, Florence, 1612.
406. – De proportionum instrumento a se invento...tractatus. 4⁰, Strassburg, 1617.
407. – Systema cosmicum. 4⁰, Strassburg, 1635.
408. Gallonio, Antonio. Vita Beati P. Philippi Nerii. 8⁰, Maintz, 1602.
409. Garcaeus, Johann. Harmonia de ratione institutionis scholasticae. 8⁰, Wittenberg, 1566.
410. Gassendi, Pierre. Institutio astronomica. 8⁰, London, 1653.
411. Gellius, Aulus. Noctes Atticae. 12⁰, Leyden, 1644.
412. Gemma, Reiner, Frisius. De principiis astronomiae & cosmographiae, deque usu globi. 8⁰, Antwerp. 1530.
413. – [Another copy.]
414. Ghetaldus, Marinus. De resolutione et compositione mathematica libri quinque. Folio, Rome, 1640.
415. – Promotus Archimedis. 4⁰, Rome, 1603.
416. Giraldus, Lilius Gregorius. Historiae poetarum tam Graecorum quam Latinorum dialogi decem. 8⁰, Basle, 1545.
417. – De re nautica libellus. 8⁰, Basle, 1540.
418. Gislenius (Angerius) Seigneur de Busbecq. Epistolae...ad Rudolphum II. 8⁰, Brussels, 1630.
419. Glanvill, Joseph. Essays on several important subjects in philosophy and religion. 4⁰, London, 1676.
420. Glassius, Salomon. Philologiae sacrae libri quinque, 4⁰, Frankfurt, 1653.
421. Glisson, Francis. Anatomia hepatis. 8⁰, London, 1654.
422. Glorioso, Giovanni Camillo. Ad theorema geometricum a nobilissimo viro propositum, J. C. G. responsum. 4⁰, Venice, 1613.
423. – Exercitationum mathematicarum decas prima. 4⁰, Naples, 1627.
424. – [Another copy.]
425. – Responsio J. C. G., ad vindicias B. Soveri. 4⁰, Naples, 1630.
426. Goclenius, Rudolph. Isagoge optica. 8⁰, Frankfurt, 1593.
427. Godino, Nicolao. Vita patris Gonzali Sylverae. 8⁰, Leyden, 1612.
428. Goelnitz, Abraham. Compendium geographicum. 12⁰, Amsterdam, 1643.
429. Goodman, John. A serious and compassionate inquiry. 8⁰, London, 1674.
430. Gothofredus, Dionysius. Auctores linguae latinae. 4⁰, Geneva, 1585.
431. Gouldman, Francis. A copious dictionary. 4⁰, London, 1664.
432. Granada, Luis de. Homiliae. 6 Vols. 8⁰, Antwerp, 1580.
433. Grandami, Jacques. Chronologia Christiana. 4⁰, Paris, 1668.
434. Great Britain. **Commissioners on revision of the Ecclesiastical Laws, 1550–**1552. Reformatio legum ecclesiasticarum. 8⁰, London, 1571.
435. Greek Anthlogy. Epigrammatum Graecorum libri VII. Ed. Jean Brodena. Folio, Basle, 1549.
436. – Florilegium diversorum epigrammatum. 8⁰, Paris, 1530.

437. Grégoire de Saint-Vincent. Opus geometricum quadraturae circuli et sectionum coni. Folio, Antwerp, 1647.
438. Gregory, **Bp of Nyssa, Saint.** Opera omnia. 2 Vols. Folio, Paris, 1615.
439. Gregory, **of Nazianzus, Saint.** Opera. (**Greek & Latin.**) 2 Vols. Folio, Paris, 1630.
440. – Orationes lectissimae XVI. 8⁰, Venice, 1516.
441. Gregory I, **Saint, the Great.** Opera. Folio, Paris, 1518.
442. Gregory, James. Exercitationes geometricae. 4⁰, London, 1668.
443. – Geometriae pars universalis. 4⁰, Padua, 1668.
444. – The great and new art of weighing vanity. 8⁰, Glasgow, 1672.
445. – Optica promota. 4⁰, London, 1663.
446. – [Another copy.]
447. – Vera circuli et hyperbolae quadratura. 4⁰, Padua, 1667.
448. Gregory, John. The works. 4⁰, London, 1665.
449. Grotius, Hugo. Annales et historiae de rebus Belgicis. 8⁰, Amsterdam, 1658.
450. – Annotata ad Vetus Testamentum. Folio, Paris, 1644.
451. Defensio fidei...adversus Faustum Socinum. 4⁰, Leyden, 1617.
452. – Excerpta ex tragoediis et comoediis Graecis. (**Greek & Latin.**) 4⁰, Paris, 1626.
453. – De imperio summarum potestatum circa sacra commentarius posthumus. 8⁰, Hague, 1661.
454. – De jure belli ac pacis libri tres. 8⁰, Amsterdam, 1650.
455. – [Another copy/edition.]
456. Ordinum Hollandiae...pietas vindicata. 4⁰, Leyden, 1613.
457. – Poemata. 8⁰, London, 1639.
458. – Votum pro pace Ecclesiastica. 8⁰, Amsterdam, 1642.
459. Guevara, Iohannis de. In Aristotelis mechanicas commentarii. 4⁰, Rome, 1627.
460. Guez, Jean Louis, Sieur de Balzac. Socrate Chrestien. 8⁰, Paris, 1652.
461. Guicciardini, Francesco. La historia d'Italia. 2 Vols. 4⁰, Geneva, 1636.
462. – Dell'istoria d'Italia...gli ultimi quattro libri. 4⁰, Venice, 1590.
463. Guicciardini, Ludovico. Omnium Belgii...descriptio. Folio, Amsterdam, 1613.
464. Guiffart, Pierre. Cor vindicatum, seu tractatus de cordis officio. 4⁰, Rouen, 1652.
465. Guillim, John. A display of heraldrie. Folio, London, 1638.
466. Gunning, Peter, **Bp of Ely.** The Paschal or Lent fast. 4⁰, London, 1662.
467. Gunter, Edmund. The description and use of the sector. 4⁰, London, 1636.
468. – The works. 4⁰, London, 1653.
469. Hacket, John. **Bp of Coventry & Lichfield.** A century of sermons. Folio, London, 1675.
470. Hales, John. Golden Remains. 4⁰, London, 1659.
471. Hall, Joseph. **Bp of Norwich.** Cases of conscience. 12⁰, London, 1654.
472. Hammond, Henry. Dissertationes quatuor. 4⁰, Oxford, 1651.
473. – A letter of resolution to six quaeres of present use in the Church of England. 12⁰, London, 1653.
474. – A Paraphrase and Annotations upon...the New Testament. Folio, London, 1653.

475. – A Practical Catechism. 8⁰, London, 1645.

476. – Of schisme. 12⁰, London, 1653.

477. – The works. (1st Vol.) Folio, London, 1674.

478. Harvey, William. Exercitationes anatomicae de circulatione sanguinis. 12⁰, London, 1649.

479. – Exercitationes de generatione animalium. 12⁰, Amsterdam, 1651.

480. Heereboord, Adrianus. Collegium ethicum. 4⁰, Leyden, 1648.

481. Heinsius, Daniel. Crepundia Siliana. 12⁰, Cambridge, 1646.

482. – Sacrarum exercitationum ad novum testamentum libri XX. 4⁰, Cambridge, 1640.

483. Heliodorus, **Bp of Tricca**. Aethiopicorum libri X. (Greek & Latin.) 8⁰, Heidelberg, 1596.

484. Helmont, Jan Baptista van. Opuscula medica inaudita. 2nd Ed. 4⁰, Amsterdam, 1648.

485. Henshaw, Nathaniel. Aero–Chalinos: or, a register for the air. 8⁰, Dublin, 1664.

486. Hérigone, Pierre. Cursus mathematicus. 6 Vols. 8⁰, Paris, 1634–44.

487. Hermogenes. Ars oratoria absolutissima et libri omnes. **(Greek & Latin.)** 8⁰, Cologne, 1614.

488. Hero of Alexandria. Belopoeca. 4⁰, Augsburg, 1616.

489. – Spiritalium liber. 4⁰, Urbino, 1575.

490. Hero of Constantinople. Liber de machinis bellicis. 4⁰, Venice, 1572.

491. Herodian. **The Historian**. Historiarum libri VIII. **(Greek & Latin.)** 8⁰, Basle, 1563.

492. – [Another edition.] 4⁰, Geneva, 1581.

493. Herodotus. Historia. **(Greek.)** Folio, Geneva, 1570.

494. – Historia libri IX. 8⁰, Basle, 1573.

495. Hesiod. Opera. **(Greek & Latin.)** 8⁰, Basle, 1542.

496. – Quae extant. **(Greek & Latin.)** 8⁰, Cambridge, 1672.

497. Heylin, Peter. Aerius Redivivus: or the history of the Presbyterians. Folio, Oxford, 1670.

498. Hierocles, **of Alexandria**. Commentarius in aurea Pythagoreorum carmina. Ed. John Pearson. **(Greek & Latin.)** 8⁰, London, 1654–55.

499. – [Another edition.] 8⁰, London, 1673.

500. Hilary, **Saint**. Opera. Folio, Paris, 1652.

501. Hippocrates. Aphorismi. **(Greek & Latin.)** 8⁰, Cambridge, 1633.

502. Hobbes, Thomas. Dialogus physicus. 4⁰, London, 1661.

503. – Elementorum philosophiae sectio tertia, de cive. 4⁰, Paris, 1642.

504. – Elementorum philosophiae sectio secunda, de homine. 4⁰, London, 1658.

505. – Examinatio & emendatio mathematicae. 4⁰, London, 1660.

506. Homer. Ilias & Odyssea. **(Greek.)**, Folio, Basle, 1541.

507. – Eustathii...in Homeri Iliadis et Odysseae. 3 Vols. Folio, Basle, 1559.

508. – Iliad. Trans. John Ogilby. Folio, London, 1660.

509. – Odyssey. Trans. John Ogilby. Folio, London, 1665.

510. – Index Homericum. 4⁰, Amsterdam, 1651.

511. – Index vocabulorum in Homeri. 4⁰, Heidelberg, 1604.

512. Hooke, Robert. Micrographia. Folio, London, 1665.

513. Hoornbeek, Joannes. Summa controversiarum religionis cum infidelibus. 8⁰, Utrecht, 1658.

514. Horatius Flaccus, Quintus. Opera. Ed. Daniel Heinsius. 8⁰, Leyden, 1612.

515. – Poemata omnia. 8⁰, Frankfurt, 1600.

516. – [Another edition.] 8⁰, London, 1614.

517. – [Another copy/edition.]

518. Hotman, François. De re numaria Populi Romani liber. 8⁰, Paris, 1585.

519. Hozyusz, Stanislaw. Confessio Catholicae Fidei Christiana. 8⁰, Antwerp, 1561.

520. – Confutatio prolegomenon Brentii. 8⁰, Paris, 1560.

521. Huygens, Christiaan. Brevis assertio Systematis Saturni sui. 4⁰, Hague, 1660.

522. – De circuli magnitudine inventa. 4⁰, Leyden, 1654.

523. – Systema Saturnium. 4⁰, Hague, 1659.

524. – Theoremata de quadratura hyperboles. 4⁰, Leyden, 1651.

525. Hyde, Edward, **1st Earl of Clarendon**. Animadversions upon a book, intituled Fanaticism fanatically imputed to the Catholic Church. 8⁰, London, 1673.

526. Hyde, Thomas. Catalogus impressorum librorum bibliothecae Bodleianae in Academia Oxoniensi. Folio, Oxford, 1674.

527. Innocent III, **Pope**. Epistolae. 8⁰, Paris, 1625.

528. Irenaeus, **Bp of Lyon, Saint**. Adversus Valentini, Folio, Geneva. 1570.

529. Isocrates. Orationes et epistolae. **(Greek & Latin.)** Folio, Geneva, 1593.

530. – Orationes duae. 8⁰, London, 1676.

531. Ivo, **Bp of Chartres**. Liber decretorum sive Panormia. 4⁰, Basle, 1499.

532. Jackson, Thomas. The Works. 5 Vols. Folio, London, 1653–54, 1673.

533. Jerome, **Saint**. Opera. 9 Vols. [in 3]. Folio, Antwerp, 1578–79.

534. – Epistolae aliquot selectae. 8⁰, Salamanca, 1572.

535. Jewel, John, **Bp of Salisbury**. Apologia Ecclesiae Anglicanae. 8⁰, London, 1562.

536. – The Works. Folio, London, 1611.

537. John, **Cassianus, Saint**. Opera omnia. 3 Vols. 8⁰, Douai, 1616.

538. John, **Chrysostom, Saint**. In...Evangelium secundum Matthaeum commentarii luculentissimi. 8⁰, Paris, 1548.

539. – Homiliae ad populum Antiochenum. **(Greek.)** 8⁰, London, 1590.

540. – [An unidentified work.]

541. Jones, Simon. A guide to the Young–gager. 12⁰, London, 1670.

542. Jonsius, Joannes. De scriptoribus historiae philosophicae libri IV. 4⁰, Frankfurt, 1659.

543. Josephus, Flavius. Opera. **(Greek.)** Folio, Basle, 1544.

544. Julianus, **the Apostate, Emperor of Rome**. Opera. 8⁰, Paris, 1583.

545. Junius, Adrianus. Nomenclator. 8⁰, Antwerp, 1567.

546. Junius, Franciscus. Ecclesiastici sive de natura et administrationibus ecclesiae Dei. 8⁰, Frankfurt, 1581.

547. – In epistolam S. Judae Apostoli...notae. 12⁰, Leyden, 1598.

548. – Sacrorum parallelorum libri tres. 8⁰, London, 1588.

549. Justel, Christophe. Codex canonum ecclesiae Africanae. 8⁰, Paris, 1614.

550. Justin, **Martyr, Saint**. Opera omnia. Folio, Paris, 1554.

551. – Opera. (**Greek & Latin.**) Folio, Heidelberg, 1593.
552. Justinian I, **Emperor of the East**. Corpus juris civilis. Folio, Geneva, 1614.
553. – Institutionum...libri IV. 12⁰, Paris, 1585.
554. – [Another edition.] Ed. A. Vinii, 8⁰, Amsterdam, 1663.
555. – [Another copy/edition.]
556. Juvenalis, Decimus Junius. J. Juvenalis & A. Persii Flacci Satyrae. 8⁰, Leyden, 1557.
557. – [Another edition.] 8⁰, London, 1612.
558. Kempis, Thomas à. De imitatione Christi. [Edition not identified.]
559. Kepler, Johann. Ad Vitellionem paralipomena. 4⁰, Frankfurt, 1604.
560. – Dioptrice. 4⁰, Augsburg, 1611.
561. – Epitome astronomiae Copernicanae. 8⁰, Frankfurt, 1635.
562. – Tabulae Rudolphinae. Folio, Ulm, 1627.
563. – [Another copy.]
564. Kircher, Athanasius. Prodromus Coptus sive Aegyptiacus. 4⁰, Rome, 1636.
565. Kircher, Konrad. Concordantiae Veteris Testamenti Graecae Ebraeis vocibus respondentes. 4⁰, Frankfurt, 1607.
566. Kirchmann, Johann. De funeribus Romanorum. 12⁰, Hamburg, 1605.
567. Knatchbull, **Sir** Norton. Animadversiones in libros Novi Testamenti paradoxae orthodoxae. 8⁰, London, 1659.
568. Knowles, John. An answer to Mr. Ferguson's book. 8⁰, London, 1668.
569. Lactantius, Lucius Coelius Firmianus. Opera. 8⁰, Leyden, 1652.
570. Laet, Johannes de. Novus orbis, seu Descriptionis Indiae Occidentalis libri XVII. Folio, Leyden, 1633.
571. La Faille, Jean Charles de. Theoremata de centro gravitatis partium circuli et ellipsis. 4⁰, Antwerp, 1632.
572. – [Another copy.]
573. Lake, Edward. Officium Eucharisticum. 8⁰, London, 1673.
574. Lalovera, Antonius. Quadratura circuli et hyperbolae segmentorum. 8⁰, Toulouse, 1651.
575. Lamy, Bernard. L'art de parler. 12⁰, Paris, 1676.
576. Lansbergen, Philippus van. Progymnasmatum astronomiae restitutae liber unus. 4⁰, Middelburg, 1610.
577. Larroque, Matthieu de. Histoire de l'Eucharistie. 8⁰, Amsterdam, 1671.
578. Laubegeois, Antoine. Graecae linguae breviarium. 8⁰, Douai, 1626.
579. Launoy, Jean de. Epistolarum pars 1a [– 8a.] 8⁰, Paris, 1667–73. (An odd volume?)
580. Leibniz, Gottfried Wilhelm von. Hypothesis physica nova. 12⁰, London, 1671.
581. Leo I, **Saint, Pope**. Opera. Folio, Cologne, 1561.
582. Leurechon, Jean. Mathematicall recreations. 8⁰, London, 1633.
583. Linacre, Thomas. De emendata structura Latini sermonis libri sex. 8⁰, Paris, 1532.
584. Linda, Lucas de. Descriptio orbis & omnium eius rerumpublicarum. 8⁰, Leyden, 1655.
585. Linton, Anthony. Newes of the complement of the art of navigation. 4⁰, London, 1609.
586. Lipsius, Justus. Admiranda, sive, de magnitudine Romana. 4⁰, Antwerp, 1599.

587. – De amphitheatro. 4⁰, Antwerp, 1584.
588. – Epistolae selectae. 12⁰, Paris, 1601.
589. – De militia Romana. 4⁰, Antwerp, 1596.
590. – Politicorum sive civilis doctrinae libri sex. 4⁰, Leyden, 1589.
591. – Saturnalium sermonum libri duo. 4⁰, Antwerp, 1585.
592. Littleton, **Sir** Thomas. Tenures. 8⁰, London, 1608.
593. Liturgies. **Greek Rite**. Lithurgiae, sive missae Sanctorum Patrum. (**Greek &
Latin**.) Folio, Paris, 1560.
594. – Rituale Graecorum. (**Greek & Latin**.) Folio, Paris, 1647.
595. – **English Rite**. Liber precum publicarum. (**Greek**) 8⁰, London, 1638.
596. – Liber precum publicarum. (**Latin**) 8⁰, London, 1560.
597. Livius, Titus. Historiarum quod extant. 3 Vols. 8⁰, Leyden, 1654.
598. Loggan, David. Oxonia illustrata. Folio, Oxford, 1675.
599. Lollius, Antonius. De oratione libri septem. Folio, Basle, 1558.
600. London, Royal College of Physicians. Pharmacopoea Londoniensis. Folio,
London, 1618.
601. Longinus, Dionysius Cassius. Liber de grandi sive sublimi genere orationis.
(**Greek & Latin**.) 8⁰, Geneva, 1612.
602. Lucanus, Marcus Annaeus. Pharsaliae libri X. 8⁰, Basle, 1550.
603. – [Another edition.] 8⁰, Leyden, 1627.
604. Lucian, **of Samosata**. Dialogi selecti. 12⁰, London, 1636.
605. Lucretius Carus, Titus. De rerum natura. 16⁰, Lyon, 1576.
606. – [Another edition.] 8⁰, Frankfurt, 1631.
607. – [Another edition.] 12⁰, Cambridge, 1675.
608. Ludolf, Hiob. Lexicon Aetiopico–latinum. 4⁰, London, 1661.
609. Lupus, Servatus, **Abbot of Ferrières**. Opera. Ed. Etienne Baluze, 8⁰, Paris,
1664.
610. Lycophron. Alexandra, sive Cassandra. 4⁰, Geneva, 1601.
611. Lye, Thomas. The Assemblies shorter catechism. 4⁰, London, 1674.
612. Lysias. Eratosthenes. Ed. Andrew Downes. (**Greek & Latin**.) 8⁰, Cambridge,
1593.
613. Macarius, **Saint, the elder**. Homiliae. (**Greek**.) 8⁰, Paris, 1559.
614. Mace, Thomas. Musick's monument. Folio, London, 1676.
615. Machiavelli, Niccolo. Discorsi. 4⁰, Rome, 1531.
616. – Historie Florentine. 4⁰, Florence, 1532.
617. – Libro della arte della guerra. 8⁰, Florence, 1521.
618. Mackenzie, **Sir** George. Religio Stoici. 8⁰, Edinburgh, 1663.
619. Macrobius, Ambrosius Theodosius. Opera. 8⁰, Leyden, 1628.
620. Maestlin, Michael. Epitome astronomiae. 8⁰, Tubingen, 1597.
621. Maldonatus, Joannes. Commentarii in quattuor Evangelistas. Folio, Leyden, 1607.
622. Manilius, Marcus. Astronomicon. Ed. J. J. Scaliger, 4⁰, Leyden, 1600.
623. Manuel, Comnenus, **Emperor of the East**. Legatio...ad Armenios. (**Greek
& Latin**.) 8⁰, Basle, 1578.
624. Marca, Pierre de. Dissertationum de concordia sacerdotii et imperii libri
octo. Ed. Etienne Baluze. Folio, Paris, 1663.
625. – [Another copy.]
626. Marcellinus, Ammianus. Rerum gestarum libri XVIII. 8⁰, Paris, 1544.

627. Mariana, Juan de. De rege et regis institutione libri III. 8⁰, Maintz, 1605.
628. Marlorat, Augustine. Enchiridion locorum communium theologicorum. 8⁰, London, 1588.
629. Marvel, Andrew. The rehearsal transpos'd. 8⁰, London, 1672.
630. Martialis, Marcus Valerius. Epigrammaton libri. 8⁰, London, 1615.
631. Martini, Martinus. Sinicae historiae decas prima. 8⁰, Amsterdam, 1659.
632. Maurolico, Francisco. Opuscula mathematica. 4⁰, Venice, 1575.
633. – Theoremata de lumine. 4⁰, Leyden, 1613.
634. Maximus, Tyrius. Dissertationes XLI. (**Greek & Latin**.) 8⁰, Leyden, 1607.
635. Meibomius, Marcus. De fabrica Triremium liber. 4⁰, Amsterdam, 1671.
636. – [Another copy.]
637. Mela, Pomponius. De situ orbis. Ed. Isaac Vossius, 4⁰, Hague, 1658.
638. Mengoli, Pietro. Geometria speciosa. 4⁰, Bologna, 1659.
639. – [Another copy.]
640. Mercator, Gerardus. Atlas sive cosmographicae meditationes de fabrica mundi et fabricati figura. Folio, Amsterdam, 1613.
641. Mercator, Nicholas. Institutiones astronomicae. 8⁰, London, 1676.
642. – Logarithmo–technia. 4⁰, London, 1668.
643. Mersenne, Marin. Cogitata physico mathematica. 2 Vols. 4⁰, Paris, 1644.
644. – Harmonicorum libri. Folio, Paris, 1635.
645. Middendorpius, Jacobus. De celebrioribus universi terrarum orbis Academiis libri duo. 8⁰, Cologne, 1567.
646. Minucius Felix, Marcus. Octavius. 12⁰, Oxford, 1627.
647. Montague, Richard. The acts and monuments of the Church before Christ Incarnate. Folio, London, 1642.
648. – Apparatus ad Origines ecclesiasticas. Folio, London, 1635.
649. – De originibus Ecclesiasticis commentationum tomus primus. Folio, London, 1636.
650. Montaigne, Michel de. Les Essais. 4⁰, Paris, 1617.
651. Montanus, Benedictus Arias. Antiquitates Iudaicae. 4⁰, Leyden, 1593.
652. Monte, Francesco Maria del, **Cardinal**. Vita...F. Xavier. 4⁰, Rome, 1622.
653. Monte, Guidubaldo del. Mechanicorum liber. Folio, Pisa, 1577.
654. – Planisphaeriorum universalium theorica. Folio, Pisa, 1579.
655. – [Another copy.]
656. Moore, **Sir** Jonas. Resolutio triplex. 4⁰, London, 1658.
657. More, Henry. A modest enquiry into the mystery of iniquity. Folio, London, 1664.
658. – An antidote against atheism. 8⁰, London, 1653.
659. – Conjectura cabbalistica. 8⁰, Cambridge, 1653.
660. – Enthusiasmus triumphatus. 8⁰, Cambridge, 1656.
661. – An explanation of the grand mystery of Godliness. Folio, London, 1660.
662. – Enchiridion ethicum. 8⁰, London, 1669.
663. – The Immortality of the soul. 8⁰, Cambridge, 1659.
664. Morland, **Sir** Samuel. Tuba stentoro-phonica. Folio, London, 1672.
665. Mornay, Philippe de. Mémoires. 4 Vols. 4⁰, Leyden, 1624–52. (An odd volume?)
666. – De veritate religionis Christianae. 8⁰, Antwerp, 1583.

667. Moufet, Thomas. Insectorum sive minimorum animalium theatrum. Folio, London. 1634.
668. Mulerius, Nicolaus. Tabulae Frisicae lunae–solares–quadruplices. 4⁰, Alcamar, 1611.
669. – [Another copy.]
670. Mydorge, Claude. Prodromi catoptricorum et diopticorum...libri quatuor priores. Folio, Paris, 1639.
671. – [Another copy.]
672. Mythology. Mythologici Latini. 8⁰, Heidelberg, 1599.
673. N., O. Dr. Stillingfleet's principles. 8⁰, Paris, 1671.
674. Naudé, Gabriel. Apologie pour tous les grands Personnages qui ont esté faussement soupçonnez de magie. 8⁰, Paris, 1625.
675. Naunton, **Sir** Robert. Fragmenta Regalia. 4⁰, London, 1641.
676. Needham, Walter. Disquisitio anatomica de formato foetu. 8⁰, London, 1667.
677. Nepos, Cornelius. Vitae excellentium imperatorum. 8⁰, Leyden, 1658.
678. Neri, Antonio. The art of glass. 8⁰, London, 1662.
679. Newman, Samuel. A concordance to the Holy Scriptures. Folio, Cambridge, 1662.
680. Newton, John. The art of practical gauging. 8⁰, London, 1669.
681. Nicephorus, **Saint, Patriarch of Constantinople.** Historia Ecclesiastica. **(Greek & Latin.)** 2 Vols. Folio, Paris, 1630.
682. Nicole, Pierre. Essais de morale. Vol. 1 [of 2], 8⁰, Amsterdam, 1672.
683. Nizolius, Marius. Thesaurus Ciceronianus. 8⁰, Basle, 1568.
684. Nonnus, **of Panoplis.** Paraphrasis Sancti secundum Ioannem Evangelii. **(Greek & Latin.)** 8⁰, Basle, 1571.
685. Nuñez, Pedro. Opera. Folio, Basle, 1592.
686. Oecumenius, **Bp of Tricca.** Commentaria...in Acta Apostolorum. **(Greek & Latin.)** 2 Vols. Folio, Paris, 1631.
687. Oppian. De Venatione lib. IV. 8⁰, Leyden, 1597.
688. Optatus, **Saint, Bp of Mela.** De schismate Donatistarum. 8⁰, London, 1631.
689. Origen. Opera. Folio, Leyden, 1536.
690. – In Sacras Scripturas commentaria. **(Greek & Latin.)** Ed. Daniel Huet. 2 Vols. Folio, Rouen, 1668.
691. Ortelius, Abraham. Thesaurus geographicus. Folio, Antwerp, 1587.
692. Osorio, Jeronimo, Da Fonseca, **Bp of Silves.** De regis institutione etdisciplina lib. VIII. 8⁰, Lisbon, 1571.
693. Ossat, Arnaud d', **Cardinal.** Lettres. 4⁰, Paris, 1624.
694. Oughtred, William. The circles of proportion. 8⁰, Oxford, 1660.
695. – Clavis mathematicae. 8⁰, Oxford, 1652.
696. Ovidius Naso, Publius. Opera. 3 Vols. 8⁰, Leipzig, 1607.
697. Owen, John. Epigrammatum libri tres. 8⁰, London, 1606.
698. Oxford University. Statuta selecta. 8⁰, Oxford, 1638.
699. Pagan, Blaise Françoise de. The Count of Pagan's method of delineating all manner of fortifications. Folio, London, 1672.
700. Palmer, John. The catholique planisphaer. 4⁰, London, 1658.
701. Panciroli, Guido. Rerum memorabilium libri duo. 4⁰, Frankfurt, 1609.

702. Panvino, Onofrio. Reipublicae Romanae commentariorum libri tres. 8⁰, Venice, 1558.
703. Pappus, **of Alexandria**. Mathematicae collectiones. Folio, Bologna, 1660.
704. [Paradies, Ignace–Gaston.] A Discourse of local motion. [By A. M.] 16⁰, London, 1670.
705. Parker, Samuel. **Bp of Oxford**. An account of the nature and extent of the divine dominion. 4⁰, London, 1666.
706. – A reproof of the rehearsal. 8⁰, London, 1673.
707. – A discourse of ecclesiastical politie. 8⁰, London, 1670.
708. – A defence and continuation of the ecclesiastical politie. 8⁰, London, 1671.
709. – A discourse in vindication of Bp. Bramhall. 8⁰, London, 1673.
710. – A free and impartial censure of the Platonick philosophie. 4⁰, Oxford, 1666.
711. Parlatorio. Il parlatorio delle monache. 12⁰, Geneva, 1650.
712. Pascal, Blaise. Lettres de Dettonville. 4⁰, Paris, 1659.
713. – Pensées. 12⁰, Paris, 1670.
714. – Les Provinciales. 4⁰, Paris, 1656.
715. Patrick, John. Reflexions upon the devotions of the Roman Church. 8⁰, London, 1674.
716. Patrick, Symon. Advice to a friend. 12⁰, London, 1673.
717. – A friendly debate. 8⁰, London, 1666.
718. Paulus, Aegineta. Opus de re medica. **(Greek & Latin.)** 2 Vols. Folio, Basle, 1538.
719. Pausanias, **the Traveller**. Graeciae descriptio. **(Greek & Latin.)** Folio, Hanau, 1613.
720. Peckham, Johannes. Perspectiva communis. 4⁰, Nuremberg, 1542.
721. Pecquet, Jean. Experimenta nova anatomica. 4⁰, Paris, 1651.
722. Pellison–Fontanier, Paul. The history of the French Academy. 8⁰, London, 1657.
723. Perez Ayala, Martin. De divinis Apostolicis atque Ecclesiasticis traditionibus. 8⁰, Cologne, 1560.
724. Perionius, Joachimus. Dialogorum de linguae Gallicae origine...libri quatuor. 8⁰, Paris, 1555.
725. Persius Flaccus, Aulus. A. Persii Flacci satirarum liber. Ed. Isaac Casaubon. 8⁰, Paris, 1605.
726. – [Another copy/edition.]
727. Petau, Denis. Opus de doctrina temporum. 2 Vols. Folio, Paris, 1627.
728. – Rationarium temporum in...libros tredecim tributum. 8⁰, Maintz, 1646.
729. – Uranologia. Folio, Paris, 1630.
730. – [Another copy.]
731. Peter, **Chrysologus, Saint, Abp of Ravenna**. Sermones. 4⁰, Bologna, 1534.
732. Petrus, **Lombardus, Bp of Paris**. Sententiarum libri IV. 8⁰, Louvain, 1568.
733. Petty, **Sir** William. The discourse made before the Royal Society, 26 Nov. 1674. 12⁰, London, 1674.
734. Peurbach, Georg. Novae theoricae planetarum. 8⁰, Venice, 1538.
735. Philo, **Judaeus**. Opera. **(Greek & Latin.)** Folio, Cologne, 1613.
736. Philosophical transactions. No. 38 (17 August 1668); No. 43 (11 January 1668/9.)

737. Photius, **Patriarch of Constantinople, Saint.** Epistolae. Folio, London, 1651.
738. – Myriobiblon, sive bibliotheca. Folio, Rouen, 1653.
739. Pindar. Olympia, Pythia, Nemea, Isthmia. **(Greek & Latin.)** 8⁰, Heidelberg, 1598.
740. – [Another edition.] 4⁰, Saumur, 1620.
741. Piscator, Johann. Commentarii in omnes libros Novi Testamenti. 4⁰, Herborn, 1621.
742. Platina, Bartholomeo. De vitis ac gestis summorum Pontificum. 8⁰, Paris, 1530.
743. Plato. De rebus divinis dialogi selecti. Ed. John North **(Greek & Latin.)** 8⁰, Cambridge, 1673.
744. – Opera. **(Greek & Latin.)** 2 Vols. Folio, Frankfurt, 1602.
745. Plautus, Titus Maccius. Opera. 8⁰, Basle, 1535.
746. – Comoediae viginti. 8⁰, Leyden, 1581.
747. – Lexicon Plautinum. Ed. Johann Philipp Pareus. 8⁰, Frankfurt, 1614.
748. Plinius Caecilius Secundus, Caius. Epistolarum libri X. 8⁰, Frankfurt, 1611.
749. – [Another edition.] 8⁰, Leyden, 1669.
750. Plutarch. Opera. **(Greek & Latin.)** 2 Vols. Folio, Paris, 1624.
751. Poetae. Poetae Graeci principes heroici carminis. Ed. H. Stephanus. Folio, Geneva, 1566.
752. Poland, Socinian Churches. Catechesis Racoviensis. 12⁰, London, 1651.
753. Pollux, Julius **(of Naucratis.)** Onomasticon: vocabularium. **(Greek.)** 4⁰, Basle, 1536.
754. Polyander A Kerckhoven, Joannes (et al.) Synopsis purioris theologiae. 8⁰, Leyden, 1625.
755. Polybius. Historiarum libri priores quinque. Folio, Basle, 1549.
756. – Historiarum libri qui supersunt. **(Greek & Latin.)** Ed. Isaac Casaubon. Folio, Paris, 1609.
757. Porphyry. De abstinentia ab animalibus necandis libri quatuor. 8⁰, Cambridge, 1655.
758. Porta, Giambattista della. Magiae naturalis libri XX. 12⁰, Leyden, 1651
759. Portius, Simon. Dictionarium Latinum Graeco–barbarum et litterale. 4⁰, Paris, 1635.
760. Portus, Aemilius. Dictionarium Doricum Graecolatinum. 8⁰, Frankfurt, 1603.
761. – Dictionarium Ionicum Graecolatinum. 8⁰, Frankfurt, 1603.
762. Portus, Franciscus. Aphthonius, Hermogenes, & Dionisius Longinus. F. Porti...Opera ...industriaque...illustrati atque expoliti. **(Greek.)** 8⁰, Geneva, 1569.
763. Posselius, Joannes. Syntaxis...Graeca. 8⁰, Cambridge, 1640.
764. Potter, Christopher. Want of charitie justly charged. 8⁰, London, 1633.
765. Powell, Vavasor. The Scripture's concordance. 8⁰, London, 1646.
766. Power, Henry. Experimental philosophy. 4⁰, London, 1664.
767. Proclus, **Diadochus.** De sphaera. 8⁰, Paris, 1536.
768. – In Platonis Theologiam libri sex. **(Greek & Latin.)** Folio, Hamburg, 1618.
769. Procopius, **of Caesarea.** Arcanae historia. **(Greek & Latin.)** Folio, Leyden, 1623.

770. – Historiarum Procopii libri VIII. (**Greek**.) Folio, Augsburg, 1607.

771. Prometheus, **pseud** [i.e., Théodore Agrippe D'Aubigné.] Les tragiques. 4⁰, Geneva, 1616.

772. Prosper, **of Aquitaine, Saint**. Opera. Folio, Lyon, 1539.

773. Prudentius Clemens, Aurelius. Opera. 8⁰, Basle, 1527.

774. Psellus, Michael. De operatione daemonum. (**Greek & Latin**.) 8⁰, Paris, 1615.

775. Ptolemaeus, Claudius. De hypothesibus planetarum liber singularis. Ed. John Bainbridge. (**Greek & Latin**.) 4⁰ London, 1620.

776. – Geographia. 4⁰, Venice, 1562.

777. – Planisphaerium. 8⁰, Basle, 1536.

778. – [Another edition.] 4⁰, Venice, 1558.

779. – Quadripartitum. (**Greek & Latin**.) 4⁰, Nuremberg, 1535.

780. Pufendorf, Samuel von. Elementorum jurisprudentiae universalis libri II. 8⁰, Jena, 1669.

781. Quintilianus, Marcus Fabius. Institutionum oratoriarum libri duodecim. 2 Vols. 8⁰, Leyden, 1665.

782. Rahn, Johann Heinrich. An introduction to algbera. Trans. Thomas Brancker, Ed. John Pell. 4⁰, London, 1668.

783. Ramus, Petrus. Arithmeticae libri duo. 4⁰, Basle, 1569.

784. – [Another copy.]

785. – Scholae mathematicae. 4⁰, Frankfurt, 1599.

786. Ravisius, Joannes, Textor. Epitheta. 8⁰, Geneva, 1622.

787. Ray, John. Catalogus plantarum Angliae et insularum adjacentium. 8⁰, London, 1670.

788. – A collection of English proverbs. 8⁰, London, 1670.

789. – A collection of English words. 8⁰, London, 1674.

790. – Observations, topographical, moral, & physiological; made in a journey through part of the Low–Countries, Germany, Italy and France. 8⁰, London, 1673.

791. Recueil de diverses pièces faites par plusieurs personnes illustres. 3 Vols. 12⁰, La Haye, 1669.

792. Regino, **Abbot of Prum**. De ecclesiasticis disciplinis et religione christiana. Ed. Etienne Baluze. 8⁰, Paris, 1671.

793. Regius, Henricus. Philosophia naturalis. 4⁰, Amsterdam, 1654.

794. Reinold, Erasmus. Prutenicae tabulae coelestium motuum. 4⁰, Tubingen, 1551.

795. Religion. La Religion des Hollandois. 8⁰, Cologne, 1673.

796. Remmelinus, Johannes Ludovicus. Structura tabularum quadratarum tali artificio. 4⁰, Augsburg, 1627.

797. Renerius, Vincentius. Tabulae Medicae. Folio, Florence, 1639.

798. – [Another copy.]

799. Riccioli, Giovanni Battista. Almagestum novum. 2 Vols. Folio, Bologna, 1651.

800. – Geographiae et hydrographiae reformatae libri duodecim. Folio, Bologna, 1661.

801. Ridley, **Sir** Thomas. A view of the civile and ecclesiasticall law. 8⁰, Oxford, 1662.

802. Risner, Friedrich. Opticae libri quatuor. 4⁰, Kassel, 1606.
803. Riverius, Lazarus. Praxis medica cum theoria. 2 Vols. 8⁰, Leyden, 1660.
804. Rivet, André. Operum Theologicorum. 3 Vols. Folio, Rotterdam, 1651–60.
805. Rohan, Henri de. Les mémoires du duc de Rohan. 2 Vols. 12⁰, Leyden, 1661.
806. Rolle, Samuel. The burning of London. 8⁰, London, 1667.
807. – Justification justified. 8⁰, London, 1674.
808. – Londons resurrection. 8⁰, London, 1668.
809. Ryff, Petrus. Quaestiones geometricae. 4⁰, Frankfurt, 1600.
810. Sabunde, Raymundus de. Theologia naturalis. 8⁰, Venice, 1581.
811. Saint Jure, Jean Baptiste. The Holy Life of Monr. De Renty. 8⁰, London, 1658.
812. Salignacus, Bernardus. Mesolabii expositio. 4⁰, Geneva, 1574.
813. Sallustius Crispus, Caius. Opera. 8⁰, Leyden, 1602.
814. Salvianus. **Massiliensis, Presbyter**. Opera. Ed. Etienne Baluze. 8⁰, Paris, 1664.
815. Sammes, Aylett. Britannia antiqua illustrata. Folio, London, 1676.
816. Sanderson, Robert, **Bp of Lincoln**. De juramenti promissorii obligatione praelectiones septem. 8⁰, London, 1661.
817. – De obligatione conscientiae praelectiones decem. 8⁰, London, 1660.
818. Sandius, Christophorus. Interpretationes paradoxae quatuor Evangeliorum. 8⁰, Amsterdam, 1669.
819. – Nucleus historiae ecclesiasticae. 8⁰, Amsterdam, 1669.
820. Sandys, **Sir** Edwin. Europae speculum. 4⁰, London, 1630.
821. Sarpi, Paolo. Interdicti Veneti historia. 4⁰, Cambridge, 1626.
822. Saumaise, Claude de. De modo usuarum liber. 8⁰, Leyden, 1639.
823. – Diatriba de mutuo, non esse alienationem. 8⁰, Leyden, 1640.
824. – Disquisitio de mutuo qua probatur non esse alienationem. 8⁰, Leyden, 1645.
825. – Dissertatio de foenore trapezitico. 8⁰, Leyden, 1640.
826. – Walonis Messalini de Episcopis et presbyteris contra D. Petavium Loiolitam dissertatio prima. 8⁰, Leyden, 1641.
827. Savile, **Sir** Henry. Praelectiones tresdecim in principium elementorum Euclidis. 4⁰, Oxford, 1621.
828. – [Another copy.]
829. Scaliger, Joseph Juste. Opus de emendatione temporum. Folio, Frankfurt, 1593.
830. – [Another edition.] Folio, Cologne, 1629.
831. – Scaligerana. 8⁰ Cologne, 1667.
832. Scaliger, Julius Caesar. De causis linguae Latinae libri tredecim. 8⁰, Heidelberg, 1580.
833. – Epistolae et orationes. 8⁰, Leyden, 1600.
834. – Exotericarum exercitationum liber quintus decimus de subtilitate. 8⁰, Frankfurt, 1576.
835. – Oratio pro M. Tullio Cicerone contra Des. Erasmum. 8⁰, Paris, 1531.
836. – Poemata. 8⁰, Heidelberg, 1574.
837. – Poetices libri septem. 8⁰, Heidelberg, 1581.
838. Schedius, Elias. De diis Germanis...syngrammata IV. 8⁰, Amsterdam, 1648.

839. Scheiner, Christophorus. Oculus; hoc est fundamentum opticum. 4⁰, London, 1652.
840. - [Another copy.]
841. Schooten, Franciscus A. Exercitationum mathematicarum libri quinque. 4⁰, Leyden, 1657.
842. Schoppe, Caspar. Grammatica philosophica. 8⁰, Amsterdam, 1659.
843. - Classicum belli sacri. 4⁰, Pavia, 1619.
844. - Collyrium Regium. 8⁰, Ingolstadt, 1611.
845. - Consilium Regium. 4⁰, Pavia, 1619.
846. - Infamiae Famiani. 12⁰, Sorö, 1658.
847. - Scorpiacum. 4⁰, Maintz, 1612.
848. - Suspectarum lectionum libri quinque. 8⁰, Nuremberg, 1597.
849. - Verisimilium libri IV. 8⁰. Nuremberg, 1596.
850. - [Another copy.]
851. Schott, Gaspar. Cursus mathematicus. 4⁰, Würtzburg, 1668.
852. - Magia universalis. 4 Vols., 4⁰, Würtzburg, 1657–59.
853. - Mathesis Caesarea. 4⁰, Würtzburg, 1662.
854. - Mechanica hydraulico–pneumatica. 4⁰, Würtzburg, 1657.
855. - Organum mathematicum. 4⁰, Würtzburg, 1668.
856. - Physica curiosa. 2 Vols., 4⁰, Würtzburg, 1662.
857. - Schola steganographica. 8⁰, Nuremberg, 1665.
858. - Technica curiosa. 2 Vols. 4⁰, Nuremberg, 1664.
859. Schrijver, Pieter. Collectanea veterum tragicorum. 8⁰, Leyden, 1620.
860. Schroeder, Johann. Pharmacopoeia medico–Chymica. 4⁰, Leyden, 1656.
861. Scot, Alexander. Universa grammatica Graeca. 8⁰, Leyden, 1605.
862. Scultetus, Joannes. Armamentarium chirurgicam. 8⁰, Hague, 1656.
863. Seinior, George. God, the King and the Church. 8⁰, London, 1670.
864. Selden, John. De dis Syris. 8⁰, London, 1617.
865. - De successionibus in bona defuncti. Folio, London, 1636.
866. Seneca, Lucius Annaeus. Annaei Senecae atque aliorum tragoediae. Ed. Thomas Farnaby. 8⁰, London, 1613.
867. - L. Annaeus Seneca Tragicus. Ed. P. Scriverius 8⁰, Leyden, 1621.
868. - [Another edition.] 8⁰, London, 1634.
869. Seneschallus, Michael. De tempore nati baptizati et mortui Christi. 4⁰, Liège, 1670.
870. Sennertus, Daniel. Epitome naturalis scientiae. 8⁰, Oxford, 1632.
871. - Institutiones medicinae. Folio, Paris, 1637.
872. Severus, Sulpicius. Opera omnia. 8⁰, Leyden, 1647.
873. Sharrock, Robert. The history of the propagation and improvement of vegetables. 8⁰, Oxford, 1660.
874. - Judicia (seu legum censurae) de variis Incontinentiae speciebus. 8⁰, Oxford, 1662.
875. - De officiis secundum Naturae Jus. 8⁰, Oxford, 1660.
876. Sheppard, William. The whole office of the country Justice of Peace. 8⁰, London, 1652.
877. Sherlock, William. A discourse concerning the knowledge of Jesus Christ. 8⁰, London, 1674.
878. Sherman, John. An account of faith. 4⁰, London, 1661.

879. Sidonius, **Apollinaris, Bp of Clermont.** Opera. 8⁰, Lyon, 1552.
880. - [Another edition.] 4⁰, Paris, 1652.
881. Sigonio, Carlo. De antiquo iure civium Romanorum libri duo. 4⁰, Venice, 1560.
882. - De Atheniensium Lacedaemoniorumque temporibus liber. 4⁰, Venice, 1564.
883. - Historiarum de regno Italiae libri quindecim. Folio, Frankfurt, 1575.
884. - De lege curiata magistratuum et imperatorum. 4⁰, Venice, 1569.
885. Silius Italicus, Caius. De bello Punico. 8⁰, Venice, 1523.
886. - [Another edition.] 4⁰, Paris, 1615.
887. Simson, Edward. Mosaica. 4⁰, Cambridge, 1636.
888. Sirmond, Jacques. Opuscula dogmatica veterum quinque scriptorum. 8⁰, Paris, 1630.
889. Sleidanus, Joannes. Commentarii de statu religionis et reipublicae Carolo Quinto Caesare. 8⁰, Strassburg, 1565.
890. Sluse, René François de. Mesolabum. 4⁰, Liège. 1668.
891. Smetius, Henricus. Prosodia. 8⁰, London, 1615.
892. Snelius, Willebrodus. Cyclometricus. 4⁰, Leyden, 1621.
893. - Descriptio cometae. 4⁰, Leyden, 1619.
894. - Eratosthenes Batavus. 4⁰, Leyden, 1617.
895. - Tiphys Batavus. 4⁰, Leyden, 1624.
896. Sophocles. Tragoediae VII. (**Greek.**) 4⁰, Geneva, 1568.
897. - Tragoediae VII. (**Greek.**) 16⁰, Antwerp, 1579.
898. - Tragoediae VII. (**Greek & Latin.**) 8⁰, Heidelberg, 1597.
899. Soverus, Bartholomaeus. Curvi ac recti proportio. 4⁰, Padua, 1630.
900. Spanheim, Friedrich, **the Elder.** Dubiorum evangelicorum pars prima. 4⁰, Geneva, 1639.
901. Sparrow, Anthony, **Bp of Norwich.** A collection of articles, injunctions, canons. 4⁰, London, 1661.
902. - [Another edition.] 4⁰, London, 1671.
903. Spenser, Edmund. The faerie queene. 4⁰, London, 1590.
904. Spinoza, Benedictus de. Tractatus Theologico-politicus. 4⁰, Amsterdam, 1670.
905. Sprat, Thomas. The History of the Royal Society of London. 4⁰, London, 1667.
906. - [Another copy.]
907. - Observations on Monsieur Sorbier's voyage into England. 8⁰, London, 1665.
908. Squire, William. The unreasonableness of the Romanists. 8⁰, London, 1670.
909. Statera (La Giusta) de' Porporati. Ricorso di Pasquino ad Apollo contra D. Olympia. 12⁰, Geneva, 1650.
910. Statius, Publius Papinius. Opera. 8⁰, Antwerp, 1595.
911. - Sylvarum libri V. Achilleidos libri II. Ed. T. Stephens. 8⁰, Cambridge, 1651.
912. Stegmann, Josua. Photinianismus. 8⁰, Frankfurt, 1626.
913. Stephanus, Henricus. Ciceronianum Lexicon Graeco-Latinum. 8⁰, Geneva, 1557.
914. - Oratorum veterum orationes. (**Greek & Latin.**) Folio, Geneva, 1575.

915. – Virtutum encomia. 16⁰, Geneva, 1573.
916. Stevin, Simon. Hypomnemata mathematica. 2 Vols. Folio, Leyden, 1605.
917. Stifel, Michael. Arithmetica. 8⁰, Nuremberg, 1544.
918. – [Another copy.]
919. Stillingfleet, Edward. **Bp of Worcester**. An answer to several late treatises, occasioned by a book entituled A discourse concerning the idolatry practised in the Church of Rome. 8⁰, London, 1673.
920. – A defence of the discourse concerning the idolatry practised in the Church of Rome. 8⁰, London, 1676.
921. – A discourse concerning the idolatry practised in the Church of Rome. 8⁰, London, 1671.
922. – A rational account of the grounds of Protestant religion. Folio, London, 1665.
923. Stobaeus, Joannes. Dicta poetarum. Ed. Hugo Grotius, 4⁰, Paris, 1623.
924. – Sententiae. **(Greek & Latin.)** Folio, Geneva, 1609.
925. Strabo. De situ orbis. [Edition not identified.]
926. Strada, Famianus. Prolusiones academicae oratoriae, historicae, poeticae. 12⁰, St. Omer, 1619.
927. Suetonius Tranquillus, Caius. Duodecim Caesraes. 8⁰, Leyden, 1647.
928. Suidas. Suidas, cuius...Latinum interpretationem, et Graeci textus emendationem Ae. Portus...conscripsit. **(Greek & Latin.)** 2 Vols. Folio, Geneva, 1619.
929. Sutcliffe, Matthew. Adversus Roberti Bellarmini de purgatorio disputationem liber unus. 4⁰, London, 1599.
930. Sydenham, Thomas. Observationes medicae. 8⁰, London, 1676.
931. Sylburgius, Fridericus. Etymologicon magnum. Folio, Heidelberg, 1594.
932. Tacitus, Publius Cornelius. Opera. 8⁰, Leyden, 1621.
933. – The ende of Nero and beginning of Galba. Trans. Henry Savile. Folio, Oxford, 1591.
934. Tacquet, Andreas. Arithmetica theoria et praxis. 8⁰, Antwerp, 1665.
935. – Cylindricorum et annularum liber quintus. 4⁰, Antwerp, 1659.
936. Talmud. [A single tractate, not identified.]
937. Tasso, Torquato. La Gierusalemme liberata. 4⁰, Parma, 1581.
938. Taylor, Jeremy, **Bp of Down and Connor**. The rule and exercises of holy living...(&...holy dying.) 8⁰, London, 1651.
939. – Of the sacred order. 4⁰, Oxford, 1642.
940. Templer, John. Idea theologiae Leviathanis. 8⁰, London, 1673.
941. Terentius, Publius. Comoediae. 8⁰, Leyden, 1644.
942. Tertulianus, Quintus Septimus, Florens. Opera. Ed. Franciscus Junius. Folio, Franecker, 1597.
943. – [Another edition.] Ed. Nicolas Rigault. Folio, Paris, 1635.
944. Thaddaeus, Joannes. Conciliatorium Biblicum. 12⁰, Amsterdam, 1648.
945. Theoderet, **Bp of Cyrus**. Opera omnia. **(Greek & Latin.)** 4 Vols. (of 5.) Folio, Paris, 1642.
946. Theodosius, **of Tripoli**. Sphaericorum. 4⁰, Paris, 1558.
947. Theon, **of Smyrna**. Theonis Smyrnaei eorum, que in mathematicis ad Platonis lectionem utilia sunt, expositio. Ed. Ismael Boulliau. **(Greek & Latin.)** 4⁰, Paris, 1644.

948. – [Another copy.]
949. Theophrastus. Characteres ethici. Ed. Isaac Casaubon. (Greek & Latin.) 8⁰, Leyden, 1592.
950. – [Another copy.]
951. – De historia plantarum libri decem. **(Greek & Latin.)** Folio, Amsterdam, 1644.
952. Theophylact, **Simocatta**. Historiae Mauricii Tiberii Imp. libri VII. **(Greek & Latin.)** 2 Vols. 4⁰, Ingolstad, 1604.
953. Thibaut, Pierre. The art of Chymistry. 8⁰, London, 1668.
954. Thorndike, Herbert. A discourse of the forberance or the penalties which a due reformation requires. 8⁰, London, 1670.
955. – A discourse of the right of the church in a Christian state. 8⁰, London, 1649.
956. – An epilogue to the tragedy of the Church of England. Folio, London, 1659.
957. – A letter concerning the present state of religion. 8⁰, London, 1656.
958. – De ratione ac jure finiendi controversias Ecclesiae disputatio. Folio, London, 1670.
959. – Of religious assemblies. 8⁰, Cambridge, 1642.
960. Thou, Jacques Auguste de. Doctorum virorum elogia Thuanea. 8⁰, London, 1671.
961. Thruston, Malachi. De respirationis usu primario, diatriba. 8⁰, London, 1670.
962. Thucydides. De bello Peloponnesiaco. [Edition not identified.]
963. Tillotson, John. **Abp of Canterbury**. The rule of faith. 8⁰, London, 1666.
964. Toletus, Franciscus. Instructio Sacerdotum ac Poenitentium. 4⁰, Cologne, 1621.
965. Tomkins, Thomas. The modern pleas for comprehension...considered. 8⁰, London, 1675.
966. Torriano, Giovanni. Select Italian proverbs. 12⁰, Cambridge, 1642.
967. Torricelli, Evangelista. Opera geometrica. 4⁰, Florence, 1644.
968. – De sphaera et solidis sphaeralibus libri duo. 4⁰, Florence, 1644.
969. Trogus Pompeius. Justini ex Trogi Pompeii historiis externis libri XLIV. 8⁰, London, 1593.
970. Tuckney, Anthony. Forty sermons. 4⁰, London, 1676.
971. Usher, James, **Abp of Armagh**. Annales Veteris Testamenti. Folio, London, 1648.
972. – Annotationes in Vetus Testamentum. 8⁰, London, 1653.
973. – Britannicarum ecclesiarum antiquitates. 4⁰, Dublin, 1639.
974. – De Macedonum et Asianorum anno solari dissertatio. 8⁰, London, 1650.
975. Usuardus, Monachus Saugermanensis. Martyrologium. 8⁰, Louvain, 1568.
976. Valerius Maximus. Dictorum factorumque memorabilium libri novem. 8⁰, Frankfurt, 1601.
977. Valerius, Lucas. De centro gravitatis solidorum libri tres. 4⁰, Rome, 1604.
978. Valla, Laurentius. De linguae Latinae elegentiae. 4⁰, Paris, 1542.
979. – [Another edition.] 8⁰, Antwerp, 1557.
980. Varenius, Bernardus. Geographia generalis. Ed. Isaac Newton. 8⁰, Cambridge, 1672.

981. Varro, Marcus Terentius. Opera. Ed. J. J. Scaliger. 8°, Geneva, 1653.
982. Vasari, Giorgio. Le vite de piu eccellenti architetti, pittori, et scultori Italiani. 4°, Florence, 1550.
983. Vegetius Renatus, Flavius. De re militari libri quatuor. Ed. Godescalcus Stewechius. 4°, Antwerp, 1585.
984. – [Another edition.] 8°, Leyden, 1592.
985. Velleius Paterculus, Marcus. Historiae Romanae. 8°, Leyden, 1594.
986. Vergilius, Polydorus. De rerum inventoribus libri VIII. 12°, Amsterdam, 1651.
987. Vermigli, Pietro Martire. Loci communes. Folio, London, 1583.
988. Viète, François. Opera mathematica. Folio, Leyden, 1646.
989. Vigerus, Franciscus. De praecipuis Graecae dictionis idiotismis. 8°, Cambridge, 1647.
990. Villegas, Alfonso di. Discorsi sopra gli Evangeli. 4°, Venice, 1603.
991. – Instructions Chrestiennes sur les évangiles de tous les dimanches de l'année. 8°, Paris, 1611.
992. Villeroy, Nicolas de Neufville, Seigneur de. Mémoires d'Estat. 4 Vols. 8°, Paris, 1622. (An odd volume?)
993. Vincent, **of Lerins, Saint**. Peregrini, id est, ut vulgo perhibetur, Vincentii Lirinensis, adversus prophanas haereses. 12°, Oxford, 1631.
994. – Pro Catholicae Fidei antiquitate et universitate adversus profanas omnium Haereseon novationes. 8°, Paris, 1554.
995. Virgilius Maro, Publius. Catalecta. Ed. Joseph Scaliger. 8°, Leyden, 1617.
996. – Opera. 8°, Paris, 1582.
997. – [Another edition.] Folio, Basle, 1586.
998. Vitruvius Pollio, Marcus. De architectura. 4°, Strassburg, 1543.
999. Vives, Joannes Ludovicus. Libri XII de disciplinis. 8°, Leyden, 1612.
1000. Viviani, Vincenzo. De maximis et minimis. Folio, Florence, 1659.
1001. Voet, Gijsbert. Exercitatio et Bibliotheca studiosi theologiae. 12°, Utrecht, 1651.
1002. Vossius, Gerardus Joannes. De arte grammatica. 4°, Amsterdam, 1635.
1003. – De artis poeticae natura ac constitutione liber. 4°, Amsterdam, 1647.
1004. – Chronologiae sacrae isagoge. 4°, Hague, 1659.
1005. – Etymologicon linguae Latinae. Folio, London, 1662.
1006. – Historiae de controversiis quas Pelagius eiusque reliquiae moverunt. 4°, Leyden, 1618.
1007. – De historicis Graecis libri IV. 4°, Leyden, 1624.
1008. – De historicis Latinis libri III. 4°, Leyden, 1627.
1009. – De imitatione, cum oratoria tum praecipue poetica; deque recitatione veterum. 4°, Leyden, 1626.
1010. – Institutiones linguae Graecae. 8°, Leyden, 1626.
1011. – Latina grammatica. 8°, Amsterdam, 1648.
1012. – De logices et rhetoricae natura et constitutione. 4°, Hague, 1658.
1013. – Oratoriarum institutionum libri sex. 4°, Leyden, 1645.
1014. – Poeticarum institutionum libri tres. 4°, Amsterdam, 1647.
1015. – De quatuor artibus popularibus, de philologia, et scientiis mathematicis. 4°, Amsterdam, 1650.

1016. – Rhetorices contractae, sive partitionum oratoriarum libri V. 8°, Leyden, 1622.
1017. – De theologia Gentili, et physiologia Christiana, sive de origine et progressu idolatriae. 2 Vols. 4°, Amsterdam, 1641.
1018. Vossius, Isaac. De lucis natura et proprietate. 4°, Amsterdam, 1662.
1019. – [Another copy.]
1020. – De Septuaginta interpretibus. 4°, Hague, 1661.
1021. Walker, Obadiah. Of education. 12°, Oxford, 1673.
1022. Walker, William. A modest plea for infants baptism. 12°, Cambridge, 1677.
1023. Wallis. Adversus M. Meibomii, de proportionibus dialogum. 4°, Oxford, 1657.
1024. – Arithmetica infinitorum. 4°, Oxford, 1656.
1025. – Commercium epistolicum. 4°, Oxford, 1658.
1026. – Due correction for Mr. Hobbes. 8°, Oxford, 1656.
1027. – Elenchus geometria Hobbianae. 8°, Oxford, 1655.
1028. – Hobbiani puncti dispunctio. 12°, Oxford, 1657.
1029. – Hobbius Heauton-timorumenos. 12°, Oxford, 1662.
1030. – Mechanica. 4°, London, 1670.
1031. – Operum mathematicorum pars prima. 4°, Oxford, 1657.
1032. – Tractatus duo, prior de Cycloide posterior Epistolaris...de Cissoide. 4°, Oxford, 1659.
1033. Ward, Seth, **Bp of Salisbury**. Astronomia geometrica. 8°, London, 1656.
1034. – Idea trigonometriae...Item praelectio de cometis. Et Inquisitio in Bullialdi astronomiae philolaicae fundamenta. 4°, Oxford, 1654.
1035. – Six sermons. 8°, London, 1672.
1036. Webster, William. Webster's tables. 8°, London, 1629.
1037. Weemers, Jacobus. Lexicon Aetiopicum. 4°, Rome, 1638.
1038. Wells, John. Sciographia, or the art of shadowes. 8°, London, 1635.
1039. Wendelin, Marcus Frederick. Christianae theologiae libri II. 12°, Amsterdam, 1653.
1040. – Institutionum politicarum lib. III. 12°, Amsterdam, 1654.
1041. Wharton, Thomas. Adenographia, sive glandularum totius corporis descriptio. 8°, London, 1656.
1042. Whitby, Daniel. An endeavour to evince the certainty of Christian faith. 8°, Oxford, 1671.
1043. White, Thomas. Chrysaspis seu scriptorum suorum in scientiis obscurioribus apologiae vice propalata tutela geometrica. 16°, London, 1659.
1044. – The grounds of Obedience and Government. 12°, London, 1655.
1045. – Institutionum Peripateticarum ad...K. Digbaei pars theorica. 12°, London, 1647.
1046. Widdrington, Roger (i.e., Thomas Preston.) Disputatio theologica. 8°, London, 1613.
1047. Wilkins, John. Mathematicall magick. 8°, London, 1648.
1048. Willis, Thomas. Cerebri anatome, cui accessit nervorum descriptio et usus. 4°, London, 1664.
1049. – Diatribae duae medico-philosophicae, quarum prior agit de fermentatione...altera de febribus. 8°, London, 1659.

1050. – Pathologiae cerebri, et nervosi generis specimen. 4⁰, Oxford, 1667.
1051. Willughby, Francis. Ornithologiae libri tres. Folio, London, 1676.
1052. Windet, James. De vita functorum statu. 8⁰, London, 1664.
1053. Wingate, Edmund. An exact abridgement of all Statutes. 8⁰, London, 1655.
1054. Winterton, Ralph (Ed.) Poetae minores Graeci. 8⁰, Cambridge, 1635.
1055. Wright, Edward. Certaine errors in navigation. 4⁰, London, 1599.
1056. Xenophon. Opera. (**Greek & Latin.**) Folio, Geneva, 1561.
1057. Zucchi, Nicola. Nova de machinis philosophia. 4⁰, Rome, 1649.
1058. – Optica Philosophia experimentalis et ratione a fundamentis constituta. 2 Vols. 4⁰, Leyden, 1652–56.

Titles not identified

1059. Ale of Alendo.
1060. Dionysius.
1061. Hist. eccl. & ritualis epist.
1062. Imperialia decreta de cultu imaginum.
1063. Oxonii delineatio.

Appendix: Books in Barrow's Library Catalogue that may have been acquired by Isaac Newton

1. Alsted, Johann Heinrich. Thesaurus chronologiae. 8⁰, Herborn, 1650.
2. Ambrose, **Saint, Bp of Milan**. Opera. 3 Vols., Folio, Basle, 1567.
3. Apollodorus, **Atheniensis**. Bibliotheces, sive de deorum origine. (**Greek & Latin.**) 8⁰, Rome, 1555.
4. Archimedes. Opera quae extant. Ed. David Rivault. (**Greek & Latin.**) Folio, Paris, 1615.
5. Bacon, Francis. Opuscula varia posthuma. 8⁰, London, 1656.
6. Basil, **Saint, Abp of Caesaria in Cappodocia**. Opera omnia. 3 Vols. (**Greek & Latin.**) Folio, Paris, 1618.
7. Bible, **Polyglot**. Biblia Sacra Polyglotta. 6 Vols. Folio, London, 1655–7.
8. – **Concordance**. Concordantiae Testamenti Novi Graecolatinae. Ed. Henricus Stephanus. Folio, Geneva, 1600.
9. Buxtorf, Johann. Lexicon Hebraicum et Chaldaicum. 8⁰, Basle, 1615.
10. Camden, William. Britannia. 8⁰, London, 1586.
11. Casaubon, Isaac. Animadversionum in Athenaei Dipnosophistas. 2 Vols. Folio, Leyden, 1600.
12. Cosin, John, **Bp of Durham**. Historia transubstantiationis papalis. 8⁰, London, 1675.
13. Courcelles, Estienne de. Opera theologica. Folio, Amsterdam, 1675.
14. Daillé, Jean. De cultibus religiosis Latinorum. 4⁰, Geneva, 1671.
15. – De imaginibus libri IV. 12⁰, Geneva, 1641.
16. – De jejuniis et quadragesima liber. 8⁰, Daventer, 1654.
17. Descartes, René. Geometria. Ed. Franciscus van Schooten. 2 Vols. 4⁰, Amsterdam, 1659–61.
18. Diophantus. Arithmeticorum libri sex. Folio, Paris, 1621.

19. Fabri, Honoré. Synopsis geometrica. 8⁰, Lyon, 1669.
20. Field, Richard. Of the Church. 4⁰, London, 1606.
21. Franckenberger, Reinhold. Chronologiae Scaligero–Petavianae breve compendium. 4⁰, Wittenberg, 1661.
22. Grandami, Jacques. Chronologia Christiana. 4⁰, Paris, 1668.
23. Gregory, **of Nazianzus, Saint**. Opera. **(Greek & Latin.)** 2 Vols. Folio, Paris, 1630.
24. Gregory, James. Exercitationes geometricae. 4⁰, London, 1668.
25. – Geometriae pars universalis. 4⁰, Padua, 1668.
26. – Vera circuli et hyperbolae quadratura. 4⁰, Padua, 1667.
27. Hyde, Thomas. Catalogus impressorum librorum bibliothecae Bodleianae in Academia Oxoniensi. Folio, Oxford, 1674.
28. Julianus, **the Apostate, Emperor of Rome**. Opera. 8⁰, Paris, 1583.
29. Mercator, Nicholas. Logarithmo–technia. 4⁰, London, 1668.
30. Montanus, Benedictus Arias. Antiquitates Iudaicae. 4⁰, Leyden, 1593.
31. More, Henry. An antidote against atheism. 8⁰, London, 1653.
32. – The Immortality of the soul. 8⁰, Cambridge, 1659.
33. Nicephorus, **Saint, Patriarch of Constantinople**. Historia Ecclesiastica. **(Greek & Latin.)** 2 Vols. Folio, Paris, 1630.
34. Optatus, **Saint, Bp of Mela**. De schismate Donatistarum. 8⁰, London, 1631.
35. Philosophical transactions. No. 38 (17 August 1668); No. 43 (11 January 1668/9.)
36. Photius, **Patriarch of Constantinople, Saint**. Epistolae. Folio, London, 1651.
37. – Myriobiblon, sive bibliotheca. Folio, Rouen, 1653.
38. Porta, Giambattista della. Magiae naturalis libri XX. 12⁰, Leyden, 1651.
39. Potter, Christopher. Want of charitie justly charged. 8⁰, London, 1633.
40. Procopius, **of Caesarea**. Arcanae historia. **(Greek & Latin.)** Folio, Leyden, 1623.
41. Riccioli, Giovanni Battista. Almagestum novum. 2 Vols. Folio, Bologna, 1651.
42. Ridley, **Sir** Thomas. A view of the civile and ecclesiasticall law. 8⁰, Oxford, 1662.
43. Sandius, Christophorus. Interpretationes paradoxae quatuor Evangeliorum. 8⁰, Amsterdam, 1669.
44. Scheiner, Christophorus. Oculus; hoc est fundamentum opticum. 4⁰, London, 1652.
45. Schooten, Franciscus A. Exercitationum mathematicarum libri quinque. 4⁰, Leyden, 1657.
46. Sluse, René François de. Mesolabum. 4⁰, Liège. 1668.
47. Stillingfleet, Edward. **Bp of Worcester**. An answer to several late treatises, occasioned by a book entituled A discourse concerning the idolatry practised in the Church of Rome. 8⁰, London, 1673.
48. – A discourse concerning the idolatry practised in the Church of Rome. 8⁰, London, 1671.
49. Suidas. Suidas, cuius... Latinum interpretationem, et Graeci textus emendationem Ae. Portus... conscripsit. **(Greek & Latin.)** 2 Vols. Folio, Geneva, 1619.
50. Sylburgius, Fridericus. Etymologicon magnum. Folio, Heidelberg, 1594.

51. Thou, Jacques Auguste de. Doctorum virorum elogia Thuanea. 8⁰, London, 1671.
52. Usher, James, **Abp of Armagh**. Annales Veteris Testamenti. Folio, London, 1650.
53. Vigerus, Franciscus. De praecipuis Graecae dictionis idiotismis. 8⁰, Cambridge, 1647.
54. Vossius, Gerardus Joannes. De theologia Gentili, et physiologia Christiana, sive de origine et progressu idolatriae. 2 Vols. 4⁰, Amsterdam, 1641.
55. Ward, Seth, **Bp of Salisbury**. Idea trigonometriae...Item praelectio de cometis. Et Inquisitio in Bullialdi astronomiae philolaicae fundamenta. 4⁰, Oxford, 1654.

INDEX

abstraction, 201-2
Accademia del Cimento, 50
Act of Uniformity, 53, 275
Adam, 308, 313
Aeschylus, 291
Aesop, 16
Aguilon, François d', 111
Akehurst, Alexander, 23, 34, 36
Aldrich, Henry, 77
Alembert, Jean le Rond d', 160
algebra, 190, 195, 196, 200-2; symbolic, 237, 239, 240, 244n
Alhazen (Ibn al-Haytham), 111, 122, 130, 134, 163
Alhazen's problem, 139, 140, 153-4, 155-6, 174n
Allen, Thomas, 47
Allestree, Richard, 23
Alsted, Johann Heinrich, 179
Ames, William, 315
anaclastic problem, 132-3, 155, 171n
analysis, 207, 238; algebraic, 183, 188-9, 195, 210, 213, 237, 239, 240; Barrow's attitude toward, 189-91, 196, 201
Apollonius of Perga, 38, 44, 45, 49, 54, 75, 76, 77, 78, 101n, 194, 206, 212, 213, 214, 218, 245n, 298, 300
Archimedean spiral, 206
Archimedes, 20, 21, 41, 44, 45, 46, 54, 61, 62, 68, 75, 77, 78, 79, 194, 206, 207, 211, 229, 238, 272, 298, 299, 300
Aristophanes, 16
Aristotle (Aristotelianism), 11-13, 15, 16, 24, 26, 28, 29, 32, 35, 58, 59, 60, 96n,

119, 194, 204, 205, 263, 264, 265, 267, 269, 271, 272, 276, 277, 278, 292, 297, 315
arithmetic, 186; relation to geometry, 186, 189
Arminianism, 5-6, 257, 259, 260, 286n; Barrow's, 37-8, 260-1, 266, 273
Arnauld, Antoine, 47
Arrowsmith, John, 23, 52, 256, 258, 273
Arrowsmith, Thomas, 26
astigmatism, 107, 108, 124, 135-6, 148, 158-9, 177n
Aston, Francis, 88, 281, 282
Atkinson, Gilbert, 337, 340n
atomism, 191, 278
Atwell, George, 36
Aubrey, John, 1, 4, 43, 51, 246
Augustine, Saint, of Hippo, 185
Auzout, Adrien, 155
Aylmer, Brabazon, 90, 304, 329n

Bacon, Sir Francis, 18, 28, 29, 35, 64, 91, 265, 267, 268
Bainbridge, John, 15, 293, 294, 295, 300
Baines, Thomas, 25
Baker, Thomas, 72, 75, 79, 90
Barclay, John, 16
Barnes, Joshua, 58
Barrough, Issac, 1
Barrough, Isaac (s. of Philip), 1
Barrough, Philip, 1
Barrough, Samuel, 1
Barrovian case, 144, 159-65
Barrow, Ann (mother), 1
Barrow, Elizabeth (half-sister), 2, 82